Michael Pollak was born in Bratislava, Czechoslovakia, on 23 October 1948 and spent most of his infancy in Bolivia. He arrived with his parents in Sydney at the age of six and, apart from stints overseas and a year in Brisbane, has lived in that city ever since. He is a journalist who has worked for a range of newspapers, including the *Sydney Morning Herald*, the *Australian* and the *Australian Financial Review*. Since 1983 he has been working as a freelancer to concentrate on book writing. His first book was *Sense and Censorship*, published in 1990, which dealt with how Australian writers over more than 200 years have suffered because of our draconian censorship and defamation laws. Pollak is also concerned about environmental issues which led to the idea for this book. He worked on *Hearts and Minds* for more than six years and it became a collaboration with Margaret MacNabb soon after their marriage in 1996.

Margaret MacNabb was born in Toronto, Canada, on 7 April 1954. She began her career as a journalist with Toronto's the *Globe and Mail* newspaper in 1982; she was made an Associate Editor of the *Globe*'s highly regarded financial magazine, *Report on Business*, in 1986. In 1988, she had the opportunity to work in China as an associate editor of the *Beijing Review*. MacNabb arrived in Australia in August 1989 and joined the *Sydney Morning Herald*. She worked for the Fairfax press for many years as sub-editor, feature writer and bureau correspondent. In 1998, during a year in Brisbane, she covered such important environmental stories as the controversial tourist development at Port Hinchinbrook, the opening of the Brisbane-Gold Coast transport corridor, and the fallout from the El Nino effect on Australian agriculture.

BRIDGET ELLIOT

HEARTS AND MINDS

Creative Australians and the Environment

MICHAEL POLLAK
MARGARET MacNABB

HALE
& IREMONGER

First published in 2000.

Typeset by
Midland Typesetters Pty Ltd,
Maryborough, Victoria 3465

Printed and bound by
Colorcraft
Hong Kong

For the publisher
Hale & Iremonger Pty Limited
PO Box 205, Alexandria, NSW 2015

National Library of Australia Cataloguing-in-Publication entry:
Pollak, Michael, 1948 Oct. 23– .
Hearts and minds: creative Australians and the environment.

Bibliography.
Includes index.
ISBN 0 86806 692 3.

1. Authors, Australian–20th century. 2. Environmental health–Australia. 3. Environmental protection in art. 4. Environmental protection in literature. 5. Motion pictures–Australia–Plots, themes, etc. 6. Nature in literature. I. MacNabb, Margaret, 1954– . II. Title.

808.80355

This project has been assisted by the Commonwealth Government through the Australia Council, its art funding and advisory body.

Printed on recycled paper

CONTENTS

Acknowledgements

Hundreds of people throughout Australia and overseas have given of their time and knowledge during research for this book. Some have participated in face-to-face interviews with one or both of the authors, others have spoken over the telephone and others have provided information by email or correspondence. Those who have helped us—and they include some of the finest writers in the land—are far too numerous to mention individually.

Special mention goes to: our computer expert, Alan Myers of Brisbane, for teaching us all we know about PCs and the Internet and for getting us out of many scrapes; Annette McGuiness and Marie-Louise Ayres of the Australian Defence Force Academy Library in Canberra for their help with the Austlit database and archival material; and Jane Adam of the Sydney Office of the National Film and Sound Archive for access to scripts and other rare material.

The authors also note the unstinting belief and supportiveness over the six years of this project to the late Freda MacNabb of Toronto, Canada, and to the late Ludevit Pollak.

Preface

THE spread of new ideas is the mightiest weapon in the fight to preserve the environment. Australians today feel far more keenly about the environment than they did thirty, twenty or even ten years ago—and this growing concern is largely because of the information and ideas they have received from a variety of sources. Literature and the other arts have also had an important role in forging our beliefs, and here for the first time this link between the creative arts and green consciousness is examined closely. Famous writers and little-known communicators are examined side by side. The works cover a dazzling variety of approaches and encompass almost every genre, each having a role in exposing the destruction of the planet and presenting solutions to turn things around.

Hearts and Minds begins with an introductory chapter which defines each category and in which the reader meets a cross-section of the writers discussed in the chapters that follow. Michael Pollak and Margaret MacNabb travelled throughout Australia to personally interview such writers as novelists Jean Bedford and Peter Corris on Coochiemudlo Island off Brisbane; venerable children's author Hesba Brinsmead at Terranora in northern New South Wales; poet Les Murray in his beloved surroundings at Bunyah in the New South Wales Great Lakes region; actor Michael Craig at his North Sydney townhouse where he told of how he wrote his angry film about urban development, *The Killing of Angel Street*; playwright John Romeril at his log cabin in the wilds near the old Victorian goldrush town of Castlemaine; and prominent magazine editor Megg Miller who in the Victorian town of Seymour talks about her influential publication, *Grass Roots*. In these and scores of other interviews we learn about each writer's environmental concerns and how these influence their work.

The chapters are as follows:

Fiction In this chapter we discuss every genre of novel with an environmental theme published over the past thirty years. Starting with Xavier Herbert's path-finding *Poor Fellow My Country* and Patrick White's *The Eye of the Storm* and *A Fringe of Leaves*, through to the most modern stories of the late 1990s, ecological concerns shine through. The genres encompass mainstream fiction, detective novels, sci-fi, futuristic tales, romances and quirky, offbeat offerings. The authors examined range from some of the most prominent names in Australian fiction—Peter Carey, Tim Winton, Thea Astley and Blanche d'Alpuget to name but a few—to lesser-known but extremely potent writers, who together chronicle the growing destruction of Australia—and our planet.

Non-fiction Australians have always treated their ecology in an exploitative and rapacious manner, and a wide cross-section of books has dramatically shown this. This chapter is divided into various specific categories which include books about nuclear testing in Australia, the effects of poisons and toxics, land degradation, coastal destruction from over-development, pollution and many other issues. It discusses the works of scientists like Tim Flannery (in his landmark *The Future Eaters*) and David Smith (*Continent in Crisis*) who have written definitive ecological histories of Australia; 'green' scholars such as Eric Rolls, whose *A Million Wild Acres* and *They All Ran Wild* have influenced generations of environmental activists and writers and conservationist struggles (not only famous battles such as the Franklin dam and Port Hinchinbrook but also scores of lesser-known but equally important movements); and historians like William J. Lines whose *Taming the Great South Land* and other books have defined Australian attitudes on ecological issues. These and many other non-fiction writers have woven a most articulate tapestry of environmental concerns.

Illustrated books Most of these books—despite their glossiness and colourfulness—provide in-depth analyses of almost every environmental issue of the past 25 years and are important reference works. For example, just on the Franklin dam dispute alone there are many dazzlingly attractive books by a range of prominent authors, such as Bob Brown and Helen Gee, featuring photographs by leading exponents of environmental camera-work (Peter Dombrovskis and others). This category also encompasses many well-known writers of other genres, such as best-selling novelist Nancy Cato and poet Mark O'Connor whose other works are discussed elsewhere. Within this vast and kaleidoscopic category are books which cover the environmental problems in regions as varied as tropical rainforests, the Great Barrier Reef, Antarctica and the oceans.

Magazines At the 'glossy' end of the spectrum is *Simply Living*, and of more basic appearance is *Grass Roots*, with a bookshelf-full of varied titles in between—all with a strong commitment to achieve a better environment. Then there are influential journals such as the Dick Smith-founded *Australian Geographic*. Several journals are specific to an area, for example *Bogong* magazine in Canberra. Each magazine is carefully examined in terms of its content, the influence it has had and how it perceives its role.

Films/Plays Films of every genre have also specifically or tangentially expressed ecological concerns: everything from the quirky cult movie *Cane*

Toads, to documentaries about the Franklin dam or Terania Creek, to international blockbusters like the Mad Max series and *Crocodile Dundee*, and diverse dramas and comedies. Other issues such as urban development are expressed in documentaries like Pat Fiske's *Rocking the Foundations* and in several dramas. A radical theatre emerged in Australia in the late 1960s and by early in the next decade the anti-Vietnam War movement and other major social issues also embraced environmental concerns. When the highly respected *Theatre Australia* magazine in 1981 listed its top twenty Australian playwrights at the time, the top five—Dorothy Hewett, Alex Buzo, Jack Hibberd, John Romeril and David Williamson—had all significantly dealt with environmental issues, as had other major figures (for example, Patrick White in his play *Big Toys*). Lesser-known playwrights such as Alison Lyssa in *The Boiling Frog* also created thought-provoking and influential material.

Songs/Poetry Just as 'La Marseillaise' inspired French revolutionaries, so did the songs of their day inspire the thousands of protesters at the Franklin and at Jabiluka. From the northern rainforests of New South Wales to the shores of the Franklin, the songs of Brenda Liddiard found voice, and her work is examined here. Many famous singers and song-writers (Peter Garrett of Midnight Oil, John Schumann of Redgum, John Williamson, Vince Jones and Eric Bogle) and less well-known names (Pat Drummond, Allan Caswell, Penelope Swales and others) are closely examined. Music from jazz to folk to rock to the classics has had a green tinge. Poets of every stripe have also expressed their environmental concerns, prominent among them being high-profile women like Judith Wright and Oodgeroo Noonuccal (Kath Walker), both of whom have also written non-fiction works on many issues. Les Murray is among countless Australian poets whose verses reveal a litany of ecological concerns. Aboriginal poets like Kevin Gilbert have included environmental destruction as one facet of their race's sufferings and injustices.

Children's According to a range of critics, Australian authors are world-class when it comes to literature for young adults and stories and picture books for the youngest reader. An increasing number of kids titles are published every year as a galaxy of 'adult' authors and specialist children's writers are instilling environmental consciousness into Australia's youngest readers. Books of the greatest possible variety are discussed, ranging from a Colin Thiele classic like *Storm Boy*, to the famous Jeannie Baker collages (*Window* and *Where the Forest Meets the Sea*), to picture books like those of Narelle Oliver and Jenny Wagner.

Each chapter vividly catalogues how hundreds of writers feel about the destruction of our ecology and what should be done. This is the only book where Peter Corris, creator of the gritty big-city Cliff Hardy detective novels, sits on a quiet island lamenting ecological vandalism; where Brian Hannant, co-writer of *Mad Max 2 (The Road Warrior)*, tells of how his work is influenced by the Gaia Hypothesis; where leading British television broadcaster Sue Lloyd-Roberts tells of how what happened at Maralinga was as devastating as the meltdown at Chernobyl; and where Allan Caswell, a singer/songwriter who usually entertains at pubs and clubs, tells of how he was persecuted because he wrote a song critical of French nuclear testing at Mururoa.

This book is dedicated to the proposition that if just one person is inspired by just one individual novel or play or song or poem or a non-fiction treatise to realise what is happening to our world, these writings deserve to be part of the national treasure.

1

Home truths, inspired dreams

THEY care about the environment, they care about their own quality of life and they care about the fate of the coming generations. They, more than any others, have a significant and enduring influence on us, and on our future. 'They' are writers, and over the past 30 years they have alerted Australians to the environmental problems we face. In novels, plays, films, poetry and song, as well as in ground-breaking non-fiction works, they have sounded the warnings but also rung the bells of hope.

Pollution has always been with us, yet for millennia there were physical limitations to humanity's capacity to destroy its life-supporting planet. However, the technological advances of the past century—accelerating dizzily with every new decade—now put the earth in peril.

As a nation, we cannot rely on politicians or corporations or conventional wisdom to face up to the challenges of survival. We certainly cannot rely on environmental organisations, be they vast or local, because these are often divided and unsure of their goals. Worst of all, the green groups are often unable to get their message across or win adherents to their worthy cause because they lack the simplest notions of public relations.

Corporations in Australia and abroad have discovered the importance of public opinion, and those wishing to plunder our nation's resources often seem to be winning the battle for hearts and minds. They have learned how to dress up what they do in fine colours; they have learned how to speak the language of environmentalists so that uranium mining, woodchipping and agribusiness can be presented as dynamic, lucrative and indeed essential activities that only a crank would object to—and a heedless crank, who cares nothing for the jobs of poor workers. Big government, big companies and big media are slick and professional while the green groups are often amateurish, relying on volunteers.

The roots of green consciousness in Australia were planted alongside the dramatic rise in cultural awareness which accompanied the advent of the Whitlam Government in 1972. Unlike the Whitlam Government, the green flowering proved enduring. From the late 1960s creative Australians of all sorts were coming into their own. Throughout this book we will discuss and analyse writers of every kind, and in this chapter we will meet some of them at home, where they discuss their fears and tell of their hopes. Some are well-known environmental campaigners, yet others may seem unlikely candidates. But their concern is emphatic, and their influence has been profound.

NOVELS

Fiction has always been a powerful influence on society. An environmental catastrophe may be reported on the TV news, but as soon as the program is over and the viewer returns to mundane tasks the local and global problems tend to be forgotten. But if a horror is described in a novel, complete with the human element and the emotional consequences, a reader is touched—and takes to heart what is at stake.

Australian novelists of every variety over the past thirty years have expressed their fears about our ecology being under strain. Classic novelists like Xavier Herbert and Patrick White have written about concerns which are shared in perhaps unexpected quarters—by the writers of crime thrillers, quirky romances and even 'pulp'. The environment is the equal of other potent issues of modern fiction like Australia's search for identity, reconciliation with the indigenous population, and feminism.

ONE of the earliest novels of the 1970s with a strong environmental theme was Xavier Herbert's *Poor Fellow My Country* (1975), which conveys a tentative grappling with ecological matters, as written by an author who had ambivalent and sometimes contradictory philosophies. Since his birth just after the turn of the century, in 1901, near Geraldton, Herbert had a close relationship with the land. His upbringing in Western Australia instilled in him what his biographer Frances de Groen describes as a sense of connection with the bush. From his earliest years he became interested in botany and science, and he would become a pharmacist. He also would develop a missionary zeal for all the things he felt strongly about.

Herbert's consciousness was honed throughout the 1950s and '60s, when he spent stretches of time on his own in the bush. 'This was a long period of introspection for Herbert as he grappled with an overpowering sense of failure,' de Groen observed in an interview in Sydney on 10 November 1997. 'He was having doubts about himself—as a man and as a writer—and he wondered if the next great work he felt stirring within himself would ever be born.'

In the late 1940s, Herbert spent days at a time in the Daintree, living on the royalties of his classic novel *Capricornia* (1938). He had also read extensively from books about Australian birds and other fauna and flora. 'A deep love of wild places developed within Herbert,' de Groen said. 'The bush became a solace from failure. "You're all right here," the bush seemed to say to him, and he saw himself as Zarathustra remade in the wilderness.'

Despite these welling conservationist instincts, in the years 1950 and 1951

Herbert was living near Cairns and making a living in commercial forestry—chopping down trees and cutting cordwood, an activity he recognised as 'blasting the bush'.

Although he was gradually realising the way in which the land was being abused, he himself tried to establish a property just like the exploitative Lily Lagoons of *Poor Fellow My Country*. He sold tantalite to the German industrial giant Krupp in the years before World War II, and throughout the late 1930s he mined for gold, tantalite and tin.

For the destruction of the Australian environment he blamed the British and their toadies, the Canberra government. But this seeming enlightenment actually stemmed from a not so enlightened source—the fascist nationalist movement in which he was peripherally involved.

But his hatred of British imperialism and the Australian cringing towards it would fuel his two major novels, *Capricornia* and *Poor Fellow My Country*. 'The white man—that is, the Briton—had raped Australia, according to Herbert,' de Groen said. 'In particular he abhorred the British atomic tests; he was appalled at how these buggers were allowed in by our government to come in and blow up our land.'

De Groen, author of *Xavier Herbert: A Biography* (1998), said this most idiosyncratic of writers had also developed a deep empathy towards Aboriginal culture and believed the original inhabitants were far more capable of tending the land in a proper manner. 'Herbert's environmental concerns flowed largely from his concerns for the Aborigines,' de Groen said, 'although there were a lot of inconsistencies in his attitudes and behaviour towards them too. Through personal experiences, the reading of anthropological books as well as Julian Huxley and H.G. Wells, he identified the nationalist imperatives of Aborigines as his own.'

All of these often confused and occasionally contradictory strands of thinking fuelled the decades-long process of completing *Poor Fellow My Country*.

'There was no consistency in Herbert's thinking,' de Groen said. 'He hated how grazing destroyed water-holes yet he was friendly with redneck graziers on their properties. He saw the sharp hooves of cattle as destructive forces yet he liked horses.'

Poor Fellow My Country was the product of Herbert's own experiences and the philosophies that emerged from myriad sources. His extensive research included gaining information on feral animals from Eric Rolls, while he read a wide selection of authors such as Percy Trezise, also a longtime friend.

As a book with an environmental theme, *Poor Fellow My Country* is described by de Groen as *of* its time rather than *ahead* of its time—something she attributes to most of Herbert's work. De Groen believes *Poor Fellow* had a huge influence in its day, especially among young readers throwing off the conservative shackles of previous generations.

'*Poor Fellow My Country* did a lot of good,' de Groen said. 'Xavier received a lot of fan mail and he was able to voice the concerns of the moment.

Many readers were prompted to join groups fighting for such diverse causes as the rights of Aborigines and republicanism.'

Herbert was *the* author for changing times, according to de Groen, and he became a darling to the growing number of readers who identified with the Australian nationalism of the early 1970s ushered in by the election of the Whitlam Government. '*Poor Fellow My Country* was the book of the moment,' his biographer contends. 'It wanted to say *everything* about Australia. As he said in 1972, *Poor Fellow My Country* would bestride the nation like a colossus and piss on it.'

Herbert's love of wilderness grew stronger throughout his life and in the years before his death he was spending long periods of time right away from civilisation. Frances de Groen notes that when Xavier Herbert died a copy of Rachel Carson's *Silent Spring* (1962) was found on his bookshelf.

Poor Fellow My Country would both stimulate and enrage its readership and pave the way for even more potent and powerful novels which took up its theme of Australia's environmental crisis.

THE ecological crisis in all its ferocity, combined with dark and ominous themes, pervades the work of Gabrielle Lord, whose sense of humour, even safely within the cosy comfort of her seaside Sydney home, has a macabre edge. 'I hope I didn't create the hole in the ozone layer by writing *Salt*,' she said in an interview at Clovelly on 26 May 1998.

Barely a kilometre away from the jagged Sydney coastline surrounding Clovelly Beach, Lord reflected on the urban horrors of the city's western extremities she depicted in the equally bleak *Salt* (1990) and *Jumbo* (1986). She spent almost two decades as a teacher and as an employment officer in that grimmest of regions on the foothills of the New South Wales Blue Mountains, and she discovered at first hand the hopelessness and squalor that afflicted young people. Above all, she experienced the oppressive heat and environmental degradation that she would portray as having worsened so horribly in the futuristic *Salt*.

'In the relentless heat I pondered what might happen if there was a thinning of the atmosphere, if the sun's radiation would go right through,' Lord recalled. 'One day I said to myself, "What would happen if it never rained again? What would we do?"'

Also affecting her were the everyday scenes at the Commonwealth Employment Service, where she would interview kids who were 'becoming more and more depressed about life's prospects'. These young people would go to the employment office, 'all enthusiastic and hopeful, but after months of disappointment, of being knocked back, would take the attitude "I don't care. I don't want your fuckin' job anyway". It was so sad to watch that process over and over again.'

Lord was surrounded by the realities she would so painfully and poignantly describe in the two novels: every day there seemed to be more smog; each day seemed to be yet another without rain; and there was less chance each time that an unemployed young person would ever find a job.

'I started to view all this hopelessness as inside a walled city surrounded by the plague,' she said, describing the genesis of *Salt*. Out in the backblocks of Sydney's westernmost extremity, from her home at Glenbrook and from the CES at Blacktown, she would look at the 'great black pall over Sydney' and wonder where everything was heading.

After selling the film rights to another of her novels, *Fortress* (1980), Lord was able to devote herself to writing full time and came 'back home' to Sydney's more fashionable and affluent eastern suburbs. In the beachside suburb of Clovelly, the buildings are the same chaotic, unplanned clutter that defaces much of Sydney's urban landscape. The view from Lord's workroom, in the attic of her surprisingly spacious terrace house, is of a jumble of red rooftops with the blue sea sparkling in the distance.

Amid the jumble of housing Lord was a jumble of contradictions. One of Australia's most inventive and powerful novelists stated repeatedly that she no longer read fiction. A flightiness in speech and manner belied a serious and thoughtful soul and a powerful intellect. While Lord perched on her chair in the half-lotus position of an accomplished yogi, her eyes darted restlessly and her fingers fiddled with tendrils of hair as she spoke.

'The idea of decaying civilisations is a recurring theme in literature,' Lord said, with a couple of quick cerebral references to Anglo-Saxon poetry. 'People have always been expecting the end of the world. But now, we're abusing things like never before, and greed and waste are combining as never before. This time our apocalyptic visions might actually come true.'

The extent of the horror as shown in *Salt* is best summed up by Lord's revelation that its publishers insisted that it be set in 2074—a further 50 years into the future from the 2024 of Lord's original manuscript—because anything closer in time would be too disturbing and unpalatable.

OTHER chroniclers of urban decay have opted to escape even further—to what seemed simpler and more tranquil places. Queensland's warmth and laidback lifestyle beckoned to two such writers—Peter Corris, Australia's most popular writer of crime novels, and his wife Jean Bedford, herself a notable practitioner in the gumshoe genre.

Coochiemudlo Island, one of the numerous islands of Moreton Bay which appear as a string of dots on the map alongside the southern Queensland coast, was home to some 400 people in 1998, among them Corris and Bedford.

'Coochie', best known for its tranquil, sandy beaches, is a pleasant place. There is but one sealed road which leads up from the main beachfront, with a few dirt roads branching off from it. Here houses are nestled, mostly inconspicuously, under sprawling trees and a clear subtropical sky. The little village is a comfortable and tight-knit community, served by a shop, a couple of cafés, and with its own bus, whose driver seems to know everyone and where they are going. There are a couple of blocks of modest holiday units and even a nine-hole golf course, which has views to make more prestigious clubs weep. Pleasant stretches of water separate Coochie from Brisbane's sprawling suburbs on the mainland and the other islands.

Corris and Bedford settled on Coochie after some time in the Illawarra region of New South Wales, but they are probably most identified with Sydney's inner west.

During an interview in their lush and sunny back garden in March 1998, Bedford and Corris told of their environmental concerns and the many influences on their novels, occasionally upstaged by the regular friendly incursions of their huge and handsome dog, Pancho.

Sydney with all its problems is the meat and drink of Corris's novels featuring his best-known creation, Cliff Hardy. Corris at first tended to be dismissive about his work, saying that his crime novels were written just to provide entertainment and escapism and were not 'from the heart'. But the more he talked about himself and his books, the clearer it became that his work was suffused by deep ecological and sociological concerns. After much prompting he acknowledged that 'it is a rare Hardy [novel] that doesn't have an aside about such things as development, gentrification, heritage at risk, pollution and aircraft noise'. And he laughingly said that it was a safe bet that none of his books would ever 'give comfort to conservative politics, organised religion or capitalist enterprises'.

Another Corris creation, Ray 'Creepy' Crawley, also analyses big environmental issues. Corris revealed how in preparing *The Japanese Job* (1992), a Crawley thriller about chicanery in Queensland, he travelled to Brisbane in search of colour and background and discovered the rampant development going on in that city. He was especially struck—and appalled—by what he termed 'a hole in the ground' being prepared for what was billed as the world's tallest building. 'I didn't feel that Brisbane needed such a thing,' said Corris, who said he incorporated his concerns into Crawley's thinking.

Corris recalled how living at Glebe had radicalised him on environmental issues after having been unconcerned by things political in his younger days. He confessed to being a 'complete swot and a nerd' at three campuses—Monash, Melbourne and the Australian National University—which at the time were hotbeds of political activism. He was at the periphery of the Moratorium movement against the Vietnam War and became 'mildly Left Labor'.

When he moved to Glebe he became interested in conservation and urban preservation, and he was especially inspired when the Glebe Estate was saved, mainly at the behest of the Federal Labor Government's Minister for Urban Development, Tom Uren. 'Had it not been for Uren, there might have been a freeway through the neighbourhood,' he said.

Jean Bedford became embroiled in controversies affecting the New South Wales South Coast, and her concerns are echoed in *To Make a Killing* (1992). This novel deals with how a neighbourhood can be irrevocably changed or destroyed forever by large-scale developments. The book reflects Bedford's own feelings about council subterfuge—and shows how difficult it is for a residents' action group to maintain its integrity because of infiltration, differing strategies and its vulnerability to manipulation by competing interests.

In the Illawarra, Bedford and Corris belonged to a citizens' group called SONS (Save Our Northern Suburbs), and they attended meetings and rallies and even joined picket lines. Unlike Corris, Bedford was in the forefront of the campus radicalism at Monash, and her concern with ecological issues developed naturally from her distaste for exploitation generally and the corruption she found everywhere.

Her concern with the environment originally stemmed from being a bush girl, brought up on Victoria's Mornington Peninsula. She likened her fight for conservation with all movements against a 'patriarchal society'.

'I'm against all sorts of exploitation, whether it be against women, or children, or the disabled, or against the earth itself—it all comes from a patriarchal, hierarchical society with no regard for the future,' said Bedford, who sees her writings as the most important way 'to do something about it'. She believes that raising consciousness is the key to changing things.

As they were joined by Pancho—at first seemingly huge and menacing, but in fact delightfully friendly and playful—and sighed at the peace and calmness about them, they both expressed 'a worry' about Coochiemudlo. Corris said there were always mutterings about developing the island—and some people stood ready to bitumen the whole place if there were a dollar to be made out of it.

But Bedford exulted at how Coochie was for the time being largely unspoiled, with its reserves of melaleuca and foreshores clear of buildings with height restrictions on all the other dwellings throughout the island: 'This is a situation unusual for Queensland. We can only try to make it last.'

Within months of this pleasant afternoon, Bedford, Corris and Pancho left Coochiemudlo—and Queensland. During the interview, when they still seemed ecstatic and positive about their subtropical island paradise, a few discordant notes had been struck. Bedford and Corris spoke about the innate conservatism of Queenslanders—a rigidity of attitudes and an absence of imaginative thought. All this as Pauline Hanson's One Nation party was in the ascendant, winning 11 seats in the Queensland State election of 13 June 1998 with an agenda based on fear, philistinism and xenophobia.

When a job offer came in October 1998 for Bedford to teach at the Southern Cross University at Lismore they grabbed the chance, and upped stakes to live in Byron Bay, another subtropical coastal paradise but with a wide cross-section of people and cultures.

A year after the original interview, Corris reflected on their Queensland experience in a telephone chat on 19 May 1999. 'We were in Queensland for almost two years, and the first year was a novelty and it was a pleasure to see our new home take shape on Coochie,' Corris said. 'But by the second year the conservatism and racism got to us. Until we lived in Queensland, we had understood "racist talk" to be just background banter, it even had an element of humour about it, but in Queensland it is ingrained and nasty. Queenslanders have serious racist attitudes, and we could never feel comfortable with it.'

NON-FICTION

The writers under the banner 'non-fiction' cover the widest categories—everyone from celebrities like Tim Flannery who have popularised environmental problems to low-key professors like Charles Birch. In this category are a biography of Bob Brown and an in-depth study of Antarctica at risk.

This genre embraces many classics of Australian literature, such as several books by Eric Rolls which have had widespread influence over several generations of readers, and more recent studies by William J. Lines, who for more than a decade has unleashed controversies that have aroused the strongest of passions.

EMBRACING a range of concerns is Sharon Beder, a prolific author who has written about poisoned waters and about the international PR battle for hearts and minds. From her cosy and cluttered office in the Science Technology Studies Department at the University of Wollongong, Sharon Beder can gaze out upon tranquil parkland. There are so many trees, and so many birds, especially water-birds which wade in the network of ponds and streams, that one has the feeling that a university has somehow sprung up in the middle of a forest. All is quiet there on a very hot afternoon between semesters.

Associate Professor Sharon Beder gained her doctorate examining the filth in Australia's waterways and oceans, so she quietly appreciates the cleanliness and creative landscaping of the university grounds.

What she does not appreciate are environmental irresponsibility, and on this she is far from quiet. She is vocal against any government or Public Service secrecy that allows or perpetrates ecological harm, and she decries the corporate publicity machines which churn out false propaganda. Whenever Beder is outraged by anything, she writes books, newspaper and magazine articles, letters to editors and talks to anybody in the media who will listen. She is the master of the sound bite and is a communications rottweiler in chewing out anybody trying to minimise or divert attention from anything she considers a pollution problem.

More than any other individual, Beder alerted residents to the scandalous state of Sydney's water pollution. Her landmark book, *Toxic Fish and Sewer Surfing* (1989), mobilised public anger against the way successive State governments had allowed Sydney's world-famous beaches to become a major health hazard. The book's sub-title, 'How deceit and collusion are destroying our great beaches', called for an end to public apathy and for decision-makers to become accountable.

As an engineer whose expertise was in the building, maintenance and politics of sewerage systems, Beder was able to counter government propaganda with indisputable facts. *Toxic Fish* was the culmination of her PhD thesis, which had been a kaleidoscopic overview of how decisions on sewer outfalls had developed since the turn of the century, and how 'engineering decision-making' had come about.

What Beder found was an unshakable culture of secrecy and manipulation, and inexorably this all led to *Global Spin* (1997) which dealt with, in the words of its sub-title, 'the corporate assault on environmentalism'. Quite appropriately, as Beder explained during an interview in her office on 14 December 1998, a book project originally looking at the sewer and muck pouring into the sea would later evolve into examining the propaganda sewer pouring out of politicians, government departments, corporations and all those organisations seeking to hide unpalatable truths from the public.

Beder—youthful-looking, fresh-faced, simply dressed and in appearance indistinguishable from most of her students—said *Global Spin* sought to expose international PR machines and the multi-million-dollar campaigns aimed at stifling legitimate green voices. She said the project stemmed from her belief that 'nobody was writing about this sort of thing'. On her previous research she had encountered one roadblock after another and she wanted readers to become aware of them.

She described a typical manipulative process. 'One time when I tried to get information from departments and instrumentalities like the State Pollution Control Commission [SPCC], they provided a very junior person to assist me; that was all very friendly but, after a while, they started to get concerned about what I was doing so they then sent out a middle-rank engineer to see what was going on. After I had been working through documents for about an hour they asked me to stop. They then wanted me to hand over my notes. I refused and put it to them that they couldn't make me hand over my notes. I was able to keep my notes but I wasn't allowed to continue looking through their documents.'

Further obstacles were placed in Beder's way as she tried to interview key engineers at the SPCC about their specific roles. 'Their guard immediately went up and I was denied access to internal memos,' she said.

During the several months that Beder was writing *Toxic Fish* she also became a spokesman for the group called Stop The Ocean Pollution, which was able to call on her as an engineer to counter the claims of the Water Board experts. 'That can be a problem with grassroots groups,' she said. 'The authorities can try to baffle everyone with science—so you need to have your own scientists, your own experts, to counter that.' Also as the book was being written, Beder provided bites of it to the media to maintain interest in the increasingly serious issue.

Another significant hurdle for Beder was the collegiate closing of ranks by engineers as documented in *Toxic Fish*. 'There is a code among engineers that they never criticise each other,' Beder said. 'While no one from the

Institution of Engineers accused me of getting my facts wrong, a senior member of the Institution questioned my attitude and threatened to make an official complaint that I was breaching their code of ethics.'

The New Zealand-born Beder, who travelled widely and became an engineer and taught at universities in Sydney before arriving in Wollongong, became a champion of clean beaches—and a hero to surfers—even though she rarely goes swimming in the sea. She became a champion of the wilderness even though she does not like bushwalking 'mainly because I have trouble with ticks'.

To Beder, fighting against pollution and preserving natural environments is 'a justice thing'. She described the way in which powerful interests had endangered public health and wellbeing during the sewerage crises as 'unjust, a desecration' and she felt the wider issue was 'a sociological rather than a technological problem'.

As we sat beneath a slow-moving fan in her tiny office, which was darkened against the heat of the day, Beder described industrial and urban pollution as among her main concerns. She said it was particularly essential to find and use new sources of energy. 'I'm trying to get a bigger picture on why as a society we are doing all this to our environment,' she said. 'Why do we allow corporations to get away with this?'

From her pleasant campus surroundings and from her home in the leafy Illawarra town of Austinmer she is the constant and compleat campaigner. At Austinmer her family is nestled between a hilly escarpment and a pleasant beach, but images of the destroyed and befouled coastlines of Sydney are never far away—the same images which brought 240,000 angry protesters to Bondi Beach in 1989 to tell their State Government that enough was enough. And now it was 1998, when for much of the year Sydney's water was unfit to drink.

Most corporations would stop at almost nothing to stifle the green voices, said Beder, who cited as a classic example that after the publication of *Global Spin* she had received a job offer from a New Zealand-based multinational to advise its chief executive on environmental matters. 'That happens,' she said. 'If you can't shut someone up you hire them, put them on your payroll, buy them off. That sort of thing tames a lot of activists.'

Global Spin is increasingly hard to get and most book shops don't stock it. Although Beder was reluctant to talk about conspiracies or undue influence, she conceded that it was 'strange' that bookshops had trouble ordering the book. 'I discovered in England that the playwright Harold Pinter had said it was his favourite book to come out that year, but I imagine he too would have had a hard time getting his hands on it,' she said.

Shortly after the conclusion of the interview, the heavens above Wollongong opened. The torrential downpour caught both interviewers and interviewee trapped in the midst of the sprawling campus. Such deluges have caused huge environmental problems throughout the Illawarra over the years, such as hillsides collapsing, dwellings submerged—and, of course, unpredictable sewage flows.

Even before the rain ceased, it can be assumed that Sharon Beder was on the phone, spelling out the facts to the media and pestering public officials.

SIMILAR torrents and harbingers of disaster were also on the mind of Percy Trezise, a legendary figure of Far North Queensland. Whereas Sharon Beder is your thoroughly modern person embracing the most up-to-date means of communication, Trezise is a crusty old soul who believes old-fashioned values and good old commonsense are needed to avert destruction. While Beder has at her disposal state-of-the art computer technology in her fight for the environment, Trezise admits to only 'the computer between my ears'.

Although in his late 70s, Trezise remains the quintessential rugged and vigorous individualist who is, among many other things, simultaneously a writer, painter, anthropologist, explorer, bushman and pilot. Speaking from Cairns, Trezise told in a telephone interview on 23 December 1998 of how the conservationist fight was a never-ending one.

Trezise has been a prolific writer, and as we spoke he was in comfortable, air-conditioned surrounds in Cairns writing some stories for children as well as doing a bit of painting. For a few months he was escaping the Big Wet surrounding his home in the Laura district, some four hours to the north-west in a four-wheel-drive along a bitumen track. He said he liked to be in Cairns at this time of the year because in Laura 'the marsh flies bite like crocodiles until April'.

The rest of the year is spent at Jowalbinna, his 40-square-kilometre property about thirty-four kilometres south-west of the Laura pub. He describes Jowalbinna as wild gorge country full of rock art and a scenic wonderland for bird-watching from where his son Stephen ran a tourist operation for 18 years.

Jowalbinna was described in evocative detail by Xavier Herbert in *Poor Fellow My Country*, and Trezise believes many of the campfire conversations between himself, Herbert and their accomplished Aboriginal friend Dick Roughsey found their way into the book and became touchstones for environmental awareness.

'Xavier Herbert was the first writer to make people aware of environmental issues,' stated Trezise in that firm and authoritative tone of voice which brooks no argument. Trezise told of the 'symbiotic' relationship between the three, who would spend 'many a Sunday arvo talking for hours over a barbecue and cheap aerated plonk' at the Cairns district of Redlynch where Herbert lived. 'Our writings reflected each other's opinions and we were great sounding boards for each other,' Trezise said.

Recalling those days of more than 25 years ago, Trezise proudly proclaimed that the environmental actions stemming from those yarns were 'very effective'. Among the achievements were rainforests throughout

North Queensland preserved under natural heritage guidelines.

Trezise admitted to many of the traits of the stubbornly distinctive North Queenslander and to a deep-rooted conservatism, and he scoffed at many of the attitudes and strategies of the green activists 'from down south'. But he also fiercely asserted that he was side by side with any 'greenies' who sought to challenge the 'greedies'. Trezise especially targetted Japanese logging interests and resort promoters.

Trezise fondly reminisced about his non-fiction book, *Last Days of a Wilderness* (1973), which appeared just before *Poor Fellow My Country*. Trezise said he, Herbert and Roughsey realised they were living in a 'beautiful part of the world' and did not want it to become like Victoria, 'much of which was almost used up'. He recalled how the creative trio had concluded that 'progress must not become destruction', that 'we all must realise the limits to growth' and, above all, every effort had to be made to preserve North Queensland rainforests.

At that time Trezise became the first chairman of the Cape York Conservation Council and he recalled one of its crowning achievements was to torpedo the first planned Japanese woodchip deal in Cairns. Trezise and his colleagues were able to convince local authorities about the harmful effects of such an industry, including the ruin of existing roads 'and just about everything else in the loggers' way'.

Trezise the adventurer summoned all of his skills to sway hearts and minds on this issue. Ten local authorities—including Cairns City Council—were involved in decisions on woodchipping, so he invited two observers from each and flew them and media representatives on a chartered 40-seater Fokker to see things for themselves.

Based on his own strategies of the early 1970s, Trezise—never short of an opinion on anything—had a few tips for the activists of the new millennium. 'To achieve change, get stuck in to your local council,' he said. 'The real power in this country resides in local government. If you're unhappy about what's happening environmentally where you live, keep changing your councillors. Local councillors have been the only people in Australia so far to be able to beat Japanese logging interests.'

If the lobbying of local government fails, Trezise boldly states, any group of citizens concerned about the fate of the forests would be quite entitled to 'buy ten 303 rifles, select some trees and just shoot them—put slugs into the trees so that loggers' saws would be broken'. He said he would be personally prepared to buy two 303s, 4000 rounds of ammunition and thus ensure no logger could cut down his chosen trees. He said he had learned about all this during World War II in Dutch New Guinea, where many trees were peppered with shrapnel and were thus of no further use to would-be exploiters.

Just as Trezise seemed all gun-happy and redneck, he was suddenly choked with emotion as he told how 'for a tree to attain the girth required by your average logger, it takes several hundreds of years' and that at some rates of exploitation in forests 'there will be no more millable timber for 500 years'.

Tourist development was of equal concern to Trezise, whose most recent fight was against the planned resort along Reservoir Road near Cairns. To Trezise, this scheme is ridiculous as well as reckless. 'The resort is planned along the last ten acres of vine scrub in the Cairns region,' he said. 'It's too wet there for sugar-growing yet they want to make it a holiday haven.'

Returning to his other pet theme, his disdain of the southern green activists, he said many of the mainstream environmental groups had 'lost touch with reality'. He contrasted their ineffectiveness with his own 'very pragmatic' Cape York group. 'Greenies should remember that they can't save everything, and so many of them are just bleeding hearts.'

Trezise—still prickly and unrepentant after all these years—again proudly paid tribute to his mate Dick Roughsey and said everybody should take heed of Aboriginal law and realise that land is all that we have to hand down to our children, and we should take care of it. Trezise said we'd all be much better off if we followed the Aboriginal maxim of 'take what you need, need what you take'.

ILLUSTRATED BOOKS

In stunningly produced and meticulously researched books that sit on coffee tables throughout Australia, the beauty of the Franklin River in Tasmania and the majesty of the Daintree or the Great Barrier Reef in Queensland is available to all. The spectacular visual presentation and detailed, often passionately written, text together have galvanised readers to preserve these precious places.

SITTING at his home in Sydney reflecting on his experiences in remote places is Alasdair McGregor, one of the finest practitioners in this genre. His books, *The Kimberley: Horizons of Stone* (1992) and *Australia's Wild Islands* (1997), both of which were collaborations with Quentin Chester, are thoroughly researched, stylishly written and beautifully presented guides to some of Australia's most magnificent wilderness areas.

McGregor says he often resents the way in which the illustrated book genre is so often not recognised as serious and worthwhile scholarship. In describing what it takes to produce one 'coffee-table' book, he told of the awesome logistics and incredible discipline needed. He reeled off just some of the attributes he needs to be able to visit and report on remote regions: he is simultaneously a writer, photographer, painter, trekker, yachtsman and organiser. He has to be able to call on all this knowledge and experience on rugged and sometimes dangerous expeditions.

There is also the small matter of being able to re-enter society smoothly, as he explained during an interview on 8 December 1998 at home in the fashionable suburb of Neutral Bay on Sydney's northside. All of the places he had travelled to are worlds away from his townhouse with its tiny walled garden. McGregor was in the process of renovating his 100-year-old house so we sat amongst elegant clutter, including some of his wondrous landscapes of Antarctica, most of which were still stacked against the walls. A few months earlier he had returned from the frozen continent and the contrast between the peace and isolation he had left behind and the hurly-burly of Sydney was so overwhelming that he fell off his garage roof after having emerged from Antarctica unscathed. 'I guess I was woolgathering,' he said, 'though I wouldn't do that on an expedition.'

'I was out of sorts for some time after returning from Antarctica,' McGregor recalled. 'I had just spent many weeks in the area around Sir Douglas Mawson's hut, in the eastern sector of Australian Antarctica, which is the windiest sea-level place on the planet and where there are constant blizzards.'

McGregor said his worst culture shock happens when he gets back to the city and realises again that the issues on the minds of most people are 'absurd and unimportant'. He said people in the cities were consumed by things which didn't matter, things that 'mean nothing when you contrast them with the big issues like the state of our oceans and climate change'.

'This planet is full of complexity and delicacy and we're destroying it before we even understand what we're doing,' he said. 'This is our life-support system, and we aren't looking after it properly.'

Inside his townhouse, McGregor was able to gather his thoughts about his constant travelling between 1993 and 1995 to research *Wild Islands*. He had already been to some of the remote areas—such as to Heard Island in 1983—and he described Heard as the 'wildest, most moving, most savage' place he knew.

McGregor was often away from home for three months at a time, spending about seven weeks in places like Heard. He said he preferred to travel around Australia 'in bites'. These bite-size chunks included such mammoth enterprises as sailing north from Cairns across the Torres Strait, then across to the West Australian coast with landfall on the mainland at Wyndham.

McGregor told of making four visits to the Kimberley before he collaborated with Chester in bringing the region to glorious life. The making of *The Kimberley* involved the occasional mishap, such as getting airsick while flying in storms during the wet season, landing on a makeshift airstrip that used to be a bauxite mine, and making a half-day trek to a creek and back just to get water—and then being drenched as the heavens opened.

During the interview, McGregor heaped one environmental concern upon another, with habitat and species loss foremost on his mind. He was worried about the state of the marine environment around the Australian mainland and its thousands of islands and he described, just as he had done in *Wild Islands*, how 'every island I've been to had heaps of stuff washed up on it . . . everything from whole fridges to thongs'. He told of medical waste being found on Christmas Island. Even Heard Island, the remotest of the remote, did not escape trashing by humans.

McGregor is concerned that within a century some of the low-lying islands may be submerged if nothing is done about global warming. Of more immediate worry is the fate of turtles amid the onrush of 'progress'. He spoke passionately about how his illustrated books—and indeed his other writings, art and photography—sought to 'heighten sensitivity and sensibility' to all of these problems.

McGregor said one of the keys to influencing people's attitudes and perceptions was to provide accurate information backed up by knowledge, to 'know what you're talking about'. In this quest, he always sought to examine the changes in perspectives about individual places, such as what Flinders, or the Dutch, thought about a particular location.

Alasdair McGregor's writing, painting and photography have become more prolific throughout the 1990s and his articles have appeared in such journals as the *Australian Financial Review*. He described this ubiquity as 'touching people, and hoping to make a difference' in how people treated their fragile environments. 'There are hundreds of issues to get across,' he said.

After the interview, McGregor took us up to his painter's loft. In this light-filled room, several large canvasses in various stages of completion stood on easels and against the walls. The contrast between McGregor's two worlds—the icy and remote majesty of Antarctica and the bustling crowds of North Sydney—could not have been more stark. Past the wonderful paintings of ice floes and rocky shores with penguins, we could see all the shoppers and buses and even the Oaks tavern, a place that always seems to gather a crowd. A faint roar of traffic was still audible through the double-glazing. It is to be hoped that some of those rushing around outside might find a similar moment of peacefulness.

MAGAZINES

Newspapers are a reader's companion one day and a recycling problem the next, whereas magazines have always had a more lasting value. In any newsagent's the magazine racks embrace every conceivable interest, and the environment holds its own in the concerns of the average Australian.

Specialised Aussie mags dedicated to influencing hearts and minds on ecological issues range from the most basic to the most glossy, and each has its worth and strength.

CHANGE oneself, see the world differently, and thereby change the world. This is the simple yet awesome philosophy that has guided *Grass Roots* magazine for just over a quarter of a century. And the guiding light for all those 26 years is Megg Miller, who has established a genre and nourished it. Megg Miller herself talks of having spent her entire 51 years 'thinking about the meaning of life'. She says she often ponders the big questions—usually late at night—and wonders, 'What are we here for?

What is it all about?' Her very name, Megg, symbolises who and what she is: 'Margaret' became 'Meg' became 'Megg' as a result of an egg flying through the air and landing on her lap without breaking as she sat near a city street corner. 'To me it seemed a miracle, a miracle of life,' Miller recalled. 'Some hoons had thrown some eggs in my direction, and another egg had broken in the gutter.' From that moment on, several years after *Grass Roots* was launched, the 'egg' in 'Megg' symbolised the meaning of life: 'The egg is something complete, something self-generating. You look at its smooth surface and it tells you nothing—but all of life is contained in it.'

Grass Roots was born in 1973, the brainchild of Megg and her then husband David whose shared idealism was turned to glorious reality in the Victorian countryside. From first Shepparton and then Euroa in central Victoria the *Grass Roots* message radiated to every corner of Australia, and from every corner the ideas flooded back, and a networking phenomenon took hold. *Grass Roots* has become more than just a magazine for its regular readers—it's a guidebook for those on the land, a source of inspiration and knowledge and, most of all, it's a source of community for those who may feel isolated, away from the props of city life.

The headquarters of *Grass Roots* magazine is now the regional centre of Seymour, just under a hundred kilometres north of Melbourne. It was in a small and nondescript coffee shop in the town's main drag that Miller met the authors on 27 May 1999 to talk about the magazine and her life, or, to put that more aptly, the magazine that *is* her life.

Seymour (population 6558 in 1996) is a charmless place, a sad and soulless jumble of buildings on a windswept plain. The district has grazing and grape-growing, and the manufacturing industries are said to include timber-milling, an agriculture engineering works, river gravel quarrying and wine production, but on the blustery main drag there are few signs of commerce or enterprise. While we waited for Miller, we walked up and down the main street for a little while, examining the displays in shop windows—tacky, shoddy merchandise, unappetising food, ugly furniture. Over lunch, Miller confirmed our feelings about the town, confessing that it 'doesn't have a heart'. She told of the difficulties in attracting *Grass Roots* staff to such a place, and she herself does not live there. Her home is on a pleasant property 'about three-quarters of an hour by car' along the Golden Valley Highway, well away from Seymour.

The main reason *Grass Roots* is based in Seymour is logistical and is a geographical compromise between Megg and her ex-husband, who maintains an advisory role in the magazine. It is a place from which parcels, mail and the magazine can be despatched and delivered with ease, and it's convenient in other ways. But Megg is not exactly happy at Seymour, and only her love and total commitment to the magazine keep her there.

Miller described Seymour as a refuge for some of Australia's more desperate citizens, such as unemployed vehicle workers relocated from down the railway line at Broadmeadows to the army houses farther north. In Seymour, those with jobs mostly commute to Melbourne every day—an hour each way by train. Seymour's refugee status was made even more poignant by the welcome signs in windows throughout the town to the recently arrived Kosovo Albanians given sanctuary at the Puckapunyal army base, eighteen kilometres to the west. We had noticed these signs, and had actually taken them for advertisements for some sort of Gaelic festival, perhaps because there was nothing foreign or exotic about any of the faces we saw in Seymour—and the language was decidedly foreign and exotic. Signs aside, it is debatable what sort of welcome the refugees would have had in Seymour. The mood in the town was sceptical at best—and among such poor people it may seem hard to be asked to share.

Miller herself has a vibrancy and an inner glow which these dismal surrounds have failed to quench—and which speak volumes about the healthiness of her lifestyle. She is a vegetarian (she says she spends too much time with animals to be able to eat them), a life-long 'sporty person', and regular farm chores have kept her very fit.

Megg Miller is well-travelled and cosmopolitan, self-confessedly more at home in the sophisticated cappuccino circuit of Melbourne than the basic café at Seymour; she is extremely well educated, being a translator of foreign languages and a trained teacher; she is an expert and worldwide authority on pure-bred chickens, regularly travelling to overseas capitals to spread and receive the latest information.

Above all, even amid a town with few prospects and a struggle to survive, Miller constantly dreams of better things—and she shares the dreams with her readers and wants the readers to share their dreams with her.

Megg Miller's odyssey which took her to Seymour passed through Melbourne and some very contrasting towns of rural Victoria. She was born in 1947 at Warrnambool, the coastal town at the western edge of the Great Ocean Road. Miller said her grassroots philosophy was forged by her upbringing. She described her family as 'very working class' and thrifty, leaning towards self-sufficiency. She said of her father that he could do anything—carpentry, shoemaking, any sort of repairs, electrical work—and she would regularly join him in the garden, learning about many of the natural wonders she would later share with her *Grass Roots* readers.

Warrnambool, a very wet and windy provincial town which often cops the full blasts from Antarctica, had a severe effect on her asthmatic brother, so the family left when Megg was aged seven. They moved to Numurkah, just north of the regional centre of Shepparton; Miller described Numurkah as a basic Victorian country town at that time populated by soldier–settlers. Miller recalled that even at high school she was part of her family's organic way of living and that 'I never tasted processed food until I left home'.

From her parents, whom she described as 'doers', Miller learned a range of skills—carpentry, how to sole a shoe, sewing, cooking and preserving. When she left home she went to Melbourne Teachers College to expand her knowledge of arts and crafts, later becoming an art teacher.

It was during her tertiary education that she met David, who would later attend agricultural college when Megg began teaching at Shepparton, further creating a 'synergy' from which *Grass Roots* things would flow. Megg and David lived a simple rural life surrounded by horses on agistment as well as about 50 of their own sheep. Megg and David, as a young courting couple, had driven around buying up coloured sheep, long before they became fashionable, because Megg liked the different coloured wools for her spinning and weaving and other craft projects. They had bought the coloured sheep very cheaply from baffled farmers who hadn't really known what to do with them. In those days, only white wool was saleable. From this start grew Megg Miller's interest in such things as gene pools and the need to sustain variety.

Megg recalled how she and David were in the vanguard of many of the social movements which would find such confident and clear voice in *Grass Roots*. She described how in the early 1970s she and David 'got in on the ground floor' of the arts-and-crafts and other movements which would radicalise young Australians on many issues. 'There was very little environmental concern in Melbourne in those days,' she said. 'It was too early to think that we'd run out of anything. Only after 1969 and the Vietnam Moratoriums did thinking youth come together and start questioning things.'

Like many of their peers, Megg and David arrived at environmental consciousness through circuitous routes. 'The draft for military service was more important than the environment back then,' Miller said. 'And to me and David issues to do with handicrafts and heritage were of more pressing importance than ecological matters.' Despite their involvement on campus in the radical Students for a Democratic Society, Megg recalled that she and David were 'not politicos in any way'.

The genesis of *Grass Roots*, therefore, was in plans to write a book about 'the old ways', such things dedicated to the notion of self-sufficiency as bread-making and how to spin and weave one's own clothes. That book never happened (there would be other books later), but *Grass Roots* did.

From Issue No 13 onwards, Megg took over design and played a more active role in determining the magazine's message.

Miller stressed the importance of reaching out beyond the narrow band of rural craft types. 'Making contact with people everywhere is most important,' she said. 'People in the big city are sustained by reading about country matters—even if they have very little direct contact with such things.' She saw such reaching out as a vital counter against the 'alienation' facing an increasing number of people today, a trend which Miller saw as 'most damaging'.

Despite all the difficulties in producing the magazine, *Grass Roots* is never short of material for publication, with Miller effusively proclaiming that at any one time she has 300 articles from which to choose. 'We have a wonderful filing cabinet full of stories—mostly from our readers, who often become regular contributors,' she said.

Miller speculated about why this was so—readers of other magazines don't become contributors en masse—and she thought it had something to do with *Grass Roots* being a 'family'. She said *Grass Roots* wanted people to feel that they had a part in the magazine, that 'it reflected their lives, their concerns, their feelings'.

'I guess the fact that they want to share their ideas with other *Grass Roots* readers means that they do feel this way,' she said. 'I hope so.'

Megg Miller feels strongly about battery hens, writing an article in *Grass Roots* strongly opposing the practice. More than anyone, she has popularised backyard poultry keeping. She proudly and eloquently spoke of how she and *Grass Roots* 'changed the face of poultry keeping' and dramatically swayed consumer attitudes towards the types of hens kept.

'My one achievement on this earth' is how Megg Miller described the changes in the keeping of backyard hens in Australia and beyond. 'I know that some of the poultry I have helped to popularise would have just died out by now—we would have lost those breeds forever.'

But every month new issues are raised in the pages of *Grass Roots*, and her achievements can in fact be multiplied many times over.

AN INNER glow, a power and a passion burn as fiercely in the very heart of Melbourne as in the Victorian bush as Chris Baxter approaches almost twenty years at the helm of his magazine, *Wild*. Surrounded by the fashionable cappuccino meccas of Toorak Road on one side and Prahran on the other, Baxter, a true child of the metropolis, turns his mind to Australia's—and the world's—wildest and remotest places.

Where Megg Miller has a wispy, quiet and unobtrusive strength, Chris Baxter is a full-on giant. Exact statistics of his height and weight are irrelevant and would not do him justice; suffice to say he fills the room not just with his man-mountain physical presence but with the grandness of his dreams and the formidableness of his achievements. Here is a guy who not only might conquer mountains, overcome geographical hurdles and tackle all of nature's elements but who does so regularly.

Since 1981 Baxter has been sharing his feelings with the readers of *Wild*. If a just-discovered wonder of nature excites him, he and his kindred band of writers and photographers tell his readers all about it; if he is appalled by an environmental outrage, his inner fury is never kept inside for very long.

The pages of *Wild* are suffused by the parallel rejoicing at nature's splendours and anger at its vandalisation that have forged Baxter's personality. The earliest influences of his upbringing and youth combined with the mounting concerns of adulthood are the stuff of *Wild*.

Baxter, born in Melbourne in 1946, spent 1961 at the exclusive Timbertop boarding school (also once attended by Prince Charles) in the high country of north-eastern Victoria. Thus, as a 15-year-old, he witnessed the start of the logging industry in the Victorian Alps, an experience that would have a lasting impact. 'I became very aware of the clear-felling on Mount Stirling, and I saw the damage done by rampant ski developments, and the roads that followed,' Baxter recalled in an interview with the authors at *Wild*'s South Yarra offices on 28 May 1999. Baxter's eyes were also opened by the increasing number of four-wheel-drives invading the area as tourist villages were opened up at resorts like Mount Buller. But the logging was the main concern for the young Chris. 'They would log on one side of a creek and then proceed to the other side,' he said, adding that while this was happening there was 'no interest in the curriculum at Timbertop about environmental issues'. Deep inside him, the schoolboy felt these stories needed to be told, and the first *Wild* seed was planted in his mind.

His inner feelings were further stirred up as he proceeded to Monash University, Australia's hotbed of student radicalism in the late 1960s. At Monash, Baxter was on the Students Representative Council at the same time as Australia's best-known student agitator Albert Langer, and he worked on the *Lot's Wife* student magazine with another key protester, Pete Steedman, later to become a controversial federal Labor MP. While he shared the anti-Vietnam and anti-Establishment stances of the Monash radicals, Baxter turned to Christianity rather than the mainstream left-wing agenda. 'At Monash, I began my life as an active Christian,' Baxter said. 'From this stemmed a further concern for the environment, as Christian teaching is to preserve and care for what's on earth. Christianity teaches that we are stewards, not owners, of the earth, and our duty is to nurture it.'

Meanwhile, Baxter's growing spirituality became linked with his love of the outdoors and a spirit of adventure. These partly grew out of being one of four boys in his family who early in life discovered the joys of bush-walking and rock-climbing.

After graduating in Arts and gaining a Diploma of Education, he taught at school for two years but his main interest remained with the outdoors—and with writing about and photographing it. Then Baxter travelled overseas, where his concerns crystallised even further. In Britain and the Continent he saw for himself the population pressures on resources and how 'there was no timber left in the British Isles'. In Europe he saw the 'degradation' of the Alps, such as the construction of ski lifts in the Dolomites, amid a development ethos which he saw as 'disgraceful'. But he also witnessed some encouraging signs, a few glimmers of hope; in Britain he was impressed at how industrial cities like Sheffield were preserving heritage areas, and how the National Trust bought up tracts of land to lock it away from developers. He saw such trends as 'a great step' for civilisation and they had an abiding influence on him.

Returning to Melbourne at the end of 1972 all fired up and wanting to break into journalism, he quickly came to realise that at age 27 he was too old for the ABC cadet intake. At that moment his father, who ran a business consultancy, brought him into the firm. For the next seven years, Chris received very hands-on training about every aspect of running a small enterprise. 'I learnt more in this time than I did at Monash,' Baxter noted, and he gained invaluable insights into how *Wild* might survive in a forbidding publishing environment which devoured most independent magazines.

Baxter took time off without pay to learn about the publishing industry, the formats of magazines, how advertising worked, about marketing and demographics. He also did research at such reference points as the US Library of Congress and the Sierra Club, and he discovered how American and British publications handled what he was seeking to do: put out a magazine with an emphasis on rucksack sports with a small environmental element to it. 'All my research showed me that the magazine I had in mind needed a high recreational element in it to survive,' he said. At the start of the 1980s, with less than $6000 in the bank and a cluttered front room at his Prahran home as his headquarters, Baxter decided to bring *Wild* to life.

In the months after September 1980, when Baxter returned from the USA, he put all he had learned about small business into practice. He had little money, was paying off a mortgage and worked out of that little room which had kitchen rather than office chairs, a bed, a piano and a labrador dog. An answering machine was the only sign of the more sophisticated times ahead. Despite the cash shortage, he had standby air tickets at the ready to enable him to fly up and down the east coast to promote the magazine, the 'expensive' pages of which were already dummied to show advertisers what was possible.

Baxter was working seven days a week, sometimes through the night, to ensure that the first issue *was* good and *looked* good. 'People could see it had promise, and we had to prove that,' Baxter recalled. The issue of June 1981 had a print run of 8,000, of which 1,000 were pre-sold 'very cheaply'.

From the first issue, environmental matters were given prominence. There was a 'green pages' section as well as regular features on ecological issues and in-depth profiles of environmentalists such as Bob Brown. Baxter also estimated that between one-third and half of all editorials in *Wild* were 'pro-conservationist'.

In pursuing such a strong environmental line, Baxter boldly risked alienating some of the rugged individuals who made up some of his readership, but he was never concerned. 'Any representative sample of letters to *Wild* express strong pro-environmental opinions,' Baxter said. 'And everyone who writes for *Wild* would be simpatico to such opinions.' Every facet of the magazine reflects Baxter's attitudes; for example, conservation organisations pay only one-third of the normal advertising rate, and Baxter insists that only 'ethical' ads be accepted. As part of this policy, Baxter does not allow ads for off-road vehicles or for any tourist organisation which he believes acts irresponsibly in wild areas. Baxter hands over a percentage of the magazine's turnover to conservation groups and to organisations like the Wilderness Society and World Vision. And as a further concrete expression of his environmental concerns, the magazine is printed on paper that is 50 per cent recycled.

'I try to make *Wild* appeal to as wide a cross-section of people as possible,' Baxter said. 'And I have tried to make it something of substance in what is often a superficial, fairy-floss-like world.'

In his now well-equipped electronic office one storey above bustling Malvern Road, a short hop from that little room with the labrador dog and an equally short hop to the café hub of Chapel Street, Baxter is the quintessential Melbourne lad but one who's concerned about ecological realities far away.

He told of his greatest concern still being the logging of native forests, especially in the Victorian Alps that were already being denuded when he was a lad at Timbertop. He said he was also 'horrified' by such things as the fate of the Queensland coast and the Daintree forest, the 'scraping of the bottom' of the sea floor, and the far too many 'environmental hot spots that cover the map of Australia'. Baxter, who was actively involved in the struggle to preserve the Franklin River, saw a similar battle looming over the Wongungarra river valley in north-eastern Victoria which he described as 'the last such place without a road running through it'. To support this cause he published a photo essay on the Wongungarra, to put increased pressure on the Victorian Government to stick to its policy of conserving this region.

Despite an increasingly bad back, Baxter still lives the life of an adventurer, often going rock-climbing throughout Australia and trekking to faraway places like the remote regions of Ethiopia. He says he can't carry a 10-day pack any more, but anything that his formidable frame is up to he will still attempt.

'I can't help feeling humbled by the bush,' he said, encapsulating in the simplest of words his ecological and spiritual philosophy. 'The bush is something special. It's something fragile. It's something that's diminishing.'

The man-mountain that is Chris Baxter has an air of indestructibility about him, as he keeps his ambitious dreams alive. Only with such strength, commitment and vigour—and a potent message such as *Wild* promulgates—can our threatened environment gather defenders who are up to the task against seemingly insurmountable odds.

FILMS

Since the 1970s, Australian films of every variety have embraced environmental themes either directly or tangentially. From fine documentaries to highly cerebral features to 'pulp' entertainments there has been a deep concern for the ecology. The *Mad Max* movies, despite being mainstream popular features, depict scenes of the apocalypse; Brian Hannant, co-writer of *Mad Max 2*, subscribes to the Gaia Hypothesis and has intense worries about the fate of the planet, concerns which he translates to the screen.

On the screen, powerful documentaries tell of Australia's subservient role in atomic weapons testing, and of the trashing of the Pacific Ocean. And films ranging from a comedy like *Crocodile Dundee*, to a heart-warming drama like *Storm Boy*, to a sad and overwhelming work like *A Street to Die* touch audiences and inspire an awareness of lost belongings.

THE environmental concerns of popular film and television actor Michael Craig go far beyond the many potent issues he incorporated into *The Killing of Angel Street* (1977), whose screenplay he co-wrote at a time when Sydney was a battleground between residents and developers.

Craig became deeply interested in the many human elements of the deadly struggle in which anti-development campaigner Juanita Nielsen was murdered, Jack Mundey and his faction of the Builders Labourers Federation were crushed for placing green bans on developments which were destroying heritage buildings, and ordinary citizens were either forced from their homes or faced the appalling consequences of staying put.

During an interview at his North Sydney home on 27 May 1998, Craig chatted amiably yet fervently and knowledgeably on many local and global environmental concerns.

He told, for instance, of how for thousands of years, since before the days of the Roman Empire, the Mediterranean coast of Libya had been a fertile grain basket. But in recent generations this land had died, the desert creep of the Sahel now extending to within ten kilometres of the sea. Craig cited several factors for this, mainly new technologies which allowed mankind's destructiveness to take effect 'easier and quicker'.

'It took 400 years to build Notre Dame; now they could throw it up in two years,' he said.

The manic zeal of unfettered development is at the core of *Angel Street*, and Craig said he tried to show the intense drama of people who get into deadly trouble by taking a stand.

Craig collected some of his material from a reporter who had covered Juanita Nielsen's disappearance, and quickly became aware of how dangerous was the subject with which he was dealing. He admitted that the 'third, final draft' for the film was a significantly watered-down version, and that what appeared on screen was a 'compromise'. At one stage the script was being read or rewritten at various points of the globe, including in Wales by popular British actress Julie Christie, who had been originally intended to play the lead female role.

Despite the final version being 'slightly blurred and softened', Craig received odd late-night phone calls warning him off. Despite the compromises (and in the end Elizabeth Alexander replacing Christie in the main role), Craig said the film was 'very political and anti-government' and graphically showed to film-goers what the Victoria Street struggle entailed.

The English-born Craig, who appeared in many British films of the 1950s and '60s, became a well-known face in Australia through TV series like *GP* (for which he wrote three scripts), other films and the theatre. He described himself as typical of the Australian acting fraternity—'that is, mainly left-wing and progressive'.

At his comfortable townhouse overlooking Lavender Bay, Craig looked out in disgust at the 'unholy mix' of buildings jammed together throughout the North Sydney area and he decried the many 'awful' buildings such as the Blues Point Tower which he said 'sits like a wart' on the tip of McMahons Point.

He spoke ruefully about the rather new townhouse in which he lives. 'Oh, it all looks very comfortable and nice here,' he said, 'but a lot of things don't work properly and the builders have been unwilling to come back and get it right.' The townhouse is smart and elegant but feels unlived-in, the atmosphere of a home that is frequently left for long periods and where things are never completely unpacked.

The longer Craig spoke, the more his concerns multiplied. He expressed alarm at how multinational corporations were now large and powerful enough to sue national governments, and invariably got their way regardless of the public interest. The wishes of these governments and a nation's citizens could be brushed aside, just as the developers of Angel Street were unconcerned about the feelings of the residents they were evicting and the homes they were obliterating.

He believes the world will have to endure decades more of gas emissions into the atmosphere because the giant oil companies, even though they have alternative sources of energy in place, want to ensure maximum profits from the current fuel supplies. He was scathing of the Australian Government's ineffectual stand on reducing fuel emissions and he described any short-term improvements as a 'bleak prospect'.

As he spread his arms to indicate a vast land with vast problems, Craig told of what he had personally witnessed on his travels throughout Australia: hillsides denuded in the Yass region near the Australian Capital Territory ('brown, useless grass on barren land where the trees are gone'); the degradation of river systems, especially the effects of cotton growing at Moree in north-western New South Wales; and, high among his concerns, the unwanted changes foisted on Sydney by the 2000 Olympics.

'Apart from some tidying up and a spruced-up image in some places, the Olympics are an unwarranted interference for something which will only last for two weeks, and financially we'll be like Montreal which has never recovered from the 1976 Games,' he snorted. 'Also, chunks of the harbour will be reserved for Olympic traffic, while Sydney's real infrastructure needs are being neglected.'

Craig said he especially regretted the Olympics because they were diverting public funds from greater priorities such as changing the current method of Sydney's water supply, whereby if 100 millimetres of rain falls over four days, 95 millimetres of it is lost straight into Sydney Harbour.

He fervently called on local government to make it mandatory that all new houses be solar-powered, and that these dwellings be allowed to have water tanks.

Craig's musings were not entirely those of despair and he was heartened by some international trends such as wildlife returning throughout England.

'Given a chance, the countryside can come back,' he said. 'Similar changes could happen here, but it may be hard to persuade many Australians to alter their thinking and practices.'

He sighed with an innermost hope that there were enough Australians who believed that, environmentally, things could not continue as they were and that there would be a backlash against such things as economic rationalism gone mad.

As we left Craig to prepare for a new film role, he lit a pipe and settled down again in stylish comfort, radiating the conviction that Australians would realise that quality of life and pleasantness of surroundings count for something.

ON THE far north coast of New South Wales, nestled in the Koonorigan Range about seventy kilometres west of Byron Bay lies the little village of Nimbin. Formerly a sleepy little dairy town, it was catapulted to fame—or notoriety—in the summer of 1973 when it hosted the Aquarius Festival in Australia's own summer of love. There have always been 'alternative' types in Nimbin, people drawn there by the warmth and sunshine of the north yet determined to stay well clear of the hostile right-wing territory associated with Joh Bjelke-Petersen's Queensland.

After Aquarius, Nimbin became the hippie capital of Australia, and so it appears to be to this day. Still a small place, with a population of only about 250, it is estimated that more than 12,000 live within a ten-kilometre radius.

Not far from Nimbin, at the end of a long, winding and bumpy road that leads up into the cool green hills, lies the elegant homestead of film-makers Jeni Kendell and her partner Paul Tait. Set on a high ridge, their original farmhouse overlooks a vast spread of fields and forest with views all the way to the Queensland border. Wildlife is all around them. A constant parade of birds can be seen through the huge windows—cockatoos, black parrots and rosellas all paid us visits the day we were there—and wallabies come daily to feed on the lawn.

Kendell and Tait can smell the lush foliage outside their comfortable home, much of which has been rehabilitated by them over the years. 'All the trees were gone when we moved up here,' Tait said. 'People had been trying to farm and had cleared all this ridge. The land was degraded and it was a real mess. But we planted seedlings and native shrubs and now look at it. The trees look like they've always been here, don't they? It does give us some hope that the earth, given half a chance, can heal herself like that.'

Inside the house there radiates a glowing warmth and comfort. Light catches coloured glass ornaments in the windows and reflects warmly on the timber walls; deep couches and rich carpets on solid timber floors add to the feeling of cosiness and of being gently sheltered from the surrounding forest. But the forest is a part of the house too—a central support of the roof is a huge tree trunk. There is such an atmosphere of permanence and stability it surprises you to remember that at any moment Kendell and Tait may be off on their next project to almost anywhere in the world.

Their life revolves around film-making and seemingly nothing can shift them from their path. Kendell said they had tried to leave it behind when they left the city. 'We'd both been at the ABC for years,' she said. 'We wanted a change. We wanted to get out of the rat-race of city life and get back to the land. We tried to grow avocados. But our land wasn't right for avocados—and people kept asking us to make films—so we started all over again.'

Kendell and Tait take great pride in knowing that their film *Give Trees a Chance* (1980) directly influenced the then Premier of New South Wales, Neville Wran, to order an end to logging in the State's rainforests. Wran would later say that this was the single most important decision of his tenure and the one of which he was most proud.

Another film, *Earth First* (1986), deals with wider issues including the fight for the Franklin River and the destruction of rainforests throughout Australia.

Both are happy about the effects of *Give Trees a Chance*, because they were up against such entrenched attitudes about how forests should be treated. As Kendell put it, 'Their idea of "selective logging" was to select a forest and log it.' But through their film they were able to dramatically change people's hearts and minds—in their own community and beyond.

In an interview at their Nimbin home on 25 March 1998, Kendell and Tait said their depiction in *Give Trees a Chance* of protesters screaming against police brutality had a 'powerful effect' throughout Australia. 'It shocked people,' Tait said. 'Australians have a pretty deep suspicion about the police anyway—and they certainly don't like to see their fellow-citizens being bashed.'

He said the film sought to reflect first principles in that 'we didn't want that forest to disappear'. As outdoor types who had ridden horses and enjoyed picnics in wild areas, 'we felt that places like Terania Creek were unreasonable places to log—these places are too beautiful to disappear'. They had a particular, personal love of that forest because they used to take their small daughter, Emma, there. She had subsequently died, and they couldn't bear the thought that the beautiful place that they associated so strongly with her should die too.

The film was an indirect result of their work at the ABC and as such they felt they would be able to sell it to a television network. But the ABC declined to screen it because it had 'too much of a point of view; it's too parochial', especially as no representative from the Forestry Commission appears.

'But we could never go down the middle,' Kendell said. 'In our filming we go where our heart tells us to go. And it seems we have been able to strike the right chord.'

The lack of TV exposure did nothing to lessen the film's impact. Another local activist, John Seed, took prints all over the world—from India to the USA—and it helped to raise international awareness to the extent that more than four hundred rainforest action networks were formed. Kendell and Tait took the film to cinemas and the lecture circuit, and it won the same kind of recognition and acceptance as David Bradbury's highly popular *Frontline*.

As a footnote to the achievements of *Give Trees a Chance*, Tait revealed that it had won the Australian Film Institute's 'Courage in Film-making' award in 1980—the first and only time such a prize was given. 'It doesn't even appear in the AFI's listings,' Tait ruefully commented, noting how 'respectable' bodies shied away from anything too controversial or not entirely politically correct. *Give Trees a Chance* later won awards in the USA and in 1989 it was screened by Channel Four in Britain.

Kendell and Tait were further vindicated when *Earth First* was not only commissioned by the ABC in 1986 but also appeared as an illustrated book which gained wide exposure through the ABC shops and their marketing machine.

Jeni Kendell's environmental consciousness began when as a child in the built-up surroundings of Sydney's western suburbs she played with butterflies. 'They were my window on nature,' she recalls now. As a teenager she wanted to 'save the world' and became part of the protest movement against the Vietnam War. From that time she felt that issues of war and the environment were inextricably linked, and she wrote poems expressing her concern about destructiveness in all its forms.

In her late teens she travelled overseas, then upon her return joined the ABC as a film-maker. Among her projects was a film about land use in

Western Australia, during which she discovered the attitude of 'progress at any cost'. She became appalled that such rampant and coercive development was possible in an area which had 'about the most fragile environment on earth'.

Her feelings about the environment 'crystallised' when she moved to Nimbin. 'Amidst all this beauty,' she said, 'you can see what's at stake—what there is to lose.'

Kendell's concerns are shared by her partner, who early in his life 'fell in love' with the Australian bush. This appreciation was heightened when Tait rediscovered Australian trees after travelling in Europe. 'Australian trees are so quirky,' he said. 'There's nothing regular about them—about their shape. European trees are so tame in comparison.'

Paul Tait said he became radicalised on environmental issues after 'becoming aware about everything that went on' while filming ABC news footage, current affairs and documentaries. 'The longer I worked on news and programs like "Four Corners" I saw that no consideration or care was given to the consequences of development, especially mining.' He said that too much of what he saw 'turned my stomach'.

Kendell and Tait admitted that the '90s have proven a difficult time for everyone in conservation battles. 'People in our generation are suffering from burn-out, a terrible tiredness and battle-fatigue,' Kendell said. 'And there seems no end to the number of issues there are to deal with—mining and development. They're all the same old issues, of course. The fights that have been fought and won but have to be fought all over again against different people.'

But they are heartened by the committed band of young people prepared to travel to places as far afield as Jabiluka in the Northern Territory and Cardwell in North Queensland to go and fight those new battles. Kendell added that she and Tait fervently hoped that whoever was filming the anti-uranium protests at Jabiluka would include the songs being sung, because these are at the heart and soul of things whenever people gather to protect their precious patch. 'There will be musicians at Jabiluka,' they said. 'We just hope someone's writing it all down.'

PLAYS

Every Australian playwright of note over the past thirty years has expressed a deep concern for the environment. From the radical theatre of the late 1960s to a series of controversial and confronting plays in every decade since, environmental disaster and loss of heritage has been acted out on stage.

Everyone—from the mainstream David Williamson to angry young men like Stephen Sewell and angry old men like Patrick White—has tackled the themes of our threatened ecology and humanity's destructive impulses.

THE ecological demise of city and bush have always been of equal concern to Dorothy Hewett. Many of her plays resound with compassion for ordinary people whose personal lives and their surroundings are shattered by forces beyond their control. Hewett's environmental concerns were directly reflected in her own life when, in the early 1990s, she and her husband Merv Lilley moved to the New South Wales Blue Mountains. Although they considered themselves big-city people, they were attracted to the peace and tranquillity around Faulconbridge, a place where at last they could have their own garden.

Faulconbridge is eighty kilometres west of Sydney, about halfway up the escarpment at 447 metres above sea level. Much has been written about the great beauty of the Blue Mountains, a place which was old when America's Grand Canyon was just beginning to form six million years ago. Generations of poets and painters have attempted to capture the magic of the mountains' mysterious blue haze and describe the breath-catching grandeur of its cliffs and chasms. Sydney people have been fleeing to the mountains almost since they were discovered and certainly since the railway opened them up in the 1870s. They come in search of peace and quiet, to be inspired by the mountains and the forest, to be refreshed by the clean, healthy mountain air and, in recent years, to be relieved a little of the cost of a home in Australia's most expensive city.

In an interview on 1 June 1998, Hewett told of five and a half years at Faulconbridge, where she was enjoying 'natural sympathy' with her immediate surroundings. Their home, built in 1836, had been a Cobb & Co coach watering place for horses toiling up the mountain's eastern slopes. The dwelling is on a three-quarter-acre lot surrounded by tall gum trees. With a studio out back, it has proved a 'dream place' for writing and contemplating how people's natural surroundings could be improved.

Faulconbridge was the result of Hewett's quest for 'peace and quiet' after her hectic years in the inner-Sydney suburb of Darlinghurst. She loved being in the hub of things in a throbbing metropolis, but in the end she became 'worn down by the noise . . . and the junkies'.

Hewett explained how her deep concern for the ecology came from her father, whose character and views are enshrined in her *The Fields of Heaven* (1982). She told of how she lived on a wheat-and-sheep farm until age twelve, and how she became concerned at an early age about how in Western Australia the salt table was very close to the soil's surface. 'I learned from my father how too much agriculture is a terrible mistake,' she recalled. 'He was ahead of his time. People would laugh at his efforts to plant trees and look after the land—but eventually all of his progressive ideas gained acceptance.'

Hewett radiated a passion for ecological commonsense despite conducting the interview from her sickbed. The authors would have been prepared to do the interview on another day, but Hewett was insistent that it go ahead. She proceeded to express her anger about domestic and global problems.

'We are living in a very bland and negative time,' she said. She lamented that Australia was a very conservative nation where it was almost impossible to sustain a radical vision—and 'no real Left exists any more'.

From her bed, Hewett pointed to the southern point of her house, from where one could see the haze of Sydney's pollution. She told of how the main arteries from the metropolis to the mountains—Parramatta Road and the Great Western Highway—were 'struggling for breath'. At the mountain's foothills was the Nepean River which she described as the 'filthiest in Australia'.

Within a few months, Dorothy Hewett would leave her sickbed far behind, resume her works-in-progress and attend several writers' festivals. She would avidly resume the creative effort, the championing of our environment, the communicating and networking which have placed her at the forefront of Australian letters.

ANOTHER playwright with environmental concerns is Alex Buzo, whose *Batman's Beach-head* (1973), an adaptation of Ibsen's *An Enemy of the People* (1882), was partly inspired by a visit to the Victorian border region adjoining the Murray River. Buzo saw for himself the pollution and over-irrigation along the Murray and learned how people on both the Victorian and New South Wales sides of the river were becoming sick from chemicals, sewage and other pollutants. The otherwise politically conservative Buzo was shocked that such an outwardly peaceful and pleasant place could be so polluted as to threaten people's lives.

Alex Buzo lives with his wife, his younger daughter and an elegant grey cat in a tall, narrow, Victorian terrace house in the prosperous inner-city Sydney suburb of Queens Park. He is a short stroll from both the broad green oasis of Centennial Park and the busy, noisy, crowded multicultural hub of Bondi Junction.

Interviewed in 1995, almost a quarter-century after he got the idea for the play, Buzo said that when *Batman's Beach-head* was staged 'it wasn't politically correct to be pro-environment'. Nevertheless, his outrage at what he had seen along the Murray was such that he incorporated into the text what could happen at a popular resort if its natural attractions were being destroyed.

Buzo's concerns were expanded in *Tom* (1972), which presciently examines the ugly business ethic that was to grow up within the next decade with the advent of Australia's corporate cowboys—the raiders and the wheeler-dealers who, in the pursuit of riches, corrupted and impoverished everything they touched.

As well as becoming concerned about the increased moral atrophy in Sydney, which Buzo said he had once loved as a laidback and egalitarian place, he was also beginning to notice things like trucks belching black smoke and other types of urban decay.

Buzo said that part of his green consciousness came from living near Peter Hamilton, whom he described as one of Australia's earliest environmental activists. Buzo said Hamilton had given him insights and expert advice. In the late 1970s, as part of this growing awareness of environmental problems, Buzo wrote a magazine article about the decaying state of the Murray River as well as salination problems on the Great Australian Bight, and then released his play *Big River* (1979). This play dealt, among other things, with drastic changes such as Australian urbanisation and painful adjustments to new values between 1890 and 1910, the decades leading up to and immediately following Federation.

One thing that remained constant on this continent before and after Federation was a culture of exploitation of its resources, and this mentality has been observed—with humour, pathos and high drama—in the plays of Alex Buzo, who amid the pleasantness of Queens Park empathises with those living in degraded surroundings.

AT THE busy, 'downtown' end of sprawling Bondi Junction lives Alison Lyssa, whose play *The Boiling Frog* (1984) raised the awareness of audiences to the dangers of nuclear proliferation. On a 15-minute walk from the centre of Bondi Junction—in the opposite direction to Queens Park—one can notice the mixed tableau of this multicultural and diverse suburb.

Along one of the main drags, Old South Head Road, derelict houses sit alongside dwellings of the highest elegance and charm; failed, boarded-up businesses stand next door to ostentatiously thriving concerns. Cracked pavement is everywhere and the constant rush of traffic is maddening. Lyssa lives at the intersecting point of several prestige suburbs and those struggling to find an acceptable standard of living.

At the outset of our interview at her home on 10 December 1998, Lyssa displayed her desire to live at all times in an environmentally and ideologically sound way as she offered a pot of organically grown lemongrass tea.

Like Patrick White, who went from ultra-conservative gent to passionate political radical, Alison Lyssa underwent a dramatic conversion based on her everyday experiences and the cultural influences upon her. During her first job as a teacher—at the exclusive Frensham girls' school in the New South Wales Southern Highlands town of Mittagong—she became interested in the bush and joined fellow teachers to stop the spraying of poisons within the school grounds. She also learned about composting—a practice to which she has adhered ever since.

But the Mittagong influence went far deeper. She had arrived at the town in the early 1970s as a member of the Liberal Party and a supporter of the Vietnam War; she had believed that the British and US empires were absolutely justified in conquering the world because their foes were 'villains . . . and inferior'; she had been what more than two decades later she described as 'an ignorant person'. The major cultural influence which changed all that was seeing British director Peter Watkins' classic anti-war film *The War Game* (1965), which chronicled the horrors of a nuclear aftermath with such realism and ferocity that it was banned by the BBC in the UK and not shown on Australian television until 1987.

'This film came as a huge shock to me,' Lyssa said. 'I realised I had a lot to learn about almost everything. I had to leave what I was doing, search for new ideas and change my life.'

Since moving to Sydney in 1973, her main mode of transport has been her bicycle. 'I discovered quite early that a bicycle is a very environmentally friendly way to travel—as are public transport and walking,' she commented. She spoke of cars as being 'antithetical to community'. Lyssa likened travelling in cars to 'locking people up on private property . . . it separates people from each other, as well as using vast amounts of the earth's resources'. Similarly, she deplored that today's car culture directly linked concepts of youthful 'freedom' with an ecologically unsound mode of transport. She suggested that the continuing proliferation of cars was making the world an increasingly unsafe place and too often drivers were urged to glorify aggression.

Lyssa, a powerful and potent force behind her petite frame, explained how the writing of *The Boiling Frog* emerged directly from her deep desire to raise environmental awareness in others. As a member of a group called Women Against Nuclear Energy she was part of a movement aimed at stopping the Ranger uranium mine in the Northern Territory—a site twenty-two kilometres from Jabiluka over which a new storm of protest was brewing as we spoke. She wrote articles on the nuclear issue but then concluded 'I was no good at that'. She did conclude that the most constructive thing she could do was write a play.

Although proud of the venture and gladdened by the recognition, Lyssa in 1998 expressed many concerns about *The Boiling Frog*. She concedes that she left herself too vulnerable to hostile critics who could point to it being 'so obviously political'.

Three productions of *The Boiling Frog* were staged in 1984 as well as a smattering of smaller ones—including in the hippie community of Byron Bay and among vegetable growers in Tasmania. Lyssa recalled with happiness the 'flurry of interest' in her play.

In 1998, Lyssa told of being busy with a range of concerns and activities, and high on her list of priorities is doing something about Sydney's public transport which to her is a symptom of 'society's failure to come to terms with unsustainable energy requirements'. She writes letters to politicians on such varied topics as seeking bike trails, improving the existing public transport network and trying to stop underground car-parking. She attends council precinct meetings because she believes 'grassroots actions are important'. Once a week she is part of a volunteer bush regeneration effort by Woollahra Council at nearby Cooper Park.

At an even more grassroots level, she began a composting system at her block of units, where the residents of all twelve apartments are urged to take their fruit and vegetable leftovers to a common bin. She regularly seeks to spread this gospel to her fellow residents and has written to Waverley Council on these issues.

And she grows the lemongrass whose flavours and aromas permeated the interview.

FAR FROM Bondi Junction and far from any crowds, in a remote cabin in the heart of rural Victoria, lives well-known playwright John Romeril. His rugged and creased features, his lumber-jack woollen shirt, his roll-your-own cigarettes and propensity to yarn make him a natural for his rustic surroundings, except talk quickly turns not to the weather or the crops or what the neighbours are doing today but to existentialism, surrealism, Ezra Pound, Eisenstein, Brecht, things Japanese, and global environmentalism.

Romeril, one of the leading and enduring figures of the New Wave of Australian playwrights spawned by the Australian Performing Group in Melbourne in the late 1960s, lives in the heart of Mount Alexander Shire about six kilometres north of Castlemaine, the famous Victorian gold-rush town which found its 19th-century aspirations to grandeur significantly diminished as the year 2000 approached. John Romeril—who more than most other Australian playwrights has dealt forcefully and comprehensively with environmental issues—has lived in the Castlemaine area since 1975, having had links with the region since his birth in 1945.

Castlemaine, 120 kilometres north-west of Melbourne, was one of the main gold towns of central Victoria. The town's founders clearly had dreams of glory when they laid out its streets. Wide boulevards stretch out from the train station and large, imposing Victorian homes and public buildings are the rule. The extensive and still lovely botanical gardens were designed by Baron von Mueller, responsible for the more famous gardens in Melbourne. Castlemaine did indeed seem to be headed for greatness until the gold ran out.

Romeril meets us at the station on the day of our visit, 2 June 1999, and drives us to his cabin along a road that is a red gash through the countryside, given to sudden curves and steep hills and valleys. But we ride in comfort and feel safe and calm as Romeril points out the sights, including the spots where friends have parted company with the road.

To any visitor, Romeril's cabin is about the quietest and most serene place on the planet—the perfect place to write, which Romeril does regularly except for the three months each year he is on the road. With a buffer of about 100 hectares between himself and the rest of the world, Romeril, twice married and the father to three stepchildren, lives on his own in the cabin while his partner is city-based. When he writes, Romeril inputs his text into a PC that is powered, like all else at the cabin, by solar energy.

Romeril built his peaceful cabin in the mid-1970s, but it seems to have always been there, perched at the crest of a hill, with views of woods stretching out towards Mount Alexander. The woods are rather thin, stringybark and box trees mostly, as the old timber was cleared and cleared again to build the town, to shore up the gold mines and to keep their fires burning. The few large trees are exceptions. The cabin is solid and comfortable, made with a lot of recycled material and with elegant stained glass in the windows. The views are stunning—though not as impressive as those from the outhouse. It is positioned a little farther up the hill and features panoramic views of the valley: a true throne. The place is so quiet that when our conversation pauses there is not a sound to be heard—a rare feat in this age of incessant traffic and human encroachment.

Over a long and free-ranging interview, Romeril described how environmentalism took over the 'leftist' political priorities of his earlier writing. He told of how Marxism was dedicated to the conquest of nature and to feats of engineering which could forge the earth according to the whims and dictates of mankind. Romeril said environmentalism was a natural extension of 'a Left critique in the early '70s' and he told of how he was influenced by such things as back-to-the-earth movements, the hippie philosophy and the ethos of Walden Pond.

The influences on Romeril have been profoundly multifaceted. At first, when he thought of becoming a poet, he was influenced by the Beat Generation, until he realised he had the skills which could have an impact on Melbourne's experimental theatre such as at La Mama. At university he studied English and politics while his job at the Victorian Department of Agriculture's library heightened his interest in all things agricultural and in what was happening to the earth.

Revealing the sense of humour which permeated all of his deep reflections about the peril the world is facing, Romeril pointed to how Australian utilities were influenced by the workings of Soviet-style bureaucracy, with both Victoria's State Electricity Commission and the Tasmanian hydro authority (which was not seriously challenged until the Franklin dam dispute) run on secrecy and by people who believed their career paths were more important than the public good.

Romeril recalled how his environmental consciousness also grew out of having lived on a pacifist commune in New Zealand in the late 1960s. He described how his environmentalism developed from the various radical social movements of his lifetime, particularly the anti-war movement and feminism.

In his plays from the Australian Performing Group days to the present,

Romeril was influenced by this century's 'great aestheticians', and he described them as quite a diverse group. In the 1930s, at the populist end of the entertainment spectrum, there was the celebrated Hollywood director Busby Berkeley, whose synchronised and complex choreography reflected the machines of American industrialisation; Russian film-maker Sergei Eisenstein also 'addressed the problems of modernity, showing the energy of different component parts creating a new energy'; and the German anti-fascist playwright Bertold Brecht who was 'one of the first to propose that the Aristotelian play was over'. Romeril said he was inspired by these great creative minds to 'drag modern life onto the stage, to depict mass life in vast metropolises'.

One of Romeril's earliest excursions into the full-scale agitprop of environmentalism was *The Radio-active Horror Show* (1977), which he described as the 'by-product of mass education' and as 'a kind of journalism'.

Discussing his many plans since then and the present day, as he rolls another cigarette and continues yarning, Romeril remains good-humoured with a sense of optimism. 'Why have a mid-life crisis when you can have a mid-life adventure?' he asks. He talks of organic farming on good ground—and of more plays which he will take from his Castlemaine cabin to international arenas.

SONGS

A stirring lyric and a vigorous refrain usually accompany the important moments of our lives. One can usually remember what song was playing on one's first date, the great moments in history usually happen to musical accompaniment—and all of Australia's environmental barricades of recent times have been etched into the nation's memories because of the words chanted and sung.

The French Revolution might not have been as stirring without 'La Marseillaise'; the American Civil War may not have been awash in so much emotion without the balladeers of both sides; World War I was fired up for the Allies by reminders of what a long way away was Tipperary; World War II had Vera Lynn; and the Vietnam protest movement exploded to the backdrop of the rock music revolution of the late 1960s.

So it has been for thirty years in the struggles for the Australian environment. When the Tasmanian Government, at the height of the Franklin blockade, brought in laws which forbade protesters to 'lurk, loiter, hide, secrete', the same protesters, in their tens of thousands, chanted and sang, 'We're gonna lurk, loiter, hide, secrete'. Similar defiant sights and sounds were seen at the battle for Terania Creek in northern New South Wales. Where there is a just cause, there usually can be found balladeers and superb musicians.

A GREAT advocate of things green is Eric Bogle. Long before environmentalism became a movement or fashionable in any way, Bogle was singing about rivers and trees. Indeed, he says the world has finally caught up with folk music, which has long been celebrating the good earth. We met up with him in the Sydney offices of Larrikin Music on 23 April 1999 which, in one of those funny coincidences that occur when you're working on a project such as this, happened to be the Friday before Anzac Day. Bogle is probably best known for 'And the Band Played Waltzing Matilda', one of the definitive songs about the Anzac experience, and he is very much associated in people's minds with that holiday.

As the jets of the Australian air force screamed in formation in the skies above the building, presumably practising for their own Anzac Day gigs, Bogle spoke softly about his music and about his grave concerns for this country. The issues of war and peace were very much on everyone's minds at this time, as refugees from Kosovo were still pouring over various borders and NATO officials were on television every day talking about what they'd managed to bomb the night before.

'It enrages me,' Bogle said. 'It sickens me. We've learned nothing in all these years and all these wars since World War I, it just keeps going on and on. I don't know what we're doing with Anzac Day, I really don't. Certainly, nothing has changed.

'When I see the events of the past couple of weeks [in the Balkans], it reminds me of all the thoughtless, careless, stupid things we're doing as a society and I think we're all heading for extinction, and maybe that wouldn't be such a bad thing.'

Bogle corrected himself at this point and said: 'No, I don't mean that—not entirely. I meet wonderful people all the time, individual people are wonderful. I meet people who are working their arses off for this country, trying to save something, trying to change things. I'm thinking of the LandCare people—I find them so uplifting when I talk to them. I was a LandCare ambassador for them a couple of years ago—I gave some talks for them, did some gigs. I like so much what they're trying to do.'

Bogle, who lives in an older, inner-city suburb of Adelaide, sees some of the biggest issues confronting Australia unfolding in his own home State. 'It's changed there, in South Australia, the whole attitude to the environment,' Bogle said. 'In the days of [reformist Labor Premier] Don Dunstan, South Australia was in the forefront of the green movement, everybody was an environmentalist, it was very popular. But now, it's probably the poorest State and greenies, as they call them, have been cast as the villains. You know the sort of thing—greenies take away our jobs, our prosperity.'

He said the biggest issue in South Australia was probably coastal development, people putting in marinas, new power stations. He said there were powerful forces that want more development and not very many who want to oppose it, but any opposition was met with great ferocity. 'Just yesterday,' he said, 'at Pelican Point, a group of mostly housewives, I guess, and only 20 of them, went along to make peaceful protest against a new power station they want to put in there. They were met by 100 policemen, some of them on horseback, all of them with sticks and things—to stop 20 women. And they arrested 15 of them. That's the kind of climate we're living in.

'I used to be more active—I used to protest. When I first came to Australia, I got involved with the Moratorium movement and marched against the Vietnam War. I don't know how much good those marches did. The war stopped, of course, but I think maybe the marches did more harm than good—because of the way they were portrayed in the media, especially the commercial media. I would go along to what was mostly—entirely—a peaceful protest, but the TV could make it look like a riot. They do the same thing now with all the green protests. They try to find a hippie up a tree, the most scraggly-looking creature they can find, and interview him as a typical protester. But if you go along in a business suit, you won't get on TV, that's for sure. They don't want anybody—any middle-class people—thinking that "maybe there's something in this protest; maybe I should get involved".'

That led to talk of uranium, which Bogle sees as another big issue for South Australia and for the nation as a whole. 'I hate uranium mining; I would stop it all right now if I could—and not give a damn how many people it put out of work,' Bogle said. 'The issue of dumping nuclear waste here is big right now—everyone is all upset about that and wondering how anybody would dare suggest that we take the world's muck and bury it in our backyard. But we've got to realise that we're part of the nuclear cycle—we're up there with the big boys, even if we're not a "nuclear power". The uranium comes from here in the first place. It's only right that we should take responsibility for it. But there are so many shades of grey, aren't there? When I was young I saw everything as black and white, right and wrong, but now it's all a lot harder. Like the power station—we need it, that's the best place for it. It's not a simple problem—nothing is.'

Then, almost wistfully, he said: 'I do try never to be cynical. It's the enemy of creativity for one thing. It's the enemy of life. If I were cynical I wouldn't be able to write songs any more, I'd turn my face to the wall. Or I'd just go sit in my rose garden and wait for the end. But I care. I believe in things—important things. I know we're all here at the whim of nature, or life, or God, whatever you want to call it—so I keep writing my songs.'

These sentiments are the inner feelings of a true troubadour, of a poet who performs and extends his visions as far afield as he can.

POETRY

When you see a poet, you see an environmentalist. Almost by definition, a writer who describes the unique flora, fauna and topography of Australia is intimately concerned with its ecology. And anyone who writes about humanity—our passions, our desires and aspirations—is opposed to forces that brutalise people, that cut them off from their heritage and exploit them.

The poets we look at are young and old, conservative and radical, city-based and rural, and they span the entire continent, with one uniting thread—concern about what is happening to the environment and a deep desire to expose it and to urge Australians to mend their ways.

FEW people care as passionately about the environment as Les Murray, despite his self-confessed 'redneck' sympathies and his open contempt for those he describes as 'greenies'. At his sprawling property at Bunyah in the Great Lakes area of northern New South Wales, Murray began an interview on 3 June 1998 by castigating all greens and those who sympathised with them—the very word 'environment' seemed to annoy him. But the more he talked the more he showed compassion for an abused land—and the many country folk who are suffering because their environment has been allowed to deteriorate.

The North Coast area is one of the State's troubled regions, where towns are dying and unemployment and the despair it brings are rife. The timber industry, long the mainstay of the region, is in perpetual conflict with the demands of environmentalists. Opinions differ on the merits of conservation versus development and exploitation, or indeed the real reasons for the industry's difficulties, but no-one disputes that jobs are being lost and families are hurting. In the towns, support industries are going bankrupt and services pulling out following the collapse in timber and the stringent cutbacks of right-wing governments. In nearby Bulahdelah, for example, there is no longer a bank.

The region has great natural beauty and in many places tourism is the only viable sector. Myall Lakes National Park, a system of interlocking lakes and waterways which stretches from Seal Rocks in the north to Hawks Nest in the south, draws many visitors to the area. The lakes, separated from the sea by a series of vast sand dunes, are partly fresh and partly salt water. The system covers an area of 225 square kilometres and is one of the most important bird-breeding regions in the country. The North Coast is heavily forested, more and more of it is coming under government protection. It is home to some of the tallest trees in New South Wales including Grandis, officially the tallest tree in the State at 87 metres.

Murray's property—where he lives with his wife Valerie and their son Alexander—is reached after a drive of about ten kilometres of winding and increasingly daunting road. One drives through lovely and typically Australian rural scenery—scrub bush with little creeks and rolling paddocks where the occasional kangaroo darts through the long grass. The road surface was slippery in the misty rain and our town-bred Ford Laser was frequently confused by cattle grids and sudden pot-holes.

Murray's homestead, set seemingly in a sea of mud on this rainy day, has an inviting cosiness about it. Bright rugs decorate the fine timber floors, and the focus of the room is a wood-burning stove, beside which this intense raconteur holds court.

The poetry of Les Murray is suffused with concern for ecological degradation, and he is a keen critic of all sides of the great environmental debates. He castigates the dogmatism of the green movement in which he believes real knowledge and understanding are sidestepped in favour of emotion and political correctness. He concedes that a better management of forests is essential, although he has quarrels with some of the conventional wisdom on that subject. Mostly, though, he is angry that 'people whose livelihoods depend on timber are being hammered by the conservationists'. He deplores many green strategies which put 'working people under attack' and he explained why in the bush 'environmentalists are the biggest enemy'.

The fight between the bushies and the greenies is a 'class thing', Murray believes, and he scoffs at how 'no environmentalist will take on someone like Kerry Packer; they prefer to oppress struggling people like millers and loggers'.

'"Redneck" is usually the first word out of the mouths of those who attack country people, who are unfairly accused of everything supposedly harmful to the environment,' he said.

Murray, who described his ancestors as having been 'dirt poor', spoke of the 'hectoring spirit' of the greenies' attacks. He said the green movement needed to answer one question: 'How will you let us make a living if we do as you want?'

He suggested that if greens were so concerned about country people's welfare they should direct their energies to the 'greed and weakness of local leaders' who were spoiling the ambience of many country towns such as Taree. For example, in Taree two shopping malls destroyed the flavour of the place, driving out local businesses from the centre of town. 'You wouldn't know it now,' he said, 'but Taree's main street once had a lovely, sunny side but that was all lost when the town turned its back on the river.'

Murray undercut some of his rhetoric by conceding that on his property he 'liked to see things left alone'. He waved out the window towards a pretty, sheltered duck pond. 'I'd rather let ducks live and eels swim,' he said. 'I don't allow animals to be killed on this property.'

He sought to dispel stereotypical notions that the bush was a place utterly without enlightenment, pointing out, for example, that in the region around his property there was far less shooting now than in previous years. 'There are today more kangaroos and bandicoots, echidna and platypus than when I was a child here,' he said. 'Pademelons had vanished when I was kid, and now they are back. All we had then were rabbits.'

Les Murray, an only child, told of how since his earliest years he was taught to regard himself and all other living things as 'a continuum', of how 'rocks, trees and the weather are all part of the same ecosystem . . . we didn't put humans above it'.

Above all Les Murray made a plea for knowledge. It was time for indisputable facts to have precedence over prejudices and entrenched attitudes. There was 'not enough knowledge' about how Australians should best use their resources and achieve economic growth—and he feared for the outcome if two polarised groups of society did not try to understand and relate better with each other.

'I often think we don't love things enough,' he said. 'Or people. Maybe especially people. But if we all loved this land more, we would find a way to take care of it better.'

Amid the tranquillity of his Bunyah home and the ponds with boisterous ducks, Murray can certainly smell the flowers and enjoy many of nature's gifts.

ALSO smelling the flowers with a zest and exuberance befitting one of Australia's leading nature poets, Mark O'Connor enjoys the restful tranquillity of his Canberra home. O'Connor, despite a reputation for being strident and argumentative in pursuit of a range of controversial ecological views, lives in a humble and unassuming suburban house that radiates love and warmth amidst the national capital's notorious chill and artificiality.

From the moment we entered his home, we realised O'Connor was not the ogre portrayed by the media and by himself. This vociferous advocate of 'ecologically sustainable' population control—with a reputation for prickliness as overwhelming as Les Murray's—in person is a gentle soul with an affable demeanour.

Mark O'Connor—big, bearded and buoyant—lives, with typical poetic licence, in the Canberra suburb of O'Connor, which adjoins Lyneham where many other writers congregate. The neighbourhood is about as close as any big city can get to the ideals of bucolic peace. The streets are shaded by tall, mature trees, quiet but for the cacophony of a rich variety of birdlife. While we waited, we saw rosellas, a galah, several cockatoos—as well as about 20 of Canberra's ever-present and almost tame magpies.

As the interview began on 22 March 1999, we were joined by the house's other friendly denizens, beginning with the delightful cavalier spaniel Doeg (as in 'Excuse me, monsieur, does your doeg bite?'), also known as Kendall. This tiny white-and-black ball of fluff—small enough to fit through the cat flap—sat between the guests, snuggling against our shoulders, eventually drifting off to sleep with snores that would not disgrace a mastiff. Simultaneously surveying the scene with a confident and proprietorial air was the grey Burmese cat, Arianna, described by O'Connor as a 'silken thing of darkness'; all this as a rainbow coalition of tropical fish went about their business, swimming through their truly impressive underwater jungle. O'Connor's wife Jan was absent, but her warm touches were evident everywhere.

O'Connor, youthful and glowing in his mid-50s, reflected how prosperous career paths had beckoned to him as a prize-winning student in Melbourne, but he chose poetry instead. To O'Connor, there are things in this world worth preserving, and poetry is his way of touching hearts and minds. He is, as he put it, 'happy to stay poor' and ride around on a bicycle in a city where the bigger cars of status tend to define who one is.

The author of such well-known poetry collections as *The Eating Tree* (1980) and *Fire-stick Farming* (1990) reflected on the contradictions of how Canberra, as a planned city and a political and administrative centre bursting with introduced vegetation, could nevertheless retain so many links with nature.

'Canberra is a livable, easy and pleasant city full of intelligent people who ensure a vigorous intellectual atmosphere,' O'Connor said. As he saw us writing down that quote, he said, with mock dismay, 'Hey, don't tell people that. We want the world to keep on thinking Canberra is horrible—so they'll leave it to us.'

He said too that the Australian Capital Territory suited writers well because it was comparatively inexpensive to live there, had fewer traffic problems than the State capitals and had a 'semi-rural look and feel to it'. He added that 'when friends move to the other side of town, well, they're still only about 15 minutes away'.

As an environmentalist, O'Connor confessed to being constantly on the alert to preserve what he admired about Canberra, especially as 'this area is coming up for development'. He noted that although there weren't as many local problems (environmentally speaking) as elsewhere in Australia, nevertheless the 'ugly face of capitalism can be seen all over Canberra'. He told of how all development applications needed to be intently scrutinised even in 'our smaller city with its keen sense of community'.

He said the calibre of so many of Canberra's citizens ensured that development applications had often been fought off, and that among officials and decision-makers 'letters to the editor are considered'. But he added: 'The problem is, though, that development is so lucrative—and so many people are trying to do it—that every time one idea is knocked on the head, about three more spring up. It's an endless battle. They've already made downtown look as much like downtown Sydney as they can.'

But even in his beloved Canberra O'Connor noticed a 'growth of plutocracy' and that democracy was becoming 'weaker every year'. He pondered on how 'rules are changed to favour the rich' and that it could be seen everywhere how 'capitalism is a very powerful machine'.

He saw writing as an antidote to such social ills, and poetry had its own unique role. He had always been aware of the stirring and emotive power of songs and verses on both real and symbolic barricades and he has written glowingly of the poems and prose of Judith Wright whose 'wonderful and analytical mind' has changed perceptions more than have major political or social movements.

Mark O'Connor said he always lived with the hope that his word pictures had the capacity to influence thinking people, and he spoke of his great satisfaction that New South Wales Premier Bob Carr had found affirmation—and had 'bravely stuck his neck out in embracing a controversial position'—regarding the perils of over-population and thoughtless expansion and growth after reading O'Connor's tract *This Tired Brown Land* (1998). But O'Connor had no delusions that for the pen to be mightier than developmental swords it had to be forceful indeed because there are 'overwhelmingly powerful pressures on development—and the biggest money in Australia is connected to real estate . . . there is strong business pressure on growth and more growth, and a constantly expanding marketplace'.

Spreading his own message as far and wide as possible is at all times the passion of O'Connor, who over the years has collaborated with leading photographers to convey the beauties of wild places. He has traversed the continent from North Queensland to the Blue Mountains to the Snowy and beyond, and written the verse for such titles as *Poetry in Pictures—the Great Barrier Reef* (1985), which featured the evocative underwater photography of Neville Coleman. O'Connor said such illustrated books were aimed for a 'larger readership . . . to break out of the poetry ghetto'.

Mark O'Connor is indefatigable and committed to instilling poetry into the souls of Australians young and old.

CHILDREN

The raising of children's consciousness on environmental issues is not kids' stuff, and the writing for young Australians has embraced venerable senior citizens like Colin Thiele and young mums seeking to relate to their own offspring.

It's much more than the power of the hand that rocks the cradle; it is the guiding light that will illuminate the hearts and minds of the next generation.

THE rolling hills and valleys around the semi-rural district of Dayboro have a gentleness and tranquillity that are at the heart of Colin Thiele's character and writings. Although only forty-five kilometres northwest of Brisbane, Thiele can survey an exquisite and peaceful landscape from the hilltop home he shares with his wife Rhonda—friendly visions such as those that for decades have been expanded into celebrated and prolific novels for children and young adults.

A rare silence prevails in this elevated location as we step out of our car, and views across lush, green valleys and 'Toytown' farms stretch for many kilometres in every direction, appearing tiny and unreal from this great height. We have just left the small township of Dayboro, which at first sight seems a typical, rural Queensland crossroads, dedicated to selling groceries and petrol and feedstock to its farming neighbours. But the pub looks a little too fancy for that, and the main street features a giftshop/art gallery, and a couple of upmarket restaurants peek through landscaped subtropical foliage. Dayboro, it seems, is on the tourism beat. It is also home to a growing number of Brisbane's elite, who come looking for peace and quiet a convenient drive from the city.

The Thieles' large house is surrounded by a huge garden, made from scratch by Rhonda, who has laboured alone on the windy hillside over several years to bring it to its current lush perfection.

An indestructible and peerlessly respected presence in Australian literature, Colin Thiele has published more than a hundred books, mostly for children, over the past 50 years. Exuding strength and energy from a slight and battered frame, the result of years of battling the crippling effects of arthritis, Thiele is a feisty survivor thinking of his next project and his aim to pass on his important messages to Australia's next generation.

It is appropriate that Thiele, born at Eudunda in South Australia in 1920, should end up in what he describes as 'a physically lovely place—unhurried, with none of the frenetic pace of suburban living'. In an interview at Dayboro in March 1998, Thiele warmed to his theme: 'People here have time to talk to you; they talk about the rain and the crops—just as they did in my boyhood. Whether at the post office or outside the local pub, people know you and talk to you.'

Colin Thiele's boyhood experiences included the natural environment of the Coorong district along the South Australian coastline; from the hillside farm where he lived he roamed the surrounding country and learned about the changes of the seasons. His personal feelings and emotions towards these places suffuse many of his most famous books, such as *Storm Boy* (1963).

'You can't change attitudes unless you're involved emotionally,' he said, describing what he felt was his unique vantage of being able to share with young and usually city-dwelling readers experiences that they would almost certainly never have known.

His love of the Coorong courses through his writing. As a university student he would camp there and would walk to the mouth of the Murray River, and as he looked back on a long life he remembered the pristine wilderness, which was 'just like paradise in the early morning'.

Storm Boy and other early novels were the result of Thiele's need to write about his many joyous feelings about the Coorong and to convey the importance of such rare and wild places. Appropriately, the living-room of his Dayboro home was replete with pelicans—pelican prints, pelican statues, pelican bookends, and even a pelican-shaped lamp base—collected over the years and emblematic of the care and concern as represented by the pelicans in *Storm Boy*.

Another potent influence was his stint in the RAAF during World War II when he discovered the plight of the Aborigines during his stays in Arnhem Land and at Melville Island. Spending time among the Tiwi people, he found them 'so honest and so attuned' to their environment compared with how the white settlers behaved.

Thiele remembered how Aborigines were for so long regarded as 'animals or unimportant' and he lamented that it took such a long time for Aboriginal welfare to become a significant national issue.

After the Air Force, Thiele began a career of teaching—first at high school then as a college lecturer—and he sees his books as having an educational role. One of his great hopes is for kids to think carefully about the issues raised in his books and to treat his stories as more than just escapism.

A welter of issues are addressed throughout Thiele's voluminous body of work: the dangers of offshore oil wells in *The Fight against Albatross Two* (1974); oil spills, their devastation and their effects on wildlife in *Pinquo* (1983); issues of Aboriginal heritage in *Coorong Captive* (1985); over-fishing and human greed in *The Ab Diver* (1988); dolphins under threat in *Speedy* (1991); the Arnhem Land heritage in *Brahminy* (1995); the misuse of water and the debilitation of bureaucratic power in *High Valley* (1996); and the under-estimated destructiveness of cotton farming in *Landslide* (1997). Aimed at younger readers are *The Sknuks* (1977), a picture book dealing with nothing less than the devastation of the planet through human negligence, and *Little Tom Little* (1981), about how a Northern Territory ranger's work is never done in protecting the environment from careless people.

'A very few people can do a lot of damage,' Thiele said, looking out over the broad expanse towards Mount Mee. 'Ninety-five per cent of people are caring, but it's the others who are of concern.'

As a result of his upbringing, Thiele has always been interested in rivers—especially the Murray—and he is appalled at what has been happening to Australia's waterways. He spoke of the need for governments to immediately spend millions of dollars to 'resuscitate' the rivers. 'Schools in the Murray-Darling Basin are teaching new ways to care for our rivers, and such efforts are our only hope,' he said.

Thiele sees human ingenuity—until now so much the cause of global crisis—as a potential saviour of the planet and he is 'cautiously optimistic' that new alternative energies will be adopted as soon as possible.

As to the influence of his books, Thiele handed over a piece of paper which summed up what he felt was a poignant achievement. The scrap of paper documented how Thiele, who has had a vast correspondence with his army of young readers, especially treasured one letter from a little boy who, having read *Storm Boy* and seen the film, wrote to him saying: 'Mr Thiele, I'll never, never, NEVER kill a pelican.'

To Thiele, such feedback from a young mind is as important as the Companion of the Order of Australia (AC) which he received in 1977 for services to literature and education.

SUCH overflowing and good-humoured humanity is also the essence of another celebrated writer of children's literature, Hesba Brinsmead, who in her advanced years has retired to what should have been the bucolic serenity of Terranora, a small community high in the hills of northern New South Wales, overlooking the Queensland border.

The various views from the windows of her stately home say it all. To the south is peaceful countryside, rolling hills, canefields and a glorious profusion of subtropical vegetation and birdlife; to the north, about half a kilometre away, are hideous housing estates, with one cul-de-sac seemingly built straight atop another in a frenzied fury of over-development. The only retail outlet in this maze of remote suburban blight is a real estate office slap-bang in the middle of it all, trying to sell more and more of the project homes. An eerie quiet lay over the raw, new neighbourhood where many of the huge and ornate ranch-style homes remained empty as we slowly drove through it on the way to Brinsmead's house. Winding streets, designed perhaps to be safe and pleasant places for children to play, echoed ominously. There was not much birdsong, as most of the large trees were gone, replaced by struggling shrubbery and wide lawns, thirsty under the baking sun.

Hesba Brinsmead, who for decades fought many conservationist fights and instilled her many novels with concern for the environment, was a frail figure and said that she was now leaving the many battles still to be fought to the younger generation.

'I've retired from everything; I have struck my blow; my nieces are very active and interested. They come and talk to me about what they're doing about Jabiluka and uranium mining,' she said in an interview at her home on 7 April 1998.

Brinsmead's home, where she moved with her husband 25 years earlier, is one of the few survivors from a more gracious era. It is large and elegant and squarely built, with broad staircases and wide windows. It is surrounded by shade trees and a pleasant garden and seems cool and quiet on the very warm autumn day. About the ugly sprawl outside, she said, 'Yes, isn't it terrible? But we're all guilty. It's our fault that it looks like that.' Her family and many others had sold their farming enterprises in the region because they had been 'rated out' by the local council. Their holdings were at Banora Point, an area that had been surrounded by 'rich and beautiful farms' but was now 'just like any other city suburb'—with the exception that there were no shops or convivial places for people to gather.

'Everywhere around here there were bean and fruit paddocks, with the produce grown on the most wonderful soil,' she recalled. 'Now every ruddy inch of land is under cement.'

Brinsmead dramatically decried the senselessness of the developments around her, pointing out that the council had created 'a swamp climate without the swamp'. The giant sport and entertainment complex Club Banora was a prime example of this developmental stupidity. Brinsmead said: 'That used to be our banana patch down there and it was always the hottest, wettest place around—and it still is. Their clubhouse is 10 degrees hotter than anywhere else in the neighbourhood.'

Before Terranora Brinsmead had lived in Melbourne and she has a sister who lives in Tasmania, so during all these years she has been concerned with conservation issues throughout Australia—especially the ill-fated fight for Lake Pedder and the subsequent monumental struggle against the Franklin dam. 'South-west Tasmania is very overpowering and its preservation is paramount,' said Brinsmead, who instilled elements of the island's struggles directly into her many celebrated books for children and young adults.

Echo in the Wilderness (1972) features her adolescent hero Clippie Nancarrow, swashbuckling pilot of light planes, and the idea for the story flowed directly from events in the Pedder struggle. The character of Clippie was loosely based on Max Price, the pilot who flew her around Tasmania. Price and his passenger, prominent Pedder campaigner Brenda Hean, died in highly suspicious circumstances when their plane disappeared on a flight to an important Canberra protest in September 1972.

Clippie espouses Brinsmead's environmental concerns—and shares her propensity to be a nonconformist who goes her own way. A subsequent Clippie novel, *The Sand Forest* (1985), deals with another of Brinsmead's concerns—how crowds can wreck anything in their path, including in a remote West Australian desert.

For an adult readership she wrote *I Will Not Say the Day Is Done* (1983), in which she chronicled the Pedder struggle which she believed was 'definitely a watershed in the conservation movement'. In this book, Brinsmead describes all the elements in Tasmanian politics and society which allowed the lake's flooding. She calls the Premiers responsible for most of the Pedder decisions Tweedledum and Tweedledee because of defamation concerns and her worry that the book might be suppressed due to its many controversial revelations.

Although Brinsmead believes that 'the full story of Lake Pedder still needs to be written', hers remains a powerful and accusatory document which powerful forces in Tasmania 'tried to brush under the carpet'. Newspaper reviews were very few, so the book failed to sell nationally as even wilderness shops ignored it. Over the years copies became increasingly hard to find. But the book's launch by far-away friends—a North

Sydney bushwalkers' club—gave Brinsmead heart, as did the realisation that Greenpeace was formed during the Pedder fight 'and now it is a major force in environmental protection'.

As Hesba Brinsmead looked out the window with the pristine and wondrous view, she voiced many of the concerns which nag away at her even in retirement. She cannot understand how Australia could be allowed to remain a great desert with some green strips at the edges. 'Those little green strips near the ocean—that's where they cement it,' she scoffed.

Hesba Brinsmead may be ailing and increasingly fragile, but she remains fervently concerned with ecological and social issues.

A RANGE of environmental concerns was swirling in the life of Jenny Wagner during the late 1980s as she was creating her *Goanna* (1989). This beautiful picture book, which depicts a feisty goanna as it vainly seeks to protect its patch against predatory developers, coincided with specific environmental issues in Queensland, as well as the wider Aboriginal land-rights movement.

'Like the goanna I wrote about, the Aboriginal people have to fight for their lands as well,' Wagner said in an interview in March 1998 at her home in Eumundi on the Sunshine Coast hinterland, a 90-minute drive north of Brisbane.

The Sunshine Coast has become one of the fastest growing regions of Australia, but Eumundi, just off the Bruce Highway about twenty kilometres inland from Noosa, is a sleepy country town with a population of about 500—most of the time. It is famous, though, for its Saturday morning market, said to be the largest in Australia. On Saturdays, one is told, the population can swell to about 10,000 as buyers and sellers crush into the quaint little main street. We arrived in town on a Wednesday when the much smaller midweek market was in progress and that was busy enough. We had hoped to find some lunch but every coffee shop in town—and there are several—seemed to be full of people with more trying to get through their doors. We decided that it didn't really matter if there was no place to eat as there was nowhere to park anyway.

But normally the main street of Eumundi is very quiet under its broad canopy of huge Moreton Bay fig trees, planted after World War I as a memorial. Most of the buildings date from the turn of the century to about the 1920s, when Eumundi was a prosperous railway and timber town.

As Wagner spoke, the peace and tranquillity of her pristine bush setting were regularly cut into by the din from various directions—from the nearby busy township, the road to Noosa with its incessant traffic and the railway bridge near the turn-off to her charming tumbledown cottage.

Wagner spoke of a need for her own home and for Eumundi generally to fight against the increasing incursions of developers and the tourist industry. She feared that the quaint and village-like charm of a close-knit community would succumb as had other towns to being 'discovered'.

The UK-born Wagner—soft of voice, gentle of manner, kindly and self-effacing—told of the fifteen years spent in Eumundi and the twenty years in Queensland. Although delighted by so much that the Sunshine State had to offer, she was appalled by the 'White Shoe Brigade' mentality which had held sway in each government, regardless of political colour, over those two decades.

Wagner's most immediate concern as she gazed upon the lush vegetation impinging upon her verandah was a move to vary Eumundi's town planning guidelines to set aside a vast tract of land to build a hotel complex, swimming pools and bowls club—which she felt would change the ambience of the region utterly. She noted that if such a scheme were adopted the population of the surrounding district would treble; this would be calamitous not only for the residents but for the rare fauna and flora which thrive around Eumundi. The local Maroochy Shire Council voted against this proposal but the developer was appealing—and this was what gave Wagner disquiet.

'This will be yet another case in Queensland where anyone with white shoes can build anything,' she said. Since the day we spoke, that particular development appears to have been quashed, but the controversy remains as nearly half of the town's residents had supported the idea strongly enough to sign a petition that the development go ahead.

Further afield in Queensland, Wagner was disturbed by what was happening at Port Hinchinbrook and she was especially appalled that all of the planned devastation around Cardwell could be officially justified because it created new jobs.

'Slashing railway seats also creates new jobs for those who have to fix them—and that surely doesn't justify such an anti-social and harmful activity,' she said.

'The Port Hinchinbrook development will needlessly destroy habitat yet again—simply to make money for one arrogant man.'

Wagner likened the pending Hinchinbrook disaster to those happenings at the time she was writing *Goanna*, when a group of Aborigines were seeking to buy the station they were working on. 'The lease had to be transferred through the State Government, but Premier Joh Bjelke-Petersen said no,' she said. 'Joh said, "That's just land rights through the back door." And he was against it on principle.'

She contrasted how justice and social progress were denied to one group while another group had carte blanche to change the tradition, culture and environment of a place forever.

Similarly, and also as she was writing *Goanna*, the State electricity authority—at that time known as SEQEB—was determined to arbitrarily cut down any trees which were deemed to interfere with its plans—even if the trees were beautiful old gums.

'When the trees came down, the animals and birds which relied on them had nowhere to go,' she remembered sadly.

As one venerable gum was felled, she noticed a goanna at the resultant rubbish tip.

'It was a proud goanna running on hind legs on top of all that rubbish,' she recalled.

Hence emerged the added passion to educate young readers about the senselessness of such ruin. She sought to instil in her audience a realisation that being at one with animals is 'part of being civilised' and that to respect them was to respect ourselves and the careful natural balance of which everyone is a part.

Wagner admitted that she needed to be 'subversive' to get such a message across to a young audience, although she sought to entertain rather than present kids with an overt message.

'Children—and, indeed, their parents—must never notice that they may be getting some propaganda,' she said.

In seeking to influence her local community, she attended an anti-timber rally in Brisbane where she held up a picture-book to dramatically illustrate that old-growth trees were not needed to produce such a book. 'The nicest paper is wood-free,' she told the crowd. 'You don't need to destroy pristine forests just to make paper.'

As an educator, Wagner conducts workshops at local libraries such as Gympie, which she described as a cultural 'oasis' for the region. Gympie, with a population of about 15,000, is often thought of as the gun capital of Australia, but sometimes the local people rally to read and to discuss art and literature. 'Many of the locals are rednecks or fundamentalist Christians, but I am generally treated with respect—even if this does not always apply to my views,' Wagner said.

Despite her disarming modesty and self-proclaimed lack of environmentalist credentials—'There's no compost bin here but there *is* a cat'—her deep convictions shine out of everything she says.

Jenny Wagner continues to assiduously 'subvert' the children of such folk—and the goanna seeing its habitat submerged by a shopping complex's car-park is a very potent symbol.

THE words and ideas of these writers have proved mightier—certainly more effective—than the actions of any politicians or the campaigns of environmental groups: global, national or local. Australia's best-known and most enduring 'green' politician Bob Brown, author of illustrated books on environmental issues, probably put it best when he discussed the significance of communicating the message which persuaded people all over Australia to stop the Franklin dam: 'The Liberal and Labor parties wanted the dam. Both Houses of [the Tasmanian]

Parliament supported the dam. The unions wanted the dam. Big business wanted the dam. All three newspapers wanted the dam. Not a squeak out of the churches! But we kept going, and a year later, the river was safe.'

This statement appeared in *Wild* magazine (in Autumn 1999), which also has been committed to spreading the environmental message. During the Franklin blockade, songs were sung by thousands of protesters who inspired each other into the glory and justice of their cause. The epic Franklin struggle, like other struggles around Australia, was portrayed in novels, poems, plays and films. And these writers have sparked us to look inside ourselves in a way politicians never can.

Writers have made us think about what we are doing. They make us care about what we are doing. Then they make us do something about what we are doing. In their individual and collective ways they affect our hearts and minds forever.

2

NOVELS
Destruction, hope and glory

THE novel has always been an effective instrument for influencing the hearts and minds of readers. Throughout history, writers have chronicled human misery, injustice, greed and folly—and helped to change the societies they wrote about. In more recent times the rape of the earth, the degradation of its environment and the horrors which grow out of this destruction have been potent themes.

The novel has always had a social conscience and over the years has dealt with many issues. One thinks, for instance, of *Uncle Tom's Cabin* (c1858), which galvanised the anti-slavery movement in the USA and helped force that nation into its bloody course of reform. There is a story told that Abraham Lincoln once met its author, Harriet Beecher Stowe, and said to her: 'So this is the little lady who started this great war.' An overstatement perhaps, but there is little doubt that her highly charged prose pricked the complacency of average people, and focused their outrage on the great crime of slavery in their midst.

Writers such as Charles Dickens, Emile Zola, Alexander Solzhenitsyn, George Orwell and Thomas Mann similarly all wrote works that held up a mirror to their societies and in doing so challenged the prevailing wisdom of their times.

Novels are able to influence readers so deeply because of the nature of the relationship between the writer and the reader of fiction. In a non-fiction work, one is persuaded by force of argument or irrefutable fact to consider and maybe change one's beliefs. But in a novel, the reader enters a created world, which he or she sees through the eyes of its creator. Novels thus appeal to readers on an emotional rather than on a reasonable level, and are much more difficult to resist.

Since the 1970s, environmental concerns have dominated many major Australian novels and thereby influenced—and changed forever—the way Australians perceive the ecology. Threads of environmentalism have always appeared in Australian novels since the time of the First Fleet, but in recent generations the matter has become far more pressing.

Creative imaginations in all corners of the land have expressed their concerns in all genres, all moods, all styles—from the most famous, best-selling novelists to the most obscure. As to why our environment is becoming degraded beyond acceptable levels, almost all the novels state that the fault for this lies within ourselves.

A common theme in novels is the destructive paradox that the more high-tech society becomes, the more degraded do people's lives become; and, the further down the social scale, the greater are the misery and sense of hopelessness. In many Australian novels of the last quarter of the 1900s, the prospects for the earth and its inhabitants are depicted as extremely bleak. Life is increasingly brutal, barbaric and inhuman as the last vestiges of civilisation struggle to survive in soaring temperatures; the human metabolism can no longer properly function; and murderous competition for food reduces people to rodent-like scavengers.

The novels paint landscapes which are barren and describe those who walk upon them as bereft of decency; and if humans cannot treat other humans with dignity and humanity, there is not much hope for their life-supporting planet either. As Albert Camus (quoted in Rosie Scott's most appropriately titled *Feral City*) so eloquently put it: 'The only means of fighting a plague is—common decency.' Where decency is lacking, the earth is doomed; and, in the collection of fiction which follows, the lack of decency is painfully shown.

The increasing capacity of novels to influence Australians on environmental issues has resulted from the development over the past thirty years of a 'middle ground' of fiction, as defined by the Professor of Australian Literature at the University of Sydney, Elizabeth Webby. In 1999, Webby described many of Australia's finest novelists as being 'middle brow'—as distinct from 'popular' on one side of the spectrum, and 'high literature' on the other. Thus fine writers and their important works have been able to reach a much wider and more aware readership on such issues as the environment than would otherwise have been the case. 'Writers of crime fiction like Peter Corris, a popular novelist like Blanche d'Alpuget and even romance writers have been able to produce intelligent novels and share their concerns,' Webby said.[1]

One example of the broadening influence of important writers, according to Webby, is the case of George Turner, whose earlier 'literature' novels were largely obscure but who then attracted a large 'middle ground' of readership and greater influence with sci-fi classics like *The Sea and Summer*.

Webby, until recently the editor of the influential literary journal *Southerly*, said this middle ground of novels presented all of the important contemporary issues, all of which were interlinked, with environmental concerns examined in the light of how humans treat each other. 'Many recent novels have been a critique of the Grand Narrative in Australian writing and they challenge earlier assumptions of how Australians should see themselves,' she said. Webby said important issues like our relationship with the land could not be seen in isolation from the experiences of indigenous people, a reassessment of the place of women in society and human rights generally.

As such issues were increasingly canvassed in Australian fiction, Webby said she regarded the later works of many established novelists as being more 'important' and 'accessible' than the earlier works which made their names.

Into this category she put Patrick White, whose later works were better for the fact that they reflected such things as his growing environmental awareness, which he shared with an expanding readership. Similarly, the later books of the until then largely neglected Thea Astley were also more accessible in their style and content and had a new-found passion. Established writers, a New Wave of novelists and even practitioners of pulp fiction have all been reaching out to make us think and care and act on what is happening to our land.[2]

This middle way in Australian fiction and its growing concern with serious issues were also acknowledged by leading Canberra-based critic and academic Bruce Bennett, who described a 'publishing explosion' over the past 30 years. This large range of writing 'cut across genres' and created novels which were simultaneously 'accessible' yet part of a 'literary stream'.

Bennett, head of the School of English at the University College of the University of New South Wales at the Australian Defence Force Academy, said many prominent Australian novelists were dealing with environmental issues not as a 'manifesto' but as 'a subtle way of changing or reinforcing prevailing attitudes'. Influential novels which dealt with specific key issues in turn helped the reader to 'deal with these issues being faced in their lives'.[3]

Important themes or concerns could 'overtake a writer's thinking', according to Bennett, and in turn these could influence and inspire a growing and increasingly educated readership. Like Webby, Bennett pointed to the importance of an established master like Patrick White who 'in his novels traced the immediate environment of where he was living, first at Castle Hill in the north-western outskirts of Sydney and in his later works in the centre of the city at Centennial Park, with whose preservation he became indelibly associated'. Similarly, the writings of a new and young novelist like Tim Winton 'encouraged or stimulated a reverence for the land' in his loving and careful descriptions of West Australian scenes.

These writings were a counter-weight to modern city attitudes which tended to 'eliminate the environment' from people's thinking, where we try to pretend 'that hurricanes don't happen and that climate can be controlled'. According to Bennett, such ill-founded notions and the complacency which breeds them can be rocked by novels that describe ecological disasters and the consequences of mankind's actions. 'Living in an air-conditioned and high-rise cocoon tends to make us oblivious to many realities, especially if one is constantly on the web,' he said.

Bennett said this urban air of unreality had been satirically blowtorched by writers like Peter Carey, whose short stories and novels ridiculed ways of life that were cut off from their natural environment. Bennett noted that Carey, born in the Victorian rural town of Bacchus Marsh, was able to 'see things in cities with the eyes of a person brought up in the country'.

Bennett said modern Australian fiction, with its emotive and descriptive powers, was often 'more powerful than demonstrations' or any other political or organised actions in igniting people's concerns on ecological issues.[4]

The battered Australian environment as a touchstone of creative effort has been recognised by overseas novelists such as Canada's Margaret Atwood. 'The Outback is to Australia [as] the sea was to Melville, and as Africa is . . . to *The Heart of Darkness*,' Atwood stated on a visit to Australia. She described Australian novelists' concern with the ecology as a 'spiritual experience' and a 'contact with a deeper reality in Nature . . . it's a place of ordeal, and vision'.[5]

One common theme in recent Australian novels is that political leaders and bureaucrats—those entrusted to be guardians of our environment—are unsuited for their role. Therefore average citizens can expect no salvation, especially when ecological activists are ineffectual or fey—or worse.

THE landmark Australian novel which first raised awareness among modern readers of environmental catastrophe was Xavier Herbert's *Poor Fellow My Country* (1975), which is prefaced as being in homage 'to my poor destructed country'. Herbert, in his most important work since the famous and best-selling *Capricornia* (first published in 1938), traces how multinational greed (as personified by the Vaisey Company) runs rough-shod over the cares and aspirations of all local inhabitants, most of whom are too cowardly or apathetic to fight off threats.

Poor Fellow My Country was hailed upon release as a publishing triumph and quickly entered the best-seller lists, with 34,000 copies in hardback and 45,000 in paperback. It is a sprawling and spectacular novel, exploring countless themes including how mankind's shortcomings are severely harming the environment, and how a pulverised landscape is hurting people who already have overwhelming human deficiencies. The novel, which took decades to germinate and nine years to write, pursues one of Herbert's recurring themes—that the occupants of Australia are unfit for nationhood as long as they continue to pursue unthinking and vandalising policies whose main result is to enrich foreigners (indeed, he entitles Book Three 'Day of Shame—A Rabble Fled the Test of Nationhood').

Throughout the novel, which Herbert began writing in 1968, he warns that the vast land has the capacity to devour the children of those who have treated it ill. He makes a clarion call that Europeans will be in peril unless they learn from Australia's original inhabitants, who for millennia had realised that interference with the environment can have grave physical and spiritual consequences.

Xavier Herbert writes that since 1788 Australia's white settlers have 'disembowelled' the land and been guilty of 'spoliation' and 'greedy exploitation'. Of the process known as 'Opening Up The Country by Civilisation', he writes with poetic ferocity of how miners, graziers, developers and others had ravaged the countryside:

> Not a creek but its banks had been stripped to the rock, the detritus and humus

> gathered over a millennium and more put through the cradle and the dry-blower, washed away, blown away, its rock split with dynamite, its trees torn down for building and for firewood, seeds and seedlings dead for ever for want of earth to root in—there was gold in that earth, to be got at only by destruction of the earth.[6]

Poor Fellow My Country, described as the longest novel published in the English language, is a powerful, relentless and savage condemnation of how Australians have insufficient strength or resolve to protect their unique environment, and thus both their personal lives and the landscape which sustains them are facing ruin. The scene and tone are set very early in Book One, appropriately entitled 'Terra Australis: Blackman's Idyll Despoiled by White Bullies, Thieves and Hypocrites': 'We live in a land the wonder of which, as damned and doomed Colonials, we've been unable to see. What wonder is in our lives is taken at second-hand, moth-eaten, only half comprehended, from our origins.'[7]

As in so much of his writings spanning more than five decades, Herbert contrasts the unthinking and uncaring treatment of the land since Europeans arrived in Australia with that of the Aborigines over many thousands of years. 'To the blackfellow, there's magic in everything,' Herbert writes. Whereas the white man has always had an exploitative relationship to everything, the original inhabitants have always had a 'reverence' for an environment which they hold unique and sacred, be it 'rock, tree, waterhole'.[8]

Herbert, one of the nation's best loved and most highly respected novelists, creates in *Poor Fellow My Country* the character Jeremy Delacy, in whom is encapsulated a conscience and outrage so tragically lacking in most of his fellow Australians. Jeremy, as Herbert's environmental alter-ego, muses to his silent listeners—that is, Australia's indigenous fauna: 'I feel guilty about you, my friends. What have I done to you, in my whiteman's arrogance and ignorance of Nature?'[9]

Early in the novel his anger is directed at Lydia Lindbrooke-Esk, the almost caricature-like companion of Lord Vaisey, arch-representative of the British marauder. Jeremy has a certain grudging respect for Lydia, but their irreconcilable difference in morality and attitude is quickly discovered. Much of their conversation would be equally relevant in the Mabo and Wik debates almost two decades after the book's release:

> [Jeremy:] Conservation and pristine landscape . . . they're not the sort of words they use around here.
> [Lydia:] Well, no . . . as a matter of fact they said you want to put the country back the way it was and give it back to the blacks.
> [Jeremy:] The fools . . . they've no idea that the destruction they do is cumulative and must catch up with them sooner or later.[10]

Delacy weeps for the senseless destruction all around him and decries that 'the rule is to cut down every native tree', to be replaced by 'alien' varieties. And everywhere are 'weeds, weeds, weeds', all as part of 'the rubbish of the earth they've brought in here' and 'every abomination known to man'. What

Eric Rolls and other writers have documented in non-fiction works, Herbert describes creatively: the importation of goats, rats and cats which had all bred profusely and were rampant throughout the countryside in their millions.[11]

Herbert, a pharmacist who throughout his life held a variety of jobs and spent several years roaming in rainforests, was among the first novelists to espouse the theme that without decency everything is doomed; from the mouth of his alter-ego Delacy comes the quite explicit observation as he ponders Anzac Day in 1937 that 'we're a people without compassion . . . What sort of nation could do what we do, to animals and simple savages?'[12]

In his thorough and breathtaking exploration of Aboriginal culture and folklore, Herbert introduces the notion of Wrong-side. This is an enduring metaphor for the desecration of a viable balance in nature which had been in place for thousands of years. Aboriginal lore and legend had always decreed that severe punishment would be meted out to those who were responsible for the negative and destructive actions of Wrong-side.

Herbert's depiction of Wrong-side was noted by literary critics, who recognised this theme as an important metaphor for senseless ecological destruction. 'Herbert's construction and presentation of the myth of what might have been, his vision of a Terra Australis del Espiritu Santo respected by its inhabitants and expressing the magical spirit of the land is the great achievement of this book,' wrote critic Ken Goodwin in his comprehensive *A History of Australian Literature* (1986). To Goodwin, *Poor Fellow My Country* has a 'tragic quality' in depicting in the most forceful and evocative manner 'the ignorance, carelessness, hollowness and cupidity of the white invaders'; it is 'an epic lament for the despoliation of the land by its white inhabitants and their failure to make use of the many opportunities available'.[13]

The latter part of *Poor Fellow My Country* deals with Herbert's concern with uranium mining and the foreign influences increasingly ready to trample Australia's sovereignty to ensure supplies of the ore. Delacy's Lily Lagoons had plentiful supplies of yellowcake, causing the author to deplore the same exploitation there which was to happen 'from the desert to the sea'. Those who sat on a 'billion dollars' worth of' supplies of uranium—or alumina, manganese, iron, copper, calcium, platinum, molybdenum—'name it, and all you had to do was pull down a hill, drain a billabong, level a forest, tear up a beach, to become a millionaire! What matter if you left a desert behind, when you didn't have to live there?'[14]

Of nuclear testing in the Outback, Herbert says these have 'rendered the localities where they took place as uninhabitable as the Moon [and] the Australian people had proved to be so unconcerned about the alienation and devastation of vast tracts of their country by foreigners as scarcely to have raised a voice in protest . . .'[15]

The book's quintessential pessimism was recognised upon publication, as in the *Sydney Morning Herald*'s review which described it as 'without doubt the Australian novel of the most epic and deeply tragic vision and scope'.[16]

Another prominent critic, Laurie Clancy, in his *A Reader's Guide to Australian Fiction* (1992) grouped Herbert with Henry Handel Richardson and Christina Stead as 'mostly pessimistic and sometimes nihilistic, viewing the world often with the suspicion that is it a kind of giant mistake or joke carried out for inscrutable purposes and having no intrinsic meaning or purpose'. In a discussion mainly of *Capricornia*, Clancy dissected Herbert's disgust at his poor destructive fellows. According to Clancy, Herbert's European 'makes war on his own kind, ravages the environment, and exploits its original inhabitants'. Furthermore, *Poor Fellow* continues the bleak scenario of *Capricornia*, where the earth is left 'stark, bewildered and distressed'.[17]

Poor Fellow My Country remains consistent across its giant canvas in that, amid a plethora of issues and problems, it begins and ends with allusions to an increasingly shattered environment which is at the core of the work.

Xavier Herbert was himself consistent throughout most of his 83 years in his highly pessimistic view of Australians. In his last recorded interview, in mid-1984 at a camp eighty kilometres from Alice Springs, he described Australia as 'a nation of liars and rogues, generally'.[18]

Yet Herbert's outlook was not entirely bleak and he saw positive outcomes in brave, united stands such as the Franklin blockade in 1982. In a telegram to the anti-dam protesters, Herbert said: 'The only hope, not only of saving Terra Australis from eternal ruin, but the terrestrial globe itself, rests with you brave hearts now going into battle. My heart and soul go with you, they will die like the heart and soul of my country if you fail.'[19]

Another of Australia's major novelists, and one whose career overlapped Herbert's, was Patrick White, who adopted a far more subtle approach to environmental issues but whose outrage nonetheless shone through in his later fiction.

Pessimism and despair pervade much of Patrick White's writings. His gloom stemmed both from the sadnesses in his personal life and having witnessed such destructive events as the London Blitz and the Spanish Civil War. In his later years he felt things deeply and was radicalised into seeking to preserve the wonders of his immediate neighbourhood—Sydney's Centennial Park.

Patrick White's early fiction reflects the precariousness of the human condition. He observed, via the character of Lieselotte in *The Aunt's Story* (1948) that 'we must destroy everything, everything, even ourselves'.[20]

From an early age White, born in England in 1912, suffered from asthma and believed the reserves of coal under his parents' property in the New South Wales Hunter Valley might be responsible. On one trip to the Hunter, in 1948, he had a 'massive' asthma attack and, according to his biographer David Marr, 'after being given oxygen and morphine, he was put on the train for Sydney where he recovered quickly'.[21]

It would take many years before his views on the ecology would crystallise, but the more he saw of degraded environments the more he would seek to expose destructive acts and to change things.

As early as 1955, in *The Tree of Man*, White would portray the environment as desolate and fractious. According to Marr, 'White and Stan Parker (the ill-fated central character in *The Tree of Man*) share a God that moves with the weather, a God of storms, floods, fire and disaster . . . Lightning is his mark in the sky.' Such forces combine to bring down someone Marr describes as one of the few 'good men' in White's fiction.[22]

White's novel *Riders in the Chariot* (1961) includes his revulsion at how a landscape can be made ugly by unfettered home-building. A year before its publication, White was becoming increasingly concerned by the housing developments springing up in the then still semi-rural suburb of Castle Hill on Sydney's north-western outskirts. As the New South Wales Housing Commission was establishing plans to buy up property near Dogwoods (the farm he shared with his long-term partner Manoly Lascaris), White wrote in a 1960 letter: 'Of course we don't know how much of [the Housing Commission's] horrible boxes will be visible through the trees . . . But usually when they start to destroy in Australia, they do things very thoroughly. The national genius is for ugliness.'[23]

Marr describes the situation at that time for White and Lascaris as 'living in an oasis under threat' and the biographer shows how this was translated into *Riders in the Chariot*, whose conclusion depicts 'the chariot itself, like a golden sunset, passes over the fibro houses, ugly yet full of life, standing on the ruins of Xanadu'.[24]

At the same time White was reading volume after volume of Manning Clark's *History of Australia*—a task which would take him about 20 years. He wrote to Clark during Christmas 1968: 'Interesting to see how we have remained the same pack of snarling mongrel dogs.' Most of White's subsequent works would with increasing ferocity reflect his views of Australia and Australians.[25]

Concern about the environment and a growing compassion for people whose needs and aspirations were being brushed aside by the system are a hallmark of White's writings throughout the 1970s. One of the major catalysts was in his own backyard where he would fight tenaciously to preserve the special tranquillity of Centennial Park. When he learned of plans to turn this area and the adjoining Moore Park into giant sport complexes (and a bid to host the 1988 Olympic Games during Australia's Bicentenary), White was radicalised utterly. The previously shy and diffident writer became a vociferous activist, appearing on television, addressing public meetings and leading thousands-strong marches.

Despite the poor health of his later years, White would fight doggedly for his environmental concerns until his death at his beloved Centennial Park house in September 1990.

In *The Eye of the Storm* (1973), White encapsulates many metaphors of decay in the aging Elizabeth Hunter and those gathered around her during her last dying days. As everything around her goes to ruin, she contemplates how houses had suddenly become ugly and she fears for the desecration of Sydney's magnificent Centennial Park, near where she resides. Elizabeth Hunter laments until her last breath how anybody could subvert the park's tranquillity and grandeur.

The fate of Mrs Hunter—just like the fate of all things once glorious—is doomed, but she nevertheless retains a steadfast belief in the resilience of all things to withstand depredations. Her speculations on her own mortality could equate to her wider views about the universe: 'But I shan't die—or anyway, not till I feel like it. I don't believe anybody dies who doesn't want to—unless by thunderbolts.'[26]

White deals with these core issues of humanity's relationship with the earth more indirectly than does Herbert, but he is just as powerful, emotive and fierce. He uses analogy and metaphors to decry the destructions and, despite an overwhelming streak of pessimism, he revels in all that is precious and worthy.

As he paints the scenes surrounding Centennial Park, White describes a metropolis in crisis:

> . . . all would be swallowed up in a great rubbery volume of traffic. If the lit windows of houses offered a human belief in permanence, the oil refineries ablaze down Botany way suggested other worlds, other more demoniac values. . . . [Mrs Hunter's nurse] Flora Manhood would have liked to destroy something tonight. She breathed deeply of the chemical air and hoped she was contracting lung cancer . . .[27]

As the many warped characters struggle against their fates, White rushes headlong with images of grimness, of 'noise', of 'traffic', of 'exhaust fumes', of 'lung cancer', and of 'chaos'. White proceeds to visions of expanding urban blight, and he bitterly questions what can one expect of a building called 'Miami Flats' located in the midst of a Sydney suburban wasteland: 'at the entrance a fluorescent light sat spitting at the top of its pole, so that all the potplants in the hanging gardens of "Miami" were flickering sickeningly.'[28] Sydney had become for White a place of 'neo-brutal towers' at the epicentre of which lay the dying Elizabeth Hunter who, after a lifetime of corruption, was a deadly metaphor for uncontrolled decay.[29]

Urban blight is an omnipresent reality for all of these complex souls tossed together by fate and by Mrs Hunter. Another of Mrs Hunter's carers, night nurse Mary de Santis, sexually repressed but potentially overflowing, is especially ground down by realities she, and hundreds of thousands of other Sydney residents, simply cannot escape: 'she was only here in her "conveniently situated flatette" to spend a few daylight hours, most of those asleep. And the rent was low: not low enough, perhaps, considering the view of visceral plumbing exposed against the wall of ox-blood brick opposite.'[30]

These vistas, although very much in the vicinity of Mrs Hunter's more opulent house, are admittedly in poorer parts of Sydney. But the more elegant and exclusive harbourside and oceanside eastern suburbs also are facing a gradual deterioration. At Watsons Bay, graffiti and neglect are impinging on the supposedly highly desirable seaside district. Elegant dining is affronted by 'oil scum', 'detritus' and 'the stilts of a bleached jetty'.[31]

Whenever hope or optimism briefly emerge, they are instantly snuffed out by White. As soon as the joys and beauties of a glorious sunny day at Watsons Bay are enjoyed, the tranquil scene is quickly shattered by disgraceful mob behaviour.[32]

Biographer David Marr describes *The Eye of the Storm* as 'the progress of ugliness and greed', with White directly confronting 'corrupt Sydney, more corrupt than ever . . . [it] was booming . . . half the city seemed to have been torn down to make way for cheap apartments and glass office blocks'. In a letter in 1971 to his cousin Peggy Garland, White spoke of 'a most horrifying wave of vulgarity . . . sweeping the land'.[33]

White was just as forthright when addressing public rallies, such as his first foray into such things at age 60, when he told a crowd at Centennial Park on 18 June 1972 that he would leave Australia if an Olympic sports complex was built on such hallowed ground.

'A park is a living living-room, in which to do anything and nothing . . . it will become increasingly important as this reckless anti-civilisation gathers force,' White told the rally. 'Your parks are your breathing spaces. Guard them, cherish them . . . Protect your parks from the pressure of political concrete.'[34]

In *The Eye of the Storm* White is utterly savage at what humanity has wrought and how people live. He is remorseless in his depiction of the corporate executive types he deems more responsible than most for blotting out any niceness left in a once lustrous and compelling landscape:

> The sun had gone in besides, behind a drift of dirty cloud. And once your vision is withdrawn from you, there remain the lapping shallows, the littered sand, one competing with the other for the sludge to which the human spirit can sense itself rendered: an aimless bobbing of corks which have served their purpose, and scum, and condoms, and rotting fruit, and rusted tins and excrement.[35]

In one flashback describing Elizabeth Hunter's once-glamorous life, the scene is of an island off the Queensland coast where everything is under threat from the mad onrush of development. An indiscriminate chopping-down of trees throughout the island is described as 'violence . . . recently exploded' and the scene is set for White's next major novel.

This novel is *A Fringe of Leaves* (1976), based on the experiences of the marooned and shipwrecked Eliza Frazer, who early last century suffered indignities and humiliations when captured by Aborigines. White's Frazer character is Ellen Roxburgh, who is shown to face an unspeakable future in an environment utterly alien to her fine British upbringing. Ellen is 'in a country designed for human torment, where even beauty flaunted a hostile radiance, and the spirits of place were not hers to conjure up'.[36]

White depicts Ellen Roxburgh as weighed down by a 'resignation to suffering' and there is a dread of an 'apocalypse' as she 'recoiled shivering' from vistas which were entirely 'colourless'.[37]

As the plot unfolds towards the ill fate of Ellen—and of all those who sailed with her on the *Bristol Maid*—White remorselessly shows how from Van Diemen's Land to Sydney and north to Queensland life everywhere in the Australian colonies was harsh. Britons' arrogance and complacency in seeking to tame this alien continent are cruelly shattered during the voyage, and on both sea and land there are constant harbingers of disaster.

White wrote *A Fringe of Leaves* as he was becoming increasingly appalled at—and would most vigorously fight against—the bulldozing away of precious heritage. According to Marr, White had visited Fraser Island and had been enchanted by its stunning and pristine rainforest and by a solitude which had disappeared from almost everywhere else in Australia.

From the moment in 1974 when he witnessed at first hand the risks being faced by the island, he incorporated his fears into *A Fringe of Leaves*. Marr is emphatic: 'From this point, the writing of *A Fringe of Leaves* and the campaign for Fraser Island ran side by side.' And, as White wrote to poet Judith Wright in late 1975: 'Any kind of artist who has been to that island will remain affected by it for ever.'[38]

To White, Fraser Island was a microcosm of outrages which already had happened or were threatening, and as part of a growing international correspondence he wrote to a Swedish literary critic, Ingmar Bjorksten, in November 1980 that the only way Australians could come to their collective senses was if they became the victims of 'invasion, bloodshed, foreign occupation and a taste of starvation'.[39]

The letter to Bjorksten was among attempts by White, a major international literary figure (who had won the Nobel Prize and been awarded the Order of Australia) to influence world opinion. He wrote open letters to world leaders promoting peace and contacted prominent Australians on a range of specific issues. Despite failing health he travelled throughout Australia promoting the cause of nuclear disarmament.

Another long-established author whose later works strongly reflected her environmental concerns is Thea Astley, whose *An Item from the Late News* (1982) sprang directly from her concerns about what she saw as nuclear madness and the insane attitudes which put the planet at risk.

A bomb shelter is at the core of the novel and its main character, Wafer, is typical of Astley heroes in being one against many—that is, a sensible and caring person confronting the stupidities of the mob.

An Item from the Late News was described by Laurie Clancy as being 'as appalling and scathing a vision of life in rural Australia as has come from any novelist since Barbara Baynton . . . The pessimism and anger are almost unrelieved'.[40]

Similar assessments were published overseas, such as in the Canadian magazine *World Literature Written in English*, where Monash University's Greg Houghton described the novel as having 'a new urgency and despair in the face of imminent nuclear destruction' which represented a more ominous phase of Astley's 'chaotic, bleak and violent fictional world'. Houghton wrote that this novel portrayed 'a surreal world without hope, at a minute to midnight'. According to Houghton, Astley is 'urging us to realise that for the human race the night is very late indeed'.[41]

Astley—who throughout her life has lived in various parts of Queensland, in Sydney and on the New South Wales South Coast—was expressing a 'real rage' against tiny-minded provincialism where 'life is stranger than fiction'. Indeed, as she said in a 1985 interview, 'some of the stories [about small-town pettiness] I heard were so awful I didn't use them'. Although believing that many rural folk are thoroughly decent and the backbone of the nation, they often are swamped by less intelligent beings, with disastrous consequences for everybody's future.'[42]

'It's the indifference that makes me rage,' protests the narrator as she observes how a small town is able to ignore survival issues because it has far more important small-town concerns to worry about, blithely ignoring how the 'radio daily brought us the constants of 20th century living—neutron bombs, ICBMs, MX missile shelters . . .'[43]

There is no such complacency by Wafer, who envisages an industrialised world criss-crossed with bomb-shelters and other means of protection against an impending holocaust.

Astley's *It's Raining in Mango* (1987) virtually takes up where Herbert left off in *Poor Fellow My Country*, by showing how the earliest settlers—both convicts and those who kept them in chains—were unsuited to sensible stewardship of their new land. There were no moral foundations for building a humane society in this strange new land, whether in the treatment of the soil or of its original inhabitants: 'Only a few decades ago [in the early 19th century], poaching in Britain meant transportation or even hanging. Here, poaching is approved by the government, and murdering the owners of the local "grouse" blinked at.'[44]

Thus the scene is set for the hacking-down of virgin rainforest in North Queensland regardless of the wishes of the local population, with ruthless developers ready to squash the protesters, who are powerless to achieve a different fate. Many of the protesters are worthy and committed—including greenies with lumps of concrete chained to their feet and others roped in the top branches of trees—but Astley equally shows others to be selfish hippies whose inherent weaknesses could never overcome a determined developer.

Another stunning Astley work is *Vanishing Points* (1992), comprising two novellas, 'The Genteel Poverty Bus Company' and 'Inventing the Weather', which with exhilarating observation, uncontained rage and scathing humour examine the shallow developers who would destroy

North Queensland towns and the motley crew of locals unable to do anything about it.

In 'The Genteel Poverty Bus Company', written when she was 67, Astley creates one of the truly inspirational characters of Australian fiction. Macintosh Hope, a retired academic of the most noble yet quixotic integrity, seeks nothing but solitude and a quiet life and is prepared to fight almost literally to the death anybody who would invade his space.

Mac is in reclusive retirement on Little Brother Island and is not prepared to meekly accept the non-stop din of the yuppie paradise being built on neighbouring Hummock Island. When Mac is affronted by disco sounds he retaliates at the dead of night with sonic booms of the finest classical music. Mac is not intimidated by either the project's developer, Clifford Truscott, or the thugs sent across the water to keep him in line or the corrupt authorities who have allowed these things to happen.

Macintosh Hope is an inspiration to anybody determined to actively resist oppression. He realises that with all the forces arrayed against him he will eventually 'sicken into some soft amorphous splodge of submission' but equally strongly agrees with his mate Gamble who told him 'you don't have the right, nor the moral right, to give in so early, so easily'.[45]

From the earliest stages of the rape of Hummock Island, Mac had realised the huge struggle needed to overcome 'the prodigious power of money backed by government'.[46]

Even earlier, Mac and his assorted collection of idiosyncratic friends had seen that 'these are the last of the good years' in North Queensland what with coastal and island resorts not to mention the plans for a Cape York launching station which would ignore the wishes of Aborigines and most other local residents and with no thought of the effects on a fragile ecology.

Alternately a thoughtful tract and a rollicking and effusive read, 'The Genteel Poverty Bus Company' ingeniously examines the attitudes among a group of completely different people towards the development madness in North Queensland. Of this group Mac is brave and resolute to the point of fanaticism, unforgiving to the point of manic zeal—and Astley suggests that such stern stuff is what is required to slow down the universal destruction.

Like Rosie Scott (discussed below), Astley laments the demise of once magnificent Pacific islands becoming 'plastic Disneyworlds, shrines to the god of hamburger' and places of 'sewage and oil-spilled ruined waters'. Mac's own island also had 'a bloody history of settlement and destruction'.[47]

Also, 'The Genteel Poverty Bus Company' is a challenging and powerful depiction of noise pollution. After the book's release, Astley told of how the 'Sound Wars' sequences sprang from her distaste for loud music being imposed on people. 'I once tried to travel from Cairns to Sydney in a bus but I had to turn my ticket in in Brisbane because I just couldn't stand the radio six inches above my head, non-stop, from six in the morning till nine at night,' Astley recalled in 1992.[48]

The 'unneighbourly dispute' as depicted in the novella is, according to critic Peter Pierce, 'a metaphor for the ailments of materialistic, development-driven, amoral Australia'.[49]

The second novella, 'Inventing the Weather', shows Clifford as observed by his increasingly disillusioned wife Julie. According to Julie, Clifford Truscott is a hollow man—and all that he stands for is immoral and hypocritical. Julie shows increasing disdain for all of Clifford's activities; 'Any moment he will be looking for more coastline to despoil,' she complains. 'Clifford, having completed the despoliation of Hummock Island is looking for fresh coast to conquer.' What especially outrages Julie is that Clifford, like all of his ilk, 'has no moral guilt about Aboriginal owners, no sensitivity for the preservation of cultures, no sympathy for the disinherited'.[50]

Also, like Rosie Scott, Astley points to AIDS—as well as other diseases and the inhumanity on Aboriginal settlements—as a microcosm of wider environmental ills: 'Half the adults [on North Queensland Aboriginal reservations] rotten with DTs or TB. Most of them with syph. We're great colonisers, aren't we? There's even AIDS at a couple of places. The town whites go out and sodomise the gins.'[51]

These novellas are important contributions to contemporary fiction and have been recognised by critics as defining the extent of environmental loss, especially in the Sunshine State. Loss was the central theme of *Sun-Herald* reviewer Susan Geason, who said the vanishing points included 'people, tribes, ways of life, a landscape . . . and along with them, innocence'. Geason saw *Vanishing Points* as pivotal: 'There at the vanishing point, with notebook in hand, stands Thea Astley, bearing witness. May she thrive—for Queensland's sake, for all our sakes.'[52]

Born in Brisbane in 1925 and living much of her life in North Queensland, Astley has been recognised for almost three decades as one of the nation's major novelists, winning three Miles Franklin awards and many other prestigious prizes.

Much of her fiction after the early 1980s was based on her belief that humanity's 'vileness' was causing major ecological destruction, such as that in the Daintree rainforest of North Queensland. She was becoming increasingly concerned about the environmental omens as they were emerging in the tropics, as she related in 1985 at the Canberra Word Festival: 'The summers are becoming more and more violent. We haven't had a proper wet since 1979; it's probably because of the French with their tests.'[53]

Astley, who taught for five years throughout country towns in Queensland after graduating from the University of Queensland and the Teachers' Training College, also showed in her novels how entrenched attitudes of proud and tradition-bound rural dwellers posed one of the major obstacles to environmental reform, and she told the Word Festival how people in country towns 'are frightened of intrusion into the calm of their ways'.[54]

Such insular attitudes and the political climate were largely responsible for Astley and her husband Jack Gregson quitting Queensland. In a 1989 interview, she lamented how the Bjelke-Petersen Government was responsible for 'tearing the trees down' and the rampant development it encouraged.[55]

Thea and Jack began their retirement in 1980 near the Queensland township of Kuranda but then in 1985 moved to the New South Wales South Coast near Nowra. Her new place—a two-hectare spread which includes a five-bedroom house, appropriately solar-powered for its energy needs—has been best described by Ray Willbanks, an American scholar who interviewed the author for his *Speaking Volumes* (1991): '[Astley lives] on a hillside overlooking bright green pastures and distant valleys, and in the far distance the Pacific is a shining blue line. Kookaburras fly through the eucalyptus trees . . . kangaroos disappear in the bush that encroaches [on their property].'[56]

Of Brisbane itself, Astley has been quite scathing, describing the 'essential ugliness' of the Queensland capital and how it had 'a sort of lacerating effect the way ivy on cottages in the Cotswold would have a soothing effect'.[57]

Despite Astley's heartfelt criticisms of Queensland she has maintained an affection for the place, especially the Cairns region where she and Jack lived for more than ten years. In 1992, after revisiting the area, she described Magnetic Island near Townsville as being overrun by Melbourne developers. 'That's what happened to Cairns,' Astley said. 'When we returned there last year, I didn't recognise the city. It was one mass of glitz. I haven't been to Hamilton Island, but the pictures of the high-rises right in the middle of the Whitsundays is one of the greatest obscenities perpetrated on the coast.'[58]

Asked by Willbanks about how she perceived the importance of 'nature' in her novels, Astley replied how environmental concerns were not only at the core of her literature but the essence of her being: 'The only sorts of regrets I have about getting old is that there are fewer days to watch leaves turn over . . . I think the notion of the whole world as a Garden of Eden is as good a way to approach the Creator as any way is. But man has messed up the Garden. I find myself lately even feeling little spasms of guilt as I pick caterpillars off the acacias and crush them.'[59]

An echo of Astley's eccentric loner Macintosh Hope is evident in Lachlan Hook, pivotal character in James McQueen's litany of destruction, *Hook's Mountain* (1982). Hook displays the greatest bravery and personal resolve in defence of his little patch of Tasmanian hillside, but all of his stupendous strength is no match for the organised and corrupt entities whose sinister power holds an apathetic population in thrall.

McQueen, who with his son Stuart was arrested in January 1983 during the blockade of the Franklin dam site, a year earlier had turned down a $5000 State literature grant in protest against the Tasmanian Government's actions. But McQueen, an enigmatic figure, has often described his writings as being above all 'pragmatic' and has regularly expressed a distaste for greenie activists. McQueen, born in 1934, has also regularly defined himself as Tasmanian to the core and since 1978 has lived at Nabowla, fifty kilometres from Launceston. The remote and tiny hamlet is the setting for *Hook's Mountain*, a novel that exposes the ravages of the forestry industry.

Just as he was prepared to risk arrest over the policies of the Hydro-Electric Commission, McQueen was equally angered by the Tasmanian Forestry Commission. 'It's a tragedy,' he said at the time of the book's release. 'The Forestry Commission is destroying the natural environment.' During the same interview, McQueen intimated that writing a book like *Hook's Mountain* was the best way he could voice his protest rather than to demonstrate or to participate in direct action. 'I'm certainly not about to grab a rifle and shoot it out,' he said. 'They might put me in prison, and then I wouldn't be able to write.'[60]

McQueen has created in Hook a character who must steel himself for loss after shattering loss, but, as the hardened old warrior that he is, he is totally committed to his lost cause:

> . . . he was already beginning to consider the problems of water, of food, of equipment, of tactics, of time, of endurance; of possibilities, of necessities, of imperatives.
>
> His mind had quietly engaged the cogs of war.[61]

Lachlan Hook is not prepared to meekly accept the inevitability of his mountain being razed. 'They can't clear the whole bloody hill,' Hook says of the faceless, nameless developers. 'Christ, it's my front yard, my window, my fucking world.'[62]

There is nothing the locals can—or want—to do about an area increasingly denuded of native forest, with major clearing-works scheduled for everywhere, mostly for the uneconomic and senseless pursuit of the woodchip export trade to Japan.

Bruce Bennett has recognised the 'conservationist ethic' of *Hook's Mountain*, and seen it as a study of 'defiant individualism' in defence of natural heritage.[63]

Among the natural heritage being guarded by McQueen is anything and everything concerning his beloved Tasmania. He is even worried about the capital, Hobart, which he feared was losing its identity and essence. In an article for a special edition of *The Bulletin* magazine on Australia Day during the Bicentennial, McQueen lamented that Hobart 'is disappearing, is being slowly submerged like some antipodean Atlantis'. McQueen wrote of Hobart becoming littered with 'architectural flotsam' and he described 'imitation *bauhaus* and pink-brick block-house and lavatory gothic'. He sadly chronicled

Hobart's past being brushed aside forever: 'The sandstone islands of the old colonial town are everywhere besieged by plastic and ferro-concrete, and the smell of tar and salt is gone from the canyons of the city.' McQueen acknowledged that, unlike Cairns and other of Australia's tourist meccas, Hobart had been spared some of the developmental excesses: 'the public buildings have not been sold, nor the police station, and no Japanese developers have made a bid for the golf course. But give it time though . . .' Furthermore, what had been a city of history and character had degenerated into a 'certain counterfeit quality ... a certain fakery, a touch of all-too-obvious window-dressing'. Indeed, Hobart was stunned by feelings of 'some mythic apocalypse'.[64]

If Hobart was thus, Launceston was hardly any better, according to McQueen when he described what made him move to Nabowla originally. In an article for *Overland* in 1985, McQueen said he and his wife Barbara moved away from Launceston, 'that most constipated of provincial cities' and felt they needed in the bush 'to find a more solid and natural basis for our lives than we had had in the city'.[65]

In the article, McQueen depicts the 'gangs of the Forestry Commission' moving indiscriminately through the hilly terrain 'like a slow razor, left to right, east to west'. His anger at what he saw would be translated into the angry novel:

> It seemed incomprehensible to me then—as it still does—that they should be stupid enough to strip this beautiful hill, particularly when it is so *visible*, such a dominant feature of the land. The act seems almost a salting of malice on the rest of the wounds they have left behind. But their stupidity seems always to have that vicious, selfish quality. In quite practical terms, it seems unbelievable that otherwise reasonably rational men should tear out thousands, millions of acres of prime valuable hardwood to plant an exotic weed that produces what must be one of the *worst* timbers in the world.

McQueen noted that such activities on Blue Hill near Nabowla had been stopped, but he warned against any complacency. 'Whether it stopped because the slope grew too steep, or because "some bastard wrote a book about it and it's on the ABC", I'm not sure. What matters is, of course, that they are certainly still clear-felling somewhere else.'[66]

ONLY six years after *Poor Fellow My Country* but a world away in its offbeat modernity is another classic best-seller, *Bliss* (1981), which established Peter Carey as a major and important Australian novelist. In its extremely idiosyncratic and wacky style, *Bliss* is equally strident and pressing in its lament of Australians' fatal shortcomings in making the earth safe for human habitation.

Peter Carey's principal literary biographer, Anthony J. Hassall, has described *Bliss* and many of his other works as dissecting the 'blacker recesses of the personal and the national psyches' of Australians, with Carey characters depicted as 'trapped in surreal psychological dungeons, or the mad machinations of a technology going terribly wrong'. Carey's stories and characters are 'on the point of collapse or revolution'.[67]

Bliss was the first popular mainstream novel to suggest that the city is an utterly rotten place and most of the foods eaten by its inhabitants are poison. Four years after its release it was adapted for the cinema by Carey and director Ray Lawrence. *Bliss* the movie is a light-hearted and breezy version of the book and lacks some of its dark and menacing qualities.

The main characters are Harry Joy's nuclear family which looks at first glance to be the bedrock of what middle-class respectability is all about. Yet with every new incident each member is seen as personally corrupt, selfish and destructive. The decencies which are required to make the family prosper are built of straw. The corruption and dissolution everywhere are creating unsustainable pressures from which a new world must emerge.

Harry's extremely corporate and extremely urban world is imbued with bucolic and fresh influences when he meets Honey Barbara, who utterly rejects everything—especially the toxin-laden food—which the city has to offer. To Honey Barbara, the city is a place 'exuding poison . . . half machine, half human'.[68]

Environmental awareness is introduced only gradually as Harry's daughter Lucy is discovered to have rejected the use of pesticides around the house; this is put down to affectation, a quirkiness and the rebelliousness of youth.

If this consciousness from such an unexpected source came from left field, so does the revelation that Harry Joy, for most of his working life in the thick of an advertising agency catering to the highest echelon of corporate clients, is extremely concerned about the state of the planet and is prepared to take great risks in that cause. It had not always been thus; until his revelation, Harry Joy, advertising agency executive, was 'never heard to criticise anyone (or for that matter anything). He exhibited a blindness towards the faults of people and the injustices of the world which should have been irritating but which seemed to have almost the opposite effect: his very blindness reassured those around him and made them feel that their fears and nightmares were nothing but the products of their own overwrought imaginations'.[69]

From such inauspicious and unlikely surroundings emerges the new Harry Joy, who has turned his back on the corporate world, from which he has earned his livelihood by manipulating—and creating—its images. Joy suddenly finds himself in 'Hell' and only then does he dramatically realise the flaws and contradictions of 'normal' life.

The Harry Joy version of Hell is not only horrible but also polluted and tainted; it has a 'hot red centre' and 'hard industrial orange' from where 'poisonous yellows' radiate. Harry has apocalyptic visions of the Hell in which he suddenly finds himself and needs a special map to navigate: 'the river below appeared as black as the Styx. Barges carried their carcinogens upriver and neon lights advertised their final formulations against a blackening sky.'[70]

This Hell has all the temptations placed in the Garden of Eden, except that the sullied apple is doubly deadly as 'gleaming fruit had DDT lying just beneath the surface of its tempting skin'.[71]

The world of advertising is one well known to Carey, who worked for several agencies as he established his writing career. He was able to find his own ethical path through the advertising world, the disdain of which would feature so prominently in his novels. For example, in 1970, while working part-time as a copywriter at the firm Masius Wynne Williams in Melbourne, he wrote the campaign for the Vietnam Moratorium.

Carey, born in 1943 at Bacchus Marsh about fifty kilometres north-west of Melbourne, has lived in such far-flung places as the Victorian capital, a commune near Brisbane, in New York City and in a waterfront house at Balmain in Sydney. *Bliss* was written at a time when he was living at Yandina, an alternative community on Queensland's Sunshine Coast near Maroochydore.

Consistent with many of Carey's own stated attitudes, Harry Joy dramatically develops a consciousness about the world's ecological plight. He comes to reject the ultra-pragmatic, cynical and uncaring attitudes of the corporate world. These attitudes tolerate the spread of cancers and people dying of toxic poisons as long as progress—and profits—can continue unfettered.

Harry Joy's actions are consistent with what Carey would have done in such a situation. Asked by Ray Willbanks about the compatibility of running an ad agency in Sydney and creative writing of conscience, Carey replied: 'We never took on any business we didn't want. We were often offered tempting things, for example a cigarette business worth $5 million. We had no hesitation in saying no to it and lots of other things as well. If you are going to be in commercial life, it is great to be able to control your own destiny. It was a pleasant feeling. We could even fire clients, which we did.' Carey also told Willbanks that books like *Bliss* had the dual purpose of presenting unpalatable realities, as well as the potential in all of us to turn things around. 'The writer has a responsibility to tell the truth, not to shy away from the world as it is,' Carey said. 'And at the same time the writer has a responsibility to celebrate the potential of the human spirit.'[72]

Bliss is also a keen insight into Australia's richest and most powerful people with whom Carey rubbed shoulders from the age of twelve as a student at the exclusive Geelong Grammar. Literary biographer Karen Lamb has also dissected the influences of the advertising industry on Carey. 'Advertising . . . has provided him with unique access to, and experience of, commerce,' Lamb writes. 'He is one of the few Australian writers who have consistently interpreted Australian attitudes towards wealth, and the mythology surrounding the cult of the entrepreneur.'[73]

Hints of what would become trademark Carey themes first emerged seven years earlier in *The Fat Man In History* (1974), a spectacularly well-received collection of short stories which launched the young writer onto the literary scene. One of the stories, 'Withdrawal', is the earliest apocalyptic vision by Carey, who—starkly and uncompromisingly—introduces themes which would be developed in *Bliss*:

> Fires ringed the city . . . burning fiercely . . . At night the horizon glowed bright red as if the city were being fried on some incredibly hot plate . . . the summer of burnt houses and blackened bodies, a summer you could believe was the beginning of the end of the world . . .[74]

Images of the carcinogens which would have such a prominent place in *Bliss* are raised in another story, 'Report on the Shadow Industry', in which 'the clouds now contain the dreadful beauty of the apocalypse'. As he would in his later novels, in this story Carey discusses the barrenness and destructiveness of thoughtless and selfish consumerism which culminate in a 'poverty of the spirit' which causes a 'feeling of emptiness, that awful despair'.[75]

Wider concerns (which would punctuate *Bliss*) are also evident in other stories of the *Fat Man* collection, such as 'Conversations with Unicorns', a tragicomic interlude depicting how man's interference in nature always produced unexpected and undesired results. This story is a litany of 'man's cruelties to man and to animals . . . the slaughter of seals, the systematic murder of sheep and cattle, the subjection of horses, the killing of lions, the establishment of zoos and circuses'.[76]

The wider concerns examine various aspects of environmental degradation and urban blight which, again, are dramatically and brazenly described in the story 'American Dreams', which also angrily speculates on their cause:

> For our own town . . . we have nothing but contempt. We have treated it badly, like a whore. We have cut down the giant shady trees in the main street to make doors for the school house and seats for the football pavilion. We have left big holes all over the countryside from which we have taken brown coal and given back nothing.[77]

The concluding title story of this collection encapsulates many of these themes and gives some striking examples of how modern economics—especially the supply of essential commodities—is utterly irrational and usually based on immediate political imperatives rather than environmental effects.

Many of these themes proceed into Carey's next collection of short stories, *War Crimes* (1979), some of whose stories are even bleaker and its characters weirder. The harrowing 'A Schoolboy Prank' is a disturbing look at the monsters within us; and another, 'The Chance', visits an environment even more desperate than that in *Fat Man*. Sci-fi and doomsday scenarios permeate stories which introduce such worrisome concepts as extra-terrestrial quarantine as a means to protect a bludgeoned ecology.

A doomsday vista is evident in 'The Journey of a Lifetime' with a train journey speeding away from a socially stratified CBD to the brutish regions beyond, to 'poor dry country which marks the edges of the Great Eastern Desert' whose environment is so harsh that only prisoners, watched over by the military, are considered for the numbing work there.[78]

The powerlessness and apathy of average citizens against the ecological madness which is destroying their life-support before their very eyes are the stuff of 'The Chance':

> . . . the Americans . . . had left us with a technology we could neither control nor understand . . . we had never grasped the technicalities the Americans sold us. Our curiosity about how things worked had atrophied to such an extent that few of us bothered with understanding such things as how the tides worked and why some trees lost their leaves in autumn. It was enough that someone somewhere understood these things.[79]

Carey speculates on how this apathy and impotence against manipulation have always been a tragedy which has stopped humanity from reaching its fullest potential. As life-support mechanisms collapse all around them, citizens are doing nothing less than fulfilling their pathetic death wish.

In the closing title story, Carey magnifies these themes to discover a world in which the best business brains are those which eliminate compassion, leaving no place for the humane environment necessary for a decent quality of life. Carey dissects the dehumanising aspects of helter-skelter industrialisation and rampant technology whose conclusion is a madness which spurs the most extreme of reactions, as in *Bliss*: 'the machinery [in the factory] was terrifying enough. The vats were huge and their sheer bulk was so unrelated to anything human that I felt my throat block off . . . without any regard for life and limb.'[80]

These modern Dickensian factories which 'belched smoke in the sky and looked beyond saving' result in 'the bitter desperate faces of disenfranchised men' who must witness finality and decay everywhere including dead fish in destroyed waters, trying to survive under the stifling heat both inside and outside the crushing factories. All life is being 'suffocated' as the latest industrial revolution devours in a maddening rush those who keep the insatiable wheels spinning.

War Crimes is suffused by what Hassall describes as an 'apocalyptic vision' in which 'despite technological change, or perhaps because of it, civilisation is entering its terminal phase, order is breaking down and the barbarians are waiting in the wings'.[81]

If *Bliss* rejoiced in the potential for moral redemption, the abyss prevails in *The Tax Inspector* (1991). The setting of this most confronting of novels is the urban wasteland of Sydney's outer western suburbs, whose sprawl breeds physical and spiritual pollution of the most frightening and devastating kind.

All is in rack and ruin in ill-fated Franklin, the epitome of all that's rotten in outer suburbia:

> The Franklin Redevelopment Region now had a hundred thousand school kids. The banks of the Wool Wash were littered with beer cans and condoms and paper cups . . . Stolen cars were abandoned here, virginities were lost . . . [the] water in Deep Creek now contained lead, dioxin and methyl mercury from the paper factory on Lantana Road . . . [Benny Catchprice] could feel the poisons clinging like invisible odour-free oil slicks . . .[82]

Just as the district is rotten, so are all members of the Catchprice family, who run the local used-car lot. Some of the Catchprices are more rotten than others as they deal with three generations of sexual child abuse and almost every other moral shortcoming. The selling of second-hand cars in drab suburbia was far from the family's original dream of breaking away from Dorrigo in northern New South Wales, whose land had been turned to aridity because of ring-barking and careless agriculture.

When we first meet Granny Catchprice, the second-hand vehicles in the car yard at Franklin are 'pumping out poison' and her outrage knows no bounds:

> 'Do you know we put concrete over perfectly good soil when we made this car yard? There's concrete underneath all the gravel in the car yard . . . But there's good soil under there, and that's what upsets me. It's like a smothered baby.'[83]

As we learn of Granny's nihilistic disdain, the wider area around Franklin is also shown to be beyond salvation: 'It was an increasingly drab second-rate landscape—service stations, car yard, drive-in bottle shops and, now, three lanes each way . . . physical desolation.'[84]

The novel ends in rubble—literally rubble—as the horror reaches its inexorable climax. The odious existences of the damaged and dirtied Catchprices are blown to smithereens. Only cataclysm can result when abuse—be it sexual, moral or environmental—is left unchallenged. According to Hassall, *The Tax Inspector* is a novel of 'apocalypse now' set in 'a feral city, which runs on corruption and abuse, and which harbours little justice and precious little love'.[85]

Many such bleak themes also permeate other well-known Carey works such as *Illywhacker* (1985) and *Oscar and Lucinda* (1988), the results of a fertile imagination which cares and worries about how an odious corporate culture and equally odious marketing strategies are ruining the earth and dashing the humanity of those upon it. As Hassall put it, the result of Carey's novels is 'a haunting diagnosis of the personal and social diseases that ravage late 20th century Australia'.[86]

In the footsteps of Carey have followed other fine writers like Blanche d'Alpuget who have tackled uncomfortable issues. In *White Eye* (1993), d'Alpuget portrays a bleak and heartless world of animal experimentation and gene tampering, and its characters show a callous disregard for pending environmental collapse. D'Alpuget's description of this is brutally stark:

> It had taken millennia to create the wilderness . . . in mere decades [it turned

> to] sterility . . . The whole planet was facing environmental holocaust, as everyone knew, for man was now a plague on his own house—and everywhere there was self-disgust.[87]

D'Alpuget, born in Sydney in 1944, has received widespread recognition as a biographer (of the former Prime Minister Bob Hawke, whom she married in 1995, and industrial judge Richard Kirby) and author of novels *Monkeys in the Dark* (1980), *Turtle Beach* (1981) and *Winter in Jerusalem* (1985). She began her writing life as a journalist, first in Australia, then in Britain. Much of her work is set in South-East Asia, where she lived for many years—in Indonesia (from 1966) and later Malaysia. From there, she continued to write for various Australian newspapers and magazines.

In *White Eye*, she shows how attitudes to the environment are compliant and unthinking, especially in a Public Service culture. In Customs, for example, the motto was virtually 'Be Safe. Obey Instructions. Leave Interpretation to Others.' Because of such turnings of blind eyes and the flagrant manner in which big money across the world results in the flouting of, for example, legislation on endangered species, there is little scope for people with a conscience to redress horrible wrongs.

The suspenseful action of the novel revolves around The Research, situated near a sleepy New South Wales country town, where unspeakable experimental atrocities are conducted on illegally imported animals. Naturally, no one at The Research is accountable to anybody and nobody is big enough or brave enough to challenge this odious status quo. Everywhere is the blanket of cover-up, and at least two murders seem to be connected to a series of unexplained events at The Research.

D'Alpuget also shows how one form of ecological vandalism is often linked to another; for example, those trafficking in rare animals are often involved in illegal logging throughout South-East Asia, these activities in turn having links with criminals, big business and shady politicians everywhere.

White Eye was recognised by reviewers throughout Australia as a significant work which examined a range of important issues. Graeme Barrow in the *Canberra Times* said the novel 'raises complex moral and ethical questions about biotechnology, animal experimentation and trafficking and wildlife'. Helen Elliott in the *Sunday Age* said d'Alpuget had 'very successfully combined an adventure with masses of information and instruction about certain ethical issues in contemporary life'.[88]

Elliott went even further in recognising *White Eye*'s importance in dealing with 'the carelessness with which we treat the landscape, the callousness with which we treat animals, and, ultimately, the callousness with which we treat one another . . . There are some scenes in *White Eye* that stop your heart with their cruelty and barbarism'.[89]

The earlier *Turtle Beach* is set largely in Kuala Lumpur in the late 1970s, and tells the tragic story of the Vietnamese boat people—those last victims of the Vietnam War.

In Kuala Lumpur, the marks of colonialism—past and present—are everywhere. It is clear that the real masters of Malaysia now are big business and the international economy. Its royal family and upper classes are portrayed as fools—one prince lives for his Lamborghini; his New Age sister, who decorates her shoes with the exotic plumage of rare birds, thinks Jesus was eaten by lions.

With the prosperity brought by one colonial master after another—we are constantly reminded that Malaysia is the richest nation in the ASEAN group—comes a greed and a blatant disregard of the beauty and richness of the land. The land is sick—sick from the ravages of colonialism and economic greed:

> The residences of the ambassadors and high commissioners were here, a place which only a hundred years earlier had been virgin jungle . . . This was now the centre of town . . . In an afternoon, a man with a chain saw could undo the work of millennia. All along the roadway to the East Coast the injured trees were cloaked with curtains of parasites.[90]

Turtle Beach shows how one small nation has been irrevocably damaged by colonialism, in spite of its apparent peace and prosperity. And it is a clarion call for compassion, for understanding and peace between all people.

A more extreme—and extremely realistic—vision of one colonial ruler imposing its destructive force on its former outpost is to be found in *Maralinga My Love* (1988), in which Dorothy Johnston closely explores many aspects of British nuclear weapons testing in South Australia throughout the 1950s. The novel takes place before, during and after the catastrophe as several hapless pawns carry out orders to make the madness happen. Continuing a theme raised by Xavier Herbert in *Poor Fellow My Country*, Johnston's environment is post-holocaust, desolate and bleak. Callously mistreated victims await their inevitably tragic fate.

Maralinga My Love grew out of Johnston's curiosity and concern after a journalist friend had begun to uncover unpalatable facts of what the British tests entailed. 'It forced me out of my head and into the outside world,' she commented a few months after its release.[91]

In the book, Johnston describes how to the British scientists and servicemen who participate in blowing this alien landscape asunder this is not a real land, it is an object they can treat with disdain. They know nothing about where they are—one of the servicemen has to ask where Melbourne is—and they treat everything they do in the abstract and care not a whit about the consequences. These Britons could do whatever they wanted; none of their actions or motives was the business of the Australian servicemen who accompanied them. Australia was simply handed over to the British authorities, who were answerable to no-one—an attitude which would remain unchanged for several decades more.

To the Brits, the good ole Aussies in this backwater count for very little and the original inhabitants, the Aborigines, count for absolutely nothing at all. So little do the Brits want to be involved with this godforsaken region that all their supplies and comforts are transported from Home. Maralinga is a place inhabited not by fellow humans but rather is no more than 'a huge open-air laboratory'. Some of the British servicemen even exult at how being so close to a bomb test was an exciting and wondrous adventure.[92]

Johnston's intensity is such that if all the revelations, the growing awareness by Australians, and the findings of royal commissions didn't elicit rage against what was done, this novel will do so, especially in how it shows the routine, mechanical and uncaring manner in which the cringingly anglophile Menzies Government neglected its own citizens. Any reader would gradually boil in anger as the Minister of Defence and other government officials as depicted in the novel give knowingly false assurances and make it almost policy to tell their gullible constituents that the tests had never even taken place.

The Australians working at the test site note the indifference of their own political leaders to what was happening and they decry the complicity of all those public servants without a conscience:

> 'What do you reckon about our Prime Minister then?' Driscoll asked. 'Reckon he'll want to come down here and see for himself what's going on?'
>
> 'Menzies'll never come here,' Charlie said shortly.
>
> 'Why not?'
>
> 'Not interested. I mean he's supremely interested in whether or not the tests are a success. But I suspect he's got a secret fear of seeing one at close range.'
>
> 'Big man with feet of clay,' Alex said.[93]

The disregard by the people in power creates a wasteland as dust strangles all the trees and all the people in the afflicted regions. And worst affected are the Aborigines, who are struck down by mysterious illnesses.

Johnston carefully conveys individual attitudes, including the disgust of a small handful of those who had been at Maralinga but who found themselves impotent to change things. The most anguished among them is Graham Falconer who, in England with his wife Deborah, recalls the wasteland left behind in the South Australian wilderness, where fragments of plutonium were carelessly left lying about. Falconer seethes as he thinks about the Australians who were to suffer horribly by coming into contact with the radioactive filth:

> The British've been so bloody careless! [Falconer says] They are, so careless. Australia really is a colony to them, a rubbish dump, a bit of desert to do what they like with. They'll do whatever they can get away with . . .[94]

But Graham Falconer, like almost all those involved in the project, has a 'glamorous' life and career to think of, so the guilty do get away with it and vast chunks of South Australia are allowed to rot.

In *Maralinga My Love*, Dorothy Johnston personalises the tragedies of the tests' victims: their various sufferings, the premature deaths and the arrogant official cover-ups. Johnston deals with people who are no longer just statistics or names who feature in extracts from a royal commission report. Johnston questions the conflict between mainstream notions of friendship and loyalty on the one hand and whistle-blowing the horrors of Maralinga on the other—the eternal struggle faced by those with a conscience.

Dorothy Johnston's environmental awareness first emerged in the Victorian industrial city of Geelong, where she was born in 1948. Educated in Geelong and Melbourne, in the early 1980s she moved to Canberra. In the national capital she became one of the Canberra Seven group of prominent writers.

Johnston's upbringing in Geelong is recalled in *One for the Master* (1997), in which she describes the Barwon River as polluted and the tract of land where the city's woollen mill used to stand as a wasteland.

The river is a centrepiece and focal point of the novel, and is shown as being at the heart of the Geelong community. It is a microcosm of the triumphs and disasters of all those at Geelong. Many in this fringe city are impoverished, and so is their river: 'The air from the river was strong with rotting leaves and reeds, and sodden steaming driftwood, and soap and wool grease from the mills.'[95]

The river keeps rolling along, as generations of the city dwellers keep struggling along, and both journeys are fraught with dangers:

> Years ago the waste water [from the Highlands woollen mill] used to be piped back into the river at night . . . Red and purple and yellow leftover dyes from the vats, waste water rich with chemicals poured out of concrete pipes into the river. The river bled and cried indigo tears.[96]

The mill is the scene of horrific accidents caused by its decrepit machinery, and of exploitation of the mill workers ground down by oppressive poverty and vice. *One for the Master*, an ironic title based on who gets the lion's share of the toil within the oppressive factory, deals with all the dehumanising aspects of industrialisation, including the all-pervasive and relentless noise pollution:

> . . . rows of weavers with their ear-plugs . . . I couldn't help wondering what it must be like to silence these machines that were never silent, and thought I understood at last what the dispossessed hand-loom weavers might have felt a century and a half ago, what might have driven them to dreams and deeds of sabotage.[97]

In the wide-ranging issues of *Maralinga* and the more down-home *One for the Master*, Johnston dissects abused environments against which the occasional brave soul has the heart and mind to rebel.

Another watershed novel—this time by one of Australia's most prominent young up-and-coming authors, Tim Winton—is *Shallows* (1984), which deals mainly with whales as victims of ecological violence. In contrast to the massiveness and majesty of the whale, the human is simply not big enough to control the senseless vandals wreaking needless pain on the oceans. Winton deals with man's inadequacies in his earliest description of the West Australian coastal city of Angelus, which no longer has 'wise men and women not afraid to speak out . . . people who led'.[98]

In this novel, Winton chronicles every atrocity perpetrated by supposedly civilised humans against the world's largest creatures. Winton, born in 1960 and educated in Perth and Albany, was only 24 when he wrote the powerful novel, for which he won the Miles Franklin award. In *Shallows*, Winton described a locale he knew intimately and cared about passionately.

Shallows is set in Angelus—and there's nothing angelic about this redneck neck of the woods—whose residents couldn't care less that in their vicinity whales are being slaughtered and dead parrots litter the roadside. They blithely accept the factory-like mechanism of the flensing works, where the blood, guts and stench of the captured whales are greeted with nonchalance. Amid his vivid descriptions of the horror of the flensing process, Winton wonders at the awesomeness of these creatures and the enormity of their unjust and needless destiny.

The burghers of Angelus don't take kindly to anyone questioning their ways, especially when it comes to whaling, a time-honoured tradition and lucrative activity. The anti-whaling protesters at the flensing station are dismissed as 'foreigners' by a vicious and jingoistic group who don't tolerate any divergence from what they regard as correct thinking.

Marian Eldridge, reviewing *Shallows* in the *Canberra Times*, saw Winton's concerns for whaling and other environmental issues as a microcosm for his greater universal and spiritual feelings. 'A major theme of *Shallows* is the importance of love for one's fellow creatures—family, spouse, neighbours—and the environment,' Eldridge wrote. 'Without it one withers . . .'[99]

The acknowledgement of a greater spirituality is taken even further by another *Canberra Times* reviewer, Veronica Sen, who in her assessment of the paperback version said 'the human race itself seems to be there with the whales' in what was 'a sobering novel which challenges complacent beliefs'. Like the whales, according to Sen, humanity as seen by Winton is a holocaust where 'masses of flesh and barnacles flounder and suffocate under their own weight on the sand'.[100]

And to *Age* reviewer David Myers the tragic fates of the whales are 'symbols of the evil and the horror that have always threatened creation'.[101]

SCIENCE fiction shows environmental destruction and change as perhaps the last great challenge for mankind. A variety of Australian novels and short stories present a conveyor belt of disasters, mostly man-made, with the fragile ecology a mute victim. Australian sci-fi, in the tradition of Jules Verne and H.G. Wells, is a vehicle for social criticism—as well as political and environmental cautionary tales.

A leading commentator on sci-fi, Van Ikin, says one very significant way in which current Australian novels differ from those of late last century or the first half of this century is that 'in the 1990s the novels are increasingly aware that the future is upon us, it has an immediacy and is not way down the track as it was for Verne and Wells'. Ikin said in 1999 that 'for the past 20 years the tide of change is upon us, and people should expect that society will be made and remade over and over again' and issues such as environmentalism have a potent urgency.[102]

Ikin, a senior lecturer in Australian literature at the University of Western Australia and one of the nation's most prominent sci-fi researchers, said the genre increasingly reflected a globalised world and many Australian writers sought to make their work fit in with the 'transatlantic mainstream' of the northern hemisphere. At the same time, according to Ikin, sci-fi Down Under uniquely reflected the Australian experience, mainly because Australia was not a major mover in technological developments. Australian sci-fi reflected uniquely Australian environmental problems and challenges rather than such things as the colonisation of Mars or the moon.

To Ikin, a writer like George Turner in his landmark novel *The Sea and Summer* is one of our most important voices in warning of an environment under immediate threat. '*The Sea and Summer*, although its setting is well into the next century, captures for the reader what's happening right now in Australia,' Ikin said. 'Turner states quite directly that the country is moving more and more in the direction that he predicts.' Ikin gave as examples the grotesque apartment blocks (called Swill towers by Turner) already scarring the skylines of every Australian city, and Turner's vision of harried shoppers needing to keep moving for fear of being crushed in supermarket aisles was already part of the modern shopping experience.

Turner is the foremost literary figure and at the cutting edge of Australian sci-fi according to Ikin, who with Russell Blackford and Sean McMullen is co-author of *Strange Constellations* (1999), the definitive history of the genre. Among the other important voices is Terry Dowling, whom Ikin describes as 'taking something we're all familiar with in Australia, such as the desert, and turning it into a different and exotic—and threatened—entity. This reordering of the familiar forces us to think about our desert, our weather, our society, differently'. His co-author McMullen, a novelist and short-story writer, speculates on what would happen if certain technological changes occurred and 'expresses sociological insights into what will happen if economic rationalists are the guardians of our future'.

Ikin also acknowledges the role of several well-known Australian writers who 'now and then drop in and visit' the genre to make important statements—such as Gabrielle Lord, a 'thriller writer who in one book, *Salt*, crossed over' and Rosie Scott, a 'mainstream realist writer' who in *Feral City* took her work into extraordinary new directions.

Sci-fi also depicts the ways in which people and technology come together and the effects these have in the worlds where we live, according to Sydney writer and academic Anne Cranny-Francis. She said one of sci-fi's significant environmental concerns was examining the 'unleashing' of technology and the question as to whether it controlled humankind or whether humankind was able to control it.

Cranny-Francis, an Associate Professor in English and Cultural Studies at Macquarie University, said in 1999 that Australian sci-fi writers since the 1970s were examining futures that were almost at hand—in some cases only 20 years hence—and asking the question 'what kind of society do we want to live in?'[103]

In a discussion on both Australian sci-fi and fiction generally, Cranny-Francis said novels were able to affect hearts and minds on environmental issues because they 'engaged their readers in more than just rational ways; characters have emotional attachments with each other and they can touch the reader emotionally and thereby provoke thoughts and feelings'. Cranny-Francis, whose book *Feminist Fiction* (1990) devoted a chapter to sci-fi, said the environment was a major theme of recent fiction, especially with the millennium so close.

'There is a concern with the land, as there is in all Australian fiction,' said Cranny-Francis, whose other books include *Popular Culture* (1994) and *The Body in the Text* (1995). 'The environmental theme is not always overt or obvious, but it is there, and it usually comes through in science fiction.'[104]

The concept of ecology as a 'frontier' is deftly defined by Sean McMullen, who says the genre has been and increasingly will be 'forced to explore some of them'. For McMullen the greenhouse effect is an 'inescapable frontier', as are all aspects of the forthcoming era of 'cheap and easy' genetic engineering. 'The characters of science fiction are already exploring those,' McMullen says in his 'Author's Footnote' to a short story in his collection *Call to the Edge* (1992).[105]

Among those whom Ikin describes as 'visitors' to science fiction is Gabrielle Lord, who in *Salt* (1990) takes us to the edge of darkness. This is Sydney in the year 2074, when daily temperatures of 57 degrees Celsius reduce the populace to savagery.

As the novel begins, salt is sterilising everything with which it comes in contact on sea and land. Humans are ravaged and animals are horribly deformed. It is, as Terry Dowling describes it, a portrait of 'humanity on the wane'.[106]

Taut and frenetic, *Salt* has a clever and menacing plot. Ordinary people are victims of a fascist system driven by a remote power elite as impersonal as their computers. The two principal characters, a helicopter pilot known simply as Sando and his lover Hedda, are being crushed by an overwhelming nihilism.

Like Herbert in *Poor Fellow My Country*, Lord rails against an anonymous, unseen force which arbitrarily wreaks senseless destruction; equally, she spits at those individuals who don't have the spine to stand up against injustice—an injustice which allowed such disasters to happen. Lord states that preservation of the environment, just like preservation of anything else, requires courage and inner strength.

Whether writing about a spate of mysterious helicopter crashes or a series of grisly incidents in the never-never world outside the walled city of Sydney, Lord creates tension and suspense. She paints images of total decay as an exploding population causes the ever-expanding metropolis to burst at the seams—a situation made even worse in the aftermath of civil war.

The tone is set almost at the beginning as the author shows how gracious living is only a memory of days gone by and 'now there wasn't much left at all . . . the Civil War of 25 years ago, the ravages of the violent new climate that afflicted most of the world, had rendered these once teeming places into desert'.[107]

Where earlier generations had enjoyed meat and fruit and vegetables, now there were subsistence ersatz edibles known as polytucker; people got their energy from something called polyprotein and by ingesting 'fibre from the algaes and seaweeds'. To feed the masses, production of artificial marine foodstuffs was becoming 'the largest industry along the edges of the coast'.[108]

And if the diet for the proles was too rotten to contemplate, that for the 'elite' was not much better; the wealthy and influential feasted on 'creditable' noodles made out of 'polytucker flour' and a meal often would consist of 'bits of rat and the last of Sydney's diseased pigeons'.[109]

Very early in the novel appears Sando, grappling with increasingly inhuman surroundings while fondly reflecting on long-gone days of joy and beauty:

> The stars shone red and unwinking, and Sando thought of the days when they had twinkled, a long, long time ago. Now they pierced the blackness with a merciless clarity, like lasers, and during the day their huge brother the sun played with the globe of the earth, drying it, killing it, continent by continent.[110]

Sando goes on to explain how his once beloved surroundings have been especially victimised, as the sun utterly destroyed 'anything left after over 300 years of soil plundering'.

This Sando has a name, David Sanderson—we have been told it more than once—but we somehow quickly forget it and it's simply Sando: anonymous, impersonal and identikit. This is merely Sando, just one more bit of human fodder who eventually will not only have life snuffed out of him but will also see all of his life-support systems—and anything with a shred of decency or meaning—equally obliterated.

Apart from Sando's woes, the reasons for the general decline around him are gradually introduced: the melting of the polar icecaps, an aftermath of generations of toxic and radioactive dumping, not to mention rising ocean levels (so that the streets alongside Circular Quay are now underwater); the perpetual leaking and spilling of oil from transport vessels, which created a pollution which was 'stagnant and constant'; and, of course, the trampling of democracy and all freedoms—only a brutal authoritarian rule could have allowed such an unequal and uncaring state of affairs.

The fascistic structure creates a situation where the poor must fend for themselves as best they can while the rich and powerful have subterranean sanctuaries—where they are safe from the sun's killer rays and hermetically sealed from the collapsing society above and beyond.

Such a controlled society can be achieved only through compliance by the people, and Lord dissects the breeding-grounds of fascism: the excesses of authority can happen only because the likes of Sando allow it to. Lord cleverly shows Sando indignantly and self-righteously defending all his actions with a typical Nuremberg response: 'I don't run Western Security,' he exclaims, describing the organisation for which he flies chopper missions. 'I do my job. I get paid for it. I get a triple A rating and I file my reports. That's where I stop.'[111]

As if all these monstrosities were not enough, Lord describes how those whom the gods destroy they first shower with useless and frivolous things. Around the degraded and decrepit city of the future, the monorail still makes its pointless circular journeys, dripping oil and going nowhere, imposed still on an apathetic people.

If *Salt* is a cautionary tale portraying a Sydney of the future, *Jumbo* (1986) is a contemporary view, specifically of the urban blight and sprawl of the outer western suburbs where Lord lived for eighteen years. She describes it as a wasteland whose potential has been suffocated by conditions similar to those of *Salt*.

The action begins in an impoverished council estate that had been built on land 'bulldozed from straggly bush'. As in *Salt*, the further from the sea—and the further from prosperity—the greater the human misery:

> A few miles from the sea the easterly breeze is no longer felt and in the summer the heat increases. Twenty-five miles from the coast the air is still and oppressive, fumed with car exhausts and the spouts of industry. Forty miles inland, under the shimmering rim of the mountains, the heat can be desperate. The mountains throb in the background, gathering up and absorbing the sun's power, radiating it back on to the plain where the houses of the outer western suburbs cluster.[112]

Throughout this vast expanse of urban neglect, the decay and horribleness are matched only by the horrendous social dislocation. Lord is scathing and fierce about the sheer routine, monotony and predictability of it all:

> As well as the heat and ugliness . . . the outer west was bursting with children and adolescents, and adolescents having more children. Fat girls in tent dresses pushed fat babies in prams through the air-conditioned glitz of the shopping complex.[113]

In a masterful and profoundly important novel, Lord describes a society at its absolute straining point; as in *Salt*, any infrastructure there ever was stopped functioning a long time ago:

> The roads were already, or still, congested. One day . . . the traffic will just stop—clunk, like a huge seized-up engine, because the last possible car has just been squeezed on to the last remaining space of road. The humidity was sickening; the road shimmered in the heat and the filthy bunting of the used-car lots hung limp as if about to melt and drip on to the tar . . . [114]

Jumbo's clever and gripping plot describes a collision course between several strata of society. An inevitable doom awaits all the victims of the parched badlands suffering under a scorching sun. Everything in *Jumbo* is out of control and anarchic; all is soulless and tending to mindless violence.

Two years after the appearance of *Salt*, with all its feral imagery, came the almost inevitably titled *Feral City* (1992), by New Zealand-born Rosie Scott, who describes the Auckland environment as 'a city broken beyond repair'.[115]

Even the first lines of the novel have an ironic tone as the author sees 'the landscape of my country slipping past'. And the leading characters—ironically and bitterly named Violet and Faith—observe how 'everyone around here's too numb, they're too busy scrabbling for survival'.[116]

The initial problem seems to be connected with soil degradation, and a terrible discovery emerges after a trip from Wellington to Auckland: '[There were] cracks going metres into the earth . . . a series of earthquakes all the way from Wairarapa to Bulls,' writes Scott, born in Wellington in 1948, raised in a working-class area of Auckland and who in later years has lived in Brisbane and Sydney with her film-maker husband Danny Vendramini.[117]

Scott echoes the critique of humanity by the other novelists, revealing that the earth's doom is directly attributable to the shortcomings of its inhabitants. She rises to a pitch of anger to state that average citizens cannot evade responsibility for what is happening around them, that they themselves and their planet can be saved only by confronting the evils which would devour them.

Scott describes a 'global sickness' about which governments have neither the morality nor the political will to do anything: ' "The seasons have all changed," Violet said with foreboding. "The rain's changed. There's no freshness to it any more." '[118]

There are horrors heaped upon monstrosities, but Scott nevertheless finds a note of optimism amid the rage and despair. '[The book] is as much about humanity's ability to preserve its spirit as it is about the ability to mess things up,' she said in a 1992 newspaper interview in which she blamed reactionary political trends like New Zealand's New Right as largely responsible for a climate of exploitation and mindless destruction, all in the name of immediate and temporary gain.[119]

As a counterpoint to life-destroying economic rationalism, Scott sees salvation in what writer and literary critic Peter Hutchings, in his review of *Feral City*, describes as 'the importance of both humanist culture and humanist values . . . many of the book's key moments depict small, personal decencies rather than the grand, spectacular indecencies of politics'.[120]

The 'death of human decency' in *Feral City* is also recognised by reviewer Claire Mills, who saw the novel reflect 'the degeneration of the communal values that bind society'. According to Mills, the novel is set in a future that 'looms alarmingly close, in distance as well as time' as Scott brings 'passion and immediacy to a novel charged with a terrible vision'.[121]

A later novel, *Lives on Fire* (1993), is set across the Tasman in Brisbane and features further instances of self-destructiveness, again much of it drug-induced; an earlier work, *Nights with Grace* (1990), deals with the effects of pesticides and several other environmental issues.

In mainstream Australian science fiction, perhaps the seminal work is George Turner's *The Sea and Summer* (1987), whose futuristic action is set among the drowned ruins of what used to be Melbourne.

The complex timeline of *The Sea and Summer* spans several decades of the 21st century. Turner's doomed characters from the future reminisce about the so-called Greenhouse era, when the ecology was compromised and the economies of the world started to fall to pieces. In this Greenhouse time were built the Swill towers, 70-storey buildings which came to house 70,000 people in about as much comfort as an anthill.

In this futuristic vision, civilisation is increasingly unable to cope with the devastations of a murdered ecology. 'We are barbarians,' writes Turner as society struggles to maintain a basic standard of living available to fewer and fewer. Turner's thesis is stark: 'With survival the only moral touchstone, we show what we are. We kill in order to live. Our final decency is the ability to see what we are and exercise some rational control over it. The world's survivors will be the ruthless, not the holy meek.'[122]

The more civilised and caring people are naturally repelled by things descending to such a state, but they recognise the need for cruelties and harsh measures such as human culls. They realise it is inevitable that many among the poorer masses will need to be sacrificed in the cause of the greater good, and their main concern is not to try to stop it but rather to ensure that it happens in the most humane way possible.

This applies to the more humane elements; for the rest, apathy reigns, as always. Despite all the portents, nobody believes disaster is looming—'not in our time' is the common catch-cry. 'Refusal to believe is our surety that disaster cannot happen—at any rate, not today,' Turner writes ironically, showing how disasters heaped on disasters are inevitable.[123]

Born in Melbourne in 1916 and identified with that city until his death at age 80 in 1997, Turner also produced such significant novels as the joint winner of the 1962 Miles Franklin Award *The Cupboard under the Stairs* (1958), *Beloved Son* (1978), *Vaneglory* (1981) and *Yesterday's Men* (1983) and was the doyen of Australian sci-fi and a major literary figure for almost forty years. He described *The Sea and Summer* as crossing the line between genre sci-fi and the mainstream novel.[124]

Although stressing that he was not a scientist, Turner told the *Australian*'s Deborah Stone that as a writer he undertook 'a special role' in dissecting how current ecological problems like the greenhouse effect would affect future societies. 'The more I collected [information for *The Sea and Summer*, which also won the International Science Policy Foundation Award for the best science fiction novel of 1987 and the prestigious Arthur C. Clarke Award], the worse it looked,' Turner said.

'I'm beginning to wonder why the real brainpower hasn't gone to work [on all these various problems]. It's like Barry Jones [the then Federal Minister for Science] says, we have to become technologically literate. We also have to become literate about the world around us, and if popular entertainers can't help do that I'm damned if I know who will.'[125]

In her article, Deborah Stone described George Turner, then 72, as an 'older writer [with] a special responsibility to try to get the world thinking about critical issues'. Turner told her: 'most certainly, we have that responsibility to the next generation. We are the generation that made all the mistakes'.[126]

Turner's role was elaborated upon by *Canberra Times* reviewer Colin Steele, who described it as being part of a 'passionate desire to overcome contemporary apathy' in the context of a 'sombre and realistic projection of present trends'.[127]

For most of his life Turner lived in boarding houses and flats in the Melbourne bayside suburb of St Kilda until he settled in 1994 in the Victorian inland town of Ballarat. His long years in Melbourne suburbia gave him a profound insight into a potential world of Swill towers, social dislocation and environmental degradation.

Turner said the main 'message' of *The Sea and Summer* was that people today need to 'pay more attention to what's going on around us and we have to develop a way of dealing with it'. He said it was undesirable—indeed impossible—to isolate Australia's problems from a wider global context.

As to the novel's bleak scenario, Turner noted that his intention was not to be a prophet of doom but rather someone warning of the consequences of civilisation not changing its behaviour. 'I've only pointed out what will happen if we don't do anything,' he told Deborah Stone. 'The human race isn't going to get wiped out; it's just going to have to pull its socks up.'[128]

As the plot of *The Sea and Summer* unfolds, in the year 2061 'the sea is rising over the beaches of the world; the coastal cities face death by drowning'.[129]

In scenes similar to Gabrielle Lord's depiction of Sydney, the future Melbourne is where:

> Day by day the water advances up the streets from the shores and rivers; our placid old Yarra was long ago forced over its banks by the rising tides. The coast roads have vanished and the lower floors of the tenements are uninhabitable.[130]

Turner writes that between 2041 and 2044 global population exceeds 10 billion, a milestone which had been passed a decade sooner than the sooth-sayers had expected, making nations 'uneasily aware of their world closing in on them'.[131]

The novel makes it clear—painfully clear, yet again—that the inhuman devastations would not have arisen in a political system of democracy, decency or fairness. The futuristic world as depicted by Turner is monstrously unfair, divided by the have Sweet and the have-not Swill (who make up 90 per cent of the population). And naturally never the twain shall meet.

The Swill towers as described by Turner are a social and architectural monstrosity which, as the *Age*'s reviewer John Hanrahan put it, 'fulfill the dreams of the Housing Commission architects of the 1960s and '70s'. Concern about such things as architecture or a quality standard of living are far from the minds of these future citizens, whose sole aim is to survive into the next day. Whatever social progress there was in the latter half of the 20th century will be obliterated and made irrelevant to people living the merest and meanest existence.[132]

For the Swill—and, indeed, the Fringe, those who tragically have lost the economic strength to remain among the Sweet—things are, have always been and always will be intolerable but there are no such things as outrage or dissent because, as Turner brutally explains, anyone who protests has his or her 'brains splashed on the footpath'.[133]

In such a world there is no place for the ethics which had once glued societies to ensure a voluntary code of acceptable behaviour. Turner takes us to the year 2051, when a political functionary, Arthur Derrick, sends a memo to Nola Parkes, an administrator and educator who can remain as one of the society's most favoured so long as she doesn't step out of line:

> '. . . Nola, Nola, idealism was for the last century, when there was still time . . . No, we're down to more primitive needs . . . The State has no time to concern itself with moral quibbles.'[134]

Nevertheless, the novel is not entirely pessimistic, as noted by Van Ikin: 'The stunning achievement of [*The Sea and Summer*] is that it does say all that can be said against our age, yet still comes out on the side of being alive.'

Describing the novel as 'an amazing achievement', Ikin acknowledged how both the international sci-fi and scientific communities had 'showered' the book with awards. 'It combines the solid documentary realism of a novel like M. Barnard Eldershaw's *Tomorrow and Tomorrow and Tomorrow* (1947) with the hard-nosed scientific speculation of (Aldous Huxley's) *Brave New World* (1932)—yet its plotting is lively and gripping'.[135]

Over-population and a crumbling social structure also feature in a later Turner novel, *Brain Child* (1991), set in the next century when people live under the yoke of Population Control Laws, when unmarried parents are subject to fines and prison and even chemical castration—all within a social system that has 'raised contraception almost to the status of a competitive art form'.[136]

Turner describes the era when the earth's ecosystem is finally about to collapse:

> [In the] first year of the Long Drought . . . the Greenhouse Effect began to turn the weather patterns to agricultural bedlam. Crops wilted, cattle had to be hand fed, what grass there was stood in tinder-dry sticks, our creeks were trickles, only a few springs feeding them from the water table . . .[137]

As in *The Sea and Summer*, *Brain Child* shows how every vestige of decency has evaporated under the scorching heat, confusion and anarchy of global warming; the quality of life once enjoyed seems a utopian memory. Amid the pell-mell of rampant development, wilderness is 'trampled' and 'the last resorts of space and quiet will have vanished under factories, pollution and people'.[138]

The fragility of human existence is a theme of a Turner short story, 'Flowering Mandrake', in which he depressingly catalogues the perils caused by 'minute changes of temperature, orbit or atmosphere composition' which together could bring an end to life as we know it.[139]

Short stories such as these are the life-blood of much of science fiction, and Australia over the past 30 years has produced a variety of talented practitioners. Among them is Sean McMullen, who writes of ruined cities and utter devastation in the *Call to the Edge* nine self-contained and quite diverse short stories. He tells of 'hot, red deserts' and an era when the profoundest expression of love is 'to lock you away and protect you from the horrors of this world' and where harshness is such that 'honour, service, loyalty and ruthless discipline held the distant society together'.[140]

Much of the futuristic landscape painted by McMullen is exotic, but the rampant march of deserts is gnawingly familiar. He tells of the 'immense expanse of sand, stunted scrub and saltpans [with] little water along the way' and, as in tales of doom like *Salt*, wars are fought over basic resources; commodities once taken for granted, like coffee, 'appeared to have the value of gold'.[141]

McMullen's stories contain heroes but, unlike in classical literature, they do not always prevail; the odds against them are far too daunting and their character flaws too burdensome. The most radical of solutions are being contemplated 'to let the earth recover'.[142]

As in *The Sea and Summer*, the society described throughout *Call to the Edge* is rigid and controlled; tampering with a superior's computer access codes is an offence punishable by firing squad. It is a society in which functionality takes on added importance, and the city of Libris—with its obeisance towards knowledge and information technology—is a State within a State.

Libris is typical of the functionality at the core of every facet of society, and everyone is bred to perform a specific and defined duty. For example, waiters 'had been bred merely to make the right drinks appear in the right place without intrusion'.[143]

The crumbling or destroyed ecology has created a landscape which suggests a stone-age society—albeit one which boasts a dazzling array of high-tech tools. Indeed, the future, however advanced, is creating mostly regression: there is random and arbitrary murder as well as very organised executions to maintain central control; torture is a constant part of the administrative machinery; and constant wars are being fought across the feral landscapes. Savagery is everywhere, with rats and dogs ready to tear to shreds any human deemed vulnerable.

Call to the Edge includes many of the ecological concerns of McMullen, born in 1948. It was his first book and included his early stories as well as stories which would form the basis of his later novels. Despite success and international recognition, McMullen, who has a diploma in Computer Science from Latrobe University, has never written full-time, having been a part-time singer and musician and then a systems planner at the Bureau of Meteorology in Melbourne. He has also written non-fiction works on scientific issues and has appeared on such ABC Radio National programs as 'Ockham's Razor' and 'The Science Show'. To this strong grounding in science was added McMullen's love of the sci-fi genre which he encountered first by reading H.G. Wells and Jules Verne in comic books.

Discussing the role of sci-fi in environmental matters, McMullen said in 1993 that he was heartened that 'the general public has recently realised that Australian wildlife is wonderful and unique'. He noted that one of his early short stories, 'An Empty Wheelhouse', dealt with wildlife and resulted in 'a lot of good feedback from American readers'. McMullen said he rated ecological issues as so important that 'financial mismanagement and environmental degradation are more of a threat to Australianness than foreign invasion'.[144]

Another who has voyaged often in the short-story format is Terry Dowling, who in *Rynosseros* (1990) takes us into the futuristic era of artificial life; this often blends uncomfortably with natural life within an environment being buttressed by many more forces than mere humanity could inflict. But much remains unchanged in a desert landscape where 'all about us was desolation . . . hot unmoving terrain'. Everywhere there is 'dust', 'restlessness' and an 'odd melancholy'.[145]

This is a landscape of trees with supernatural powers and perceptions, great desert cities and of sandships which skim between them. Human-like land creatures, sea creatures, creatures of indeterminate origin, and entities of artificial intelligence all coalesce into a blur. And, almost inevitably, a central authority keeps tabs on every living thing through satellites, implants and a host of other future-tech monitoring devices.

Rynosseros is a complete novel of eight self-contained parts, some of which had appeared as separate entities in Australian sci-fi magazines. Much of it is pretentious and cryptic and many of the early pages are quite impenetrable, but there is also much poignant writing and eerie forecasting such as in 'The Only Bird in Her Name' whose hero, Tom Rynosseros, takes upon himself a quixotic crusade to save a creature, the Forgetty, a mysterious humanoid whose bite causes immediate memory to disappear.

Dowling—a Sydney teacher, writer, book reviewer, regular guest performer on the ABC TV children's show 'Mr Squiggle' between 1978 and 1986, and sci-fi personality born in 1947—exquisitely relates the arrogant, god-like attitude of the 'higher' existences towards 'lower' creatures, and even the comparatively enlightened Tom is not immune from ignorant, destructive prejudices. Tom is prepared to fight for the Forgetties for only so long as they are 'viable, true tangentials and not some short-term mutation' but he, like other superior intelligences, is prepared to 'draw our line there' and thereby arbitrarily and capriciously decide who will live and who will not.[146]

With his depiction of the Forgetty and the fragile earth's other constantly evolving creatures, Dowling provides a very broad concept of adaptability, with fading distinctions between animal, human and machine. As always, there is nothing just or logical about how the environment is treated.

In the book's most heart-warming episode, a robot marvels at a small piece of forest as a glorious, living treasure, whereas the all-powerful (human?) rulers have no capacity for such appreciation and fail to understand the robot's attitude. As ever, humankind's inhumanity and stupidity are the overriding determinants to the fate of the ecology.

The concept of various forms of extra-terrestrial life mingling with those of earth is further explored by Dowling in *Wormwood* (1991), seven self-contained stories which reveal how mankind is powerless against the myriad of post-apocalyptic forces.

Dowling continues his ecological themes in *Twilight Beach* (1993), another collection of linked short stories, the first of which, 'Shatterwrack at Breaklight', is the product of an overflowing and lush imagination. It is about a ghost femme fatale, an extremely realistic illusion which tauntingly conceals all hints of the reality beyond. This story merges into 'The Babel Ships', which continues the themes in Dowling's other works that humanity (as it was known in the distant 1990s) is being constantly mingled and diluted as well as being affected by new incarnations, including artificial intelligence, extra-terrestrials and entities which had adapted within a destroyed earth. In addition to other unspeakable ravages, the earth is suffering 'sensory holocaust'.[147]

All of Dowling's frightening scenarios lead to the mind-blowing question: if we can't properly care for our terrestrial environment, how will we ever cope with a more complex, inter-galactic future?

CRIME fiction has had a social conscience ever since sleuths appeared in popular novels—and in the past thirty years this concern has extended to environmental issues. This genre is quite influential in Australian culture, with one major retail chain, Dymocks, in 1993 reporting that more than ten per cent of its fiction sales fell under the category of 'crime'. In that year another major Sydney retailer, Abbeys Bookshop, carried at any one time 6000 titles in that category. The influence of the genre is further heightened by another statistic revealing that most 'crime' books in Australia are written by women, have female detectives as the main characters and are read mostly by women.[148]

'Crime grows directly from the environments in which people live—and this has always been reflected in the genre,' said an expert in crime fiction, John Reid.[149]

Reid, who wrote his own environmentally aware crime novels under the pseudonym Tom Howard, noted that Australian crime fiction had a proud history of dissecting vital social issues. Reid cited such important works as Fergus Hume's *The Mystery of a Hansom Cab* (1887), which commented on the vice, poverty and environmental degradations in the Melbourne of its day. Reid said all Australian crime writers since Hume had been influenced by the 'context' in which their characters found themselves and that the detective creations of Hume right through to Peter Corris were 'relevant to real places and having real concerns'. This view is echoed by crime-fiction expert Stuart Coupe, who noted that the best of the genre can provide 'the most pertinent commentary on contemporary society'.[150]

Coupe, editor/publisher of the influential *Mean Streets* magazine between 1990 and 1996, said private eyes in books could make observations about environments and social conditions in a way other characters usually could not. 'A detective thriller can become boring if it can't deal with the important issues of the day,' Coupe said.

He cited crime writers like Corris, J.R. Carroll and Dave Warner as responsible for important 'urban driven' descriptions of contemporary society. Coupe said many practitioners of the genre were deeply concerned about urban quality of life issues.

According to Coupe, the importance of the crime genre was summed up to him by American rock singer Warren Zevon, who had told him: 'I'm leaving for Sydney in a few weeks so I'm reading a bunch of Peter Corris books so I know what I'm in for.' Coupe commented in 1992: 'Zevon is not alone. There's a good argument for suggesting that any visitor to an Australian city is better off reading a selection of contemporary crime fiction than relying on a street directory and tourist brochures to get around.'[151]

This was expressed in similar terms by *Canberra Times* critic Peter Fuller, who said Corris had the ability to 'turn his love of Sydney into atmosphere [whereby] you can almost smell the city'.[152]

This 'love' of Sydney is also recognised by critic Peter Pierce, who wrote in the *Bulletin* that Corris's Cliff Hardy novels contained 'an inclusive, nostalgic sentiment, equally alert to the charms of plutocrat harbourside mansions, unsullied northern beaches, inner-city squalor'. He accords Corris the ultimate accolade of likening him with the masters of mystery: 'Fondly, we think of Holmes' London. Hardy's Sydney has the same mass, and mystery.'[153]

More than ten years earlier, Pierce had described Corris as having a 'remarkable sensitivity' in his depictions of Sydney which showed 'a responsiveness towards human uses and misuses of place'. According to Pierce, Corris also detailed the 'moral atrophy beneath the seedy dissemblings of contemporary Australians'.[154]

John Reid believes that crime fiction, although not entirely accepted as 'serious literature' by academics and mainstream critics, raises issues which cannot be raised elsewhere. Reid said the modern emphasis in crime fiction had turned more towards social responsibility, and the conflicts depicted in the novels reflected key concerns such as 'the battle lines which have emerged between greenies and farmers'.

To Reid, environmentalism has now become respectable—with its own hero figures like Dr Bob Brown—and this reality has become a part of crime fiction. He noted that major crime writers like Corris, despite his claims to be concerned only with entertainment and not with issues, were at the forefront of defining society's ills.[155]

Peter Fuller has noted Corris's depth of compassion for and understanding of society's victims, partly stemming from Cliff Hardy's creator having written his MA thesis on Aborigines and his doctorate on Kanakas. 'His years in the groves have left him with a great store of historical and sociological knowledge which finds its way into his books,' Fuller wrote.[156]

The quality and status of thriller writing is best defined by one of its most distinguished practitioners, R.F. Brissenden, who from 1953 until his retirement in 1985 taught English at the Australian National University. Brissenden—who in 1964 became the first literary editor of the national newspaper the *Australian* after having been associate editor of the literary magazine *Meanjin* for five years—said good thriller writers were intimately concerned with such issues as how 'the distinction between morality and legality tends to get blurred the bigger the stakes'. To Brissenden, detective novels must 'embody the basic principles of fiction—they have to tell a story, and keep the reader entertained . . . any profound implication will come through only because the reader is entertained.'[157]

In the crime genre, the Australian geography is extensively covered—from the mean streets of Sydney and Melbourne through to North Queensland, Brisbane, the New South Wales South Coast, Tasmania and all over the continent.

Many of the vital issues are tackled by Corris, Australia's most popular modern crime writer, not only in his prolific Cliff Hardy series but also in novels like *The Japanese Job* (1992). In this novel's introduction, a group called the Loyal Alliance, replete with anti-democratic and secretive tendencies, resolves to destabilise the elected post-Fitzgerald (that is, post-Joh) Government and institutions of Queensland.

The Japanese Job is one of the Ray Crawley novels, whose hero is an operative of a government body called the Federal Security Agency who finds himself in Brisbane to investigate what looks like the ritual murder of a Japanese mogul. The victim has been encased in concrete, on which it has been written that 'Australia is not for sale'. Crawley is dispatched amid official concern and panic about such threats against senior Japanese businessmen and their investment in Australia. Corris acknowledges the advice of journalist Brian Toohey, who would have provided invaluable help about security operations and how foreign business operates.

In Brisbane there are bizarre plans for the world's tallest building and Crawley is bothered by the 'march of greed'. He discovers how Australia is earmarked for the next place for Japanese exploitation because it is 'remote, politically stable, naïve'.[158]

The novel revolves around the death of Ohira Yasuhiro, the fifth-richest man in the world and head of the omnivorous Yasuhiro Corporation with its extensive involvement in Australian sand-mining, wetland draining and canal development—not to mention a spate of suspicious and illegal activities.

The Japanese Job embraces the best aspects of the US-style crime genre, as well as having a social conscience and the added insights of how extremism grows whenever economic opportunity is diminished or unfairness prevails.

Corris's better-known detective hero, Cliff Hardy, expresses his environmental concerns about a big metropolis like Sydney in novels like *The Washington Club* (1997). Hardy regularly ruminates on such matters as aircraft noise and car-park space impinging on once green patches. And how waterways such as the Cooks River are being increasingly and wantonly polluted: 'There was no activity on the river. Old-timers recall swimming in it, catching fish fit to eat in it, kids playing on its sandy banks, but that's all long past . . .'[159] Cliff Hardy's Sydney is a place of 'concrete' and 'brick pillars' which dot a suburbia already covered in ugly carports.[160]

In such surroundings, as in the typically ramshackle outer suburb of Earlwood, Hardy must confront the human equivalent of such rottenness, a pimp called Noel Henderson. The encounter highlights the moral decay that breeds in such a wasteland: as the knocked-about and bleeding pimp asks Hardy to 'let me wash this . . . I could get an infection', Hardy replies, 'You *are* an infection.'[161]

Corris sees urban blight throughout Sydney's outer west, and Hardy's description of Rooty Hill is especially scathing:

> The housing is a mixture of good but ugly and not so good and more ugly . . . Noel's hideaway was a fibro structure that might once have served some useful purpose as a works depot or warehouse but now it was weed-choked and derelict-looking with a sagging roof, stained walls and slumped foundations . . .[162]

Cliff Hardy displays an ongoing care about urban planning, such as high-density apartment blocks in the inner Sydney suburb of Glebe, and he wonders how specks of housing can be allowed on the fringes of what are supposed to be national parks.

In the next Cliff Hardy novel, *The Reward* (1997), Glebe is described as a place of 'foul air and . . . traffic . . . potholes and dog shit'.[163] In this novel, Hardy decries that for every improvement aimed at greening the city there is a significant negative, such as the cutting-off of public access to waterways.

Through the Hardy persona, Corris shares the feelings of any serious and committed urban resident worried about what to do against the ceaseless assaults on the quality of life. 'I wondered, not for the first time, if I shouldn't think about moving,' the exasperated private eye comments. Hardy is increasingly cynical and pessimistic about any change for the better: 'I didn't need a three-bedroom house with planes flying overhead, but I couldn't think of anywhere else I'd like to be except Bondi, and they were sure to start changing that soon.'[164]

Hardy sees himself as not just a gumshoe but as a concerned and responsible citizen as well, and as such attends residents' action meetings. He becomes increasingly concerned at how Glebe's ambience is being ruined:

> I was unsettled by some of the changes going on around here—the flight path, the Paceway, the development in Ross Street where a hectare or so of warehouses had come down, the Glebe Island Bridge for god's sake [and] the plan to build a marina on Blackwattle Bay . . .[165]

Hardy, an increasingly thoughtful detective-as-philosopher, asks himself what can one expect—environmentally or otherwise—of a nation that could advance no further from the Menzies era and the Vietnam involvement than to a society with a stupendous drug problem, ills like kiddie porn and abuse, and where nobody seems concerned about such potential disasters as the hole in the ozone layer.

At the start of the 1990s, Cliff Hardy was concerned by the general decay, deterioration and unthinking uglification of Sydney, as depicted in *Wet Graves* (1991):

> These days I feel like closing my eyes when I walk through the city. You can't rely on a building you visited on Friday afternoon still being there on Monday morning, and most of the new stuff going up looks as if it has been designed by architects who stalled at cubism.[166]

Hardy describes much of the rampant development in Sydney as a 'violation', especially the 'strangely disconcerting' tunnel-building under the Harbour. As just one more job as sleuth-for-all-seasons, Hardy organises a dead-of-night scuba venture under the Sydney Harbour Bridge to find clues into how some people disappeared, and even while doing this the ecological decline of Australia's biggest city is highlighted:

> Ray rubbed a towel over his head but oil and grease from the harbour water remained. He looked at the towel. 'Harbour's filthy. Weird, isn't it? The water was probably clean back then [in the late 1920s and early '30s, when the bridge was being built], but they treated workers like shit. Now the workers get a fair go and the environment's a great big toilet.'[167]

Environmental concerns also abound in *To Make a Killing* (1992), featuring Jean Bedford's private-eye creation, Anna Southwood. This female PI gets enmeshed in a murder investigation with suggestions that mining companies and developers in the Shoalhaven region of the New South Wales South Coast are prepared to take the ultimate umbrage against anybody who thwarts their projects. Anna Southwood has been hired by Hilda Trelawney, who doesn't accept that her husband Mark, a leading conservationist in the vanguard of local residents' action, died accidentally as the result of falling from a cliff while bushwalking. Mark Trelawney, a highly placed functionary in the Aboriginal and Torres Strait Islander Commission, was an active and voluble campaigner against Japanese resorts, marinas and many other planned developments, targetting the council-approved vandals who 'find a beautiful unspoilt place and wreck it for everyone'.[168]

Bedford laments the apathy of the average citizen, a neglect which has evolved since earliest colonial days. This apathy allowed despoliation of the forests, where trees withered from dieback. Because development was allowed to continue unfettered in most of the area centred on Nowra, 'the cattle were grazing away what goodness was left [in the soil] and the native trees were dying'. Bedford also lists, in the manner of Eric Rolls, many of the unwise agrarian practices of immigrants to the Antipodes. To the outrages of generations ago were added new ones, everything from unthinking fishing-out of all the local waters to motorised vessels which scare away the bird life.[169]

The parents of Hilda Trelawney are prime examples of the all-conquering apathy, the not-wanting-to-get-involved mentality, the belief that questioning and protesting are merely undesirable ratbaggery or stirring, even when the matter in question is the death of their son-in-law. When confronted by investigator Anna, the parents became 'uneasy' and dismiss Hilda's concerns as 'distorted'.[170]

A combination of apathy and cynical manipulation of its victims is personalised in Denis Mathews, local real-estate agent in the Shoalhaven, whose attitude would have been duly abhorred by 'madmen' like Harry Joy in Peter Carey's *Bliss*:

> 'We can't be too worried about the environment, can we? . . . I mean, human civilisation's been stuffing up the environment for thousands of years, hasn't it? We always find a way to sort it out in the end.'[171]

To Make a Killing examines the insalubrious consequences of housing developments as breeding grounds for every type of the worst social ills, alienation of the young, boredom and hopelessness, the 'silent scream' housewife syndrome and drug dependence of every variety.

Jean Bedford also suggests the self-deception of so many of the stereotypical activists in the local citizens' groups, as well as noting the other type: those who have separate political agendas. The narrative is contemptuous of local activist John Blair, who is 'a political animal . . . it's all politics with him, not really the human considerations'. Simultaneously, Bedford recognises the difficulties—indeed, sometimes the contradictions—of alliances between disparate types supposedly united in common cause: 'Still, we're grateful he's on our side—he's bloody good. And we need people who can manipulate the system . . .'[172]

Bedford cleverly recognises all facets of how the worthy idealism of many activists must be balanced by some of the tainted realities of fighting for a cause, including 'the usual ennui of meetings', and of how the average concerned citizen, being human, is driven by many forces. These forces sometimes include demons which are entangled with love and other aspirations quite separate from purely green issues.

To Make a Killing is a low-key and occasionally even dour exposition of the crime genre and thoroughly dissects many green issues. Bedford has acknowledged the educational role of her novels through which she hopes to 'build a better world of some sort' and to warn her readers about being manipulated or exploited.[173]

An equally dramatic offering is Ian Moffitt's *Gilt Edge* (1991), which from its opening pages instantly evokes the seedy side of Cairns. The descriptions of North Queensland in decay are alternately poetic, tough, mean and angry. The initial action revolves around 70-year-old Alexander Kolnachov, an idiosyncratic individualist and relic of a bygone age. He becomes a floating corpse near some proposed developments he was investigating.

Even the new hotel in town is a symptom of the malaise, 'a new disease which was creeping towards the cane: progress. The Japanese, especially, were buying up big' and it was all part of 'a dangerous dead-end into reality'.[174]

The novel creates an eerie feeling of alienation and starkly contemplates the inevitability of developers railroading local sensibilities with their projects.

Moffitt, born in Sydney in 1929 and brought up in the Manning River region of northern New South Wales, was a highly regarded veteran journalist when he turned to full-time writing in 1978. His other works, including the novel *Death Adder* (1988) and a collection of short stories, *The Electric Jungle* (1993), are full of savage condemnation of how Australians treat their land.

His disgust begins with memories of his childhood home in the Taree region: 'We lived in a lush coastal dairy district . . . but we believed (when our teachers told us) that ours was a harsh sunburnt land. We never *loved* it, as the English cherish their land; we were exiles in the wings of a secondary stage, Australia; if English history was green and blood-red, ours was burnt black and off-white, a fibrous root dangling beneath the rich foliage back home. So we loved England, then, and (with mocking reservation) admired America. [But not] the dairy paddocks, the mangrove swamps, the ghostly ringbarked gums outside our doors . . .'[175]

Moffitt deplored Australians' cringing lack of identity and he expressed a growing compassion towards this continent's dispossessed—whether the thieves were Australia's European invaders or the overseas aggressors. 'Nothing was ever ours, and we knew it then,' he wrote of his schooldays, 'our gaolers had stolen it from the Aborigines, and tossed it to us by default. We parroted patriotism, but it was not for our land. Not for Taree.'[176]

A far more over-the-top and outrageous approach to ecological matters is by Steve Wright, who in *A Drop in the Ocean* (1991) provides full-on and uncompromising descriptions of rancid environments. Like many others born in England, Wright gravitated towards the splendid beach culture of Bondi when arriving in Australia, and his concern for the ecological problems of this Sydney landmark is reflected in this novel. Born in London in 1951, Wright has been a prolific writer of plays, film scripts and magazine articles. The realist edge evident in *A Drop in the Ocean* stemmed from information gleaned from his journalism.

The novel begins at Bondi Beach, which is depicted as a cesspool of over-development compounded by physical and moral decay. Every nice aspect of Bondi is overshadowed by horrid emblems of a seedy, violent and dangerous place. As for swimming in the hallowed surf, forget it. So discovers Barry Donovan, an English immigrant and private eye, when he suggests a swim in the ocean to his girlfriend Gabrielle:

> Gabrielle reacted as though I'd just suggested a sightseeing tour of a vivisectionist's. 'I'm not going in!'
>
> 'Why not?'
>
> 'Because I don't want to get a mouthful of turds.'[177]

Every bit of the early plotline exudes despoliation and excrescence. Bondi Beach, in and out of the surf, is an utter cesspit which even the hard-bitten and cynical Donovan cannot ignore.

Yet Donovan is hardly a radical greenie—indeed, as he puts it, the only thing he has been known to recycle is his own cynicism—yet he accepts an assignment with the environmentalist group SLUDGE (Sea Life Under Danger of Gradual Extinction). The head of SLUDGE believes it is being infiltrated by powerful forces seeking to avoid tighter controls on water pollution.

The awfulness is not restricted to Bondi Beach. Wright's characters cast their net far and wide even to Sydney's North Head sewage 'treatment' works, the overwhelming stench of which partly convinces Barry that the greenies with whom he has been put into contact by Gabrielle must have a point.

Sharon Beder in *Toxic Fish* (see Chapter Three) could not have described more starkly the discoveries of this private eye:

> They've done tests on the fish caught around here and the mercury deposits are five times as high as the recommended level. And that's the *official* report. God knows what the unofficial one would be . . . the ocean had become a toxic waste dump . . .[178]

To this landscape of poison is added urban blight: 'By the time we hit Malabar, housing commission flats had joined the architectural stew and the suburban wasteland was complete, a bit like South London but with more sun.'[179]

When the head of SLUDGE, Rodney Collins, meets an untimely end—in a stormwater drain—Barry investigates the circumstances and in the process discovers the grisly secrets of all involved in the environmental struggle. Gradually he attains a rage and outrage against the many things which are causing Sydney's demise. His cynicism and sense of irony sum up the path to destruction: 'The [rotting] ferry was called *Sydney* and it seemed to sum up the city's attitude to the old and decrepit: if it couldn't be successfully renovated, pull it down or chuck it in the harbour and let it rot . . .'[180]

This stark (albeit irreverent) realism is evident in the two other books in Wright's Barry Donovan trilogy—its predecessor *Love, Avalon* (1990) and its successor *A Break in the Traffic* (1992). The most recent of Wright's novels is *Life of Riley* (1998), dedicated to the notion that deeply rooted philosophies like environmentalism begin at home.

Recalling the writing of *A Drop in the Ocean* ten years after the event, Wright said in 1999 that the environmental theme stemmed from the era in which it was written, the ecology being 'a hot issue at the time, what with Graham Richardson being the federal minister for the environment and specific problems like water pollution being constantly in the news'. He described his novel as being a 'reflection of the time in which it was set, as is the case with all good detective fiction'.[181]

To further his understanding of the key issues in *A Drop in the Ocean*, Wright went to the ocean sewage outfalls. He recalled being concerned at all of the reports about ocean pollution, as well as the air and noise pollution which were increasingly blighting what the two Englishmen—that is, Wright and his creation Donovan—believed was otherwise a great metropolis.

In 1999, Wright said his environmental concerns were much the same as when he had sat down to write *A Drop in the Ocean*. Wright, whose front door in the inner Sydney suburb of Balmain has a Greenpeace sticker, said that as the father of three children he now had greater obligations and a heightened sense of responsibility. In the house that he shares with his film producer wife Marian Macgowan and their young children Billy, Calum and Adelaide the family strives to recycle as much as they can and they constantly 'weigh up what is the best nappy option'.[182]

Wright's environmental consciousness developed in several parallel ways after he arrived in Australia in 1972, having spent more than a year travelling around Europe and Asia. In the late 1970s he lived in an alternative community in north-eastern Tasmania at the time that the Franklin dispute was hotting up. During what he called his 'Thoreau period' he lived in the Lone Star Valley near Golconda, where he enjoyed life in the most basic ways and saw 'pristine nature at work'. He also witnessed the same logging on nearby hills as described by another resident of the area, James McQueen, in *Hook's Mountain*. His concern for preserving wilderness was also heightened in 1999, when he and his family saw the World Heritage-listed spectacular sights of Lord Howe Island—the ocean, the reefs and the 'incredible collection of fish', all of which needed to be protected for the next generation.

In *A Drop in the Ocean*, Steve Wright's creation Barry Donovan—literally—gets among the muck, slime, filth and degradation at the heart of a big city as he discovers many unpalatable and incongruous truths.

Similar truths and even more unpalatable realities are at the core of *Red Nights* (1997), an in-yer-face thriller by playwright Louis Nowra (see Chapter Six). *Red Nights* shows Sydney at its rawest and most brutal and nothing about the Harbour City is shown as appealing, for example the Opera House is of a design that had been 'inspired by a piece of orange peel unwound on a table' and Fort Denison is nothing more than a 'sandstone pimple' in the middle of the harbour.[183]

The city is horrible, and so are its inhabitants. As AIDS cuts a swathe through several sectors of society, the illness is part of a wider malaise. 'There's a sort of decay, terminal decay to this era,' Nowra writes. This is a city of the utmost squalid origins where the trendiest bar had once been a lavatory.[184]

Red Nights describes the last frenzied, harrowing and increasingly menaced 24 hours in the life of mogul Nelson Taylor, a major player and mover and shaker in the corporate excesses that made Sydney. As Nelson finds himself in unstoppable decline, he sees much the same happening to the city he once controlled:

> Nelson . . . gazed out at the squat world. Only a few kilometres out of the centre of the city and he felt as if he were in another universe. Dumpy houses with red tiled roofs. It was being gentrified and now it was Japanese nouvelle cuisine restaurants and Aboriginal clothes designers. Still it would always look ugly, no matter how much they renovated the houses . . . There were so many noises, the roaring cars, the music, that voice like boots on gravel, all of it like some shrill white noise.[185]

Nowra observes how those on their expensive yachts in Sydney Harbour are quite prepared to trash their world-renowned waterway, throwing almost anything overboard. And on shore, the glittering and opulent buildings are a façade—they can rise and fall on very temporary and fleeting financial fortunes, as Nelson Taylor observes finding that his wealth, and his very existence, are at the point of extinction.

R.F. Brissenden's *Poor Boy* (1987), partly enacted in Tasmania, has as its focus the murderous international drug trade. Brissenden has a close knowledge of how politics operates. He creates an Opposition MP—with supposedly the most impeccable of green credentials, having been an activist in the fight to preserve rainforests—whose rise towards parliamentary prominence is accompanied by involvement with the heaviest of drug traders and captains of industry (in some instances these are one and the same).

Brissenden thereby explores the forbidding challenges of those in high places seeking to prevent environmental destruction. After detours of international intrigue, the action arrives in Hobart, where Brissenden reminds us of Tasmania's cruel and bloody past:

> The museum walls were hung with harpoons, flensing knives, spades and axes that had butchered the right whales in the [Derwent] estuary and the sperm whales in the ocean and had made Hobart rich. Scrimshaw, whale oil and corsets seemed a small reward for the waste . . . pictures of whales in their death agonies smashing boats and tipping broken oars and men into the waves.[186]

Poor Boy, the first of a planned trilogy, denounces the thoroughness with which the current crop of developers emulate their exploiting European forebears who upon arrival pushed aside the Aborigines, 'turned their beach into a garbage dump from the criminal refuse of London and Liverpool, and zapped the woods'.[187]

Brissenden laments a lack of morality citing the hedonism accompanying the carnival end of the Sydney–Hobart yacht race, when the revellers cannot find one second in which to reflect on Tasmania's previous and current brutalities.

His alter ego is Tom Caxton, a cliched hard-boiled foreign correspondent who can mix it with the heaviest players yet remain true to a personal code of decency. Caxton ridicules the dismissive stance of those developers and their apologists who claim that 'there must still be vast tracts of relatively untouched wilderness'. Caxton's rejoinder is withering: 'Relatively untouched . . . How can a wilderness be relatively untouched? . . . it's like being slightly pregnant.'[188]

Caxton fumes at how a majestic eagle, one of nature's proudest and most impressive forms, can be reduced by the corporate world to nothing more than a 'stylised logo . . . with extended wings grasping the globe in its talons'. And typical of the corporate raiders is one who seeks to replace wilderness with an artificial paradise.[189]

The second in the planned trilogy, *Wildcat* (1991), expands on many of Brissenden's environmental concerns. This time the action starts in the Outback, where rednecks wear T-shirts saying 'Fertilise the forests—

bulldoze a Greenie today!' Here is the overwhelming sparseness and sterility near the Queensland–Northern Territory border where 'some of the smaller trees were violently twisted, bent and uprooted'.[190]

In this wilderness Caxton discovers that 'everyone is an amateur geologist' (and unconcerned about the consequences of their actions) with the sole aim of participating in the bonanza of mineral wealth. The corpos planning to exploit the area have contacts with government and the governmental agencies which are supposed to protect the environment. As the plot thickens, any computer hackers, committed greenies, financial and software experts following the money trail who get too close to home truths find themselves missing or dead.

Despite its setting in the depth of the Outback, *Wildcat* has a harder internationalist edge and a much stronger sense of responsibility and conscience than did *Poor Boy*: 'The global program of ecological disaster—famine, desertification, annihilation of forests, erosion of irreplaceable soils, war—is driven by usury, by the banks' insatiable hunger for interest . . .'.[191]

Wildcat describes how Australia can never be protected against global influences. According to the *Sun-Herald*'s Judith White, it 'goes to the heart of government and business corruption in Australia . . . it goes close to the bone on some of the most sensitive issues in federal politics'.[192]

Brissenden, who was born in Sydney's western outskirts in 1928 and had a rural upbringing, was actively concerned throughout his life with environmental issues. In 1989 he joined other prominent writers in undersigning a letter to the editor of the *Sydney Morning Herald* to protest against woodchipping. The letter was a follow-up to that by 88 'prominent' Australians calling on the Federal Government to re-examine its woodchipping policies.

'Intensive logging [of the National Estate forests of south-eastern New South Wales] is continuing . . . and their heritage values are being destroyed,' the letter said. 'We call, too, on the Federal and State Governments to end logging in the National Estate and to give proper protection to these magnificent areas. The Federal Government should use its export powers which stopped sand-mining on Fraser Island.'[193]

In line with Brissenden's enduring concerns about nuclear technology, *Wildcat* debunks many of the myths about 'safe clean nuclear energy'.

Brissenden, who died in 1991 aged 63 after a long battle against Parkinson's disease, was recognised as a writer and poet of distinction and great variety (see Chapter Seven).

A POTPOURRI of novels—offbeat, unconventional, humorous and zany—have also dealt with the ecology. For example, the themes of white man's desecrations throughout the vast continent recur constantly in Australian novels published in the years after 1970. Every region of the country is covered in novels which embrace all moods and themes.

Among the most confronting and dramatic—and, for many reasons, extremely controversial—books is the trilogy by B. Wongar, a writer of Yugoslav descent who adopts the Aboriginal persona in narrating the tragedies of that race. In Wongar, fine writing about environmental destruction outweighs reservations about his true identity.

In *Walg* (1983), Wongar decries how the Australian environment was shaped by the demands of European invaders. 'For so long life had gone the way the whites wanted it,' writes Wongar, simply. With the advent of European civilisation, everybody and everything became brutalised:

> [Europeans] set the whole bush ablaze. From one end of the sky to the other, instead of your country you see a mountain of flame.
>
> Dadanga scratches his head. 'Why would they burn the bush?'
>
> 'To expose the rocks, kill the plants and animals—everything that is in the white man's path,' explains Wagudi.[194]

Wongar is by most accounts a Yugoslav immigrant, Sreten Bozic, although this remains shrouded in self-asserted mystery. The most reliable information suggests that he was born in the 1930s to a Yugoslav father; he has never revealed where exactly he was born nor who his mother was. It has been revealed that after he grew up in a small village in Yugoslavia, in 1960 he emigrated to Australia and spent ten years with Aboriginal tribes in the Northern Territory.

Writer Robert Drewe, who 'exposed' Wongar in 1981, noted three years later that he had a problem with an author who was 'a reincarnated Arnhem Land Aboriginal inside a Yugoslav body residing in St Kilda'. Nevertheless, in discussing *Walg*, Drewe conceded that Wongar was sometimes able to 'achieve a curious and beautiful alchemy' in a novel 'often shattering, lyrical and deeply affecting'.[195]

Wongar shows in *Walg* and many of his other writings how man's inhumanity is a sentence of death to all of the world's living things. Aboriginal rights, feelings and sensibilities count for nothing in the frenzied extraction of uranium, and Wongar describes the Europeans' readiness to mine Aboriginal reserves right down to 'the last boulder'. And, years after uranium mining had seemingly finished, whites thought nothing of spraying a whole region with toxins.[196]

The uranium miners and other white intruders leave nothing standing:

> They have stripped the top off the mountain. All the trees, scrub, even the mosses have been scraped off, leaving behind the bare rock . . . From the small plateau at Marain Cave right down to the shore of Warngi Inlet, there is not a single tree or bush to be seen; nothing but piles of stumps, topsoil and crushed branches . . .[197]

Walg, although set in such a mire of desperation, degradation and despondency, ends with a hint of optimism, with hope placed in Djumala's unborn child, who will overcome the stupidities and injustices which have been murdering the land.

The novel was widely praised by American reviewers largely unaware of the controversies surrounding Wongar's identity. Evan S. Connell, in *The New York Times Book Review*, took Wongar's work at face value as a reflection of Aboriginality and described *Walg* as 'a provocative novel'.

The second in the Wongar trilogy, *Karan* (1985), continues many of these themes as well as introducing the wider problem of British nuclear testing. Wongar estimates that the testing affected almost one-third of the total area of South Australia, well beyond the epicentre at Maralinga, as well as affecting 'scores of tribes [who] had lived [in this wider area] for about 2,000 generations'. Not only were the tests conducted arrogantly and in the utmost secrecy, but Wongar also notes how 'when the British began their nuclear testing in Central Australia, the local tribesmen, many of whom had never seen a white man before, were rounded up in the bush and removed from the area to be confined in government camps, thousands of kilometres away from their tribal country'.[198]

Karan won greater acclaim internationally than it did in Australia, including in the USA where *Boston Herald* reviewer Sue Standing described it as having 'a mythic force, a transforming power that few contemporary novels achieve'. Standing, although conceding that Wongar wrote 'under a pseudonym', said *Karan* in its 'bleak' vision of the Australian landscape had kept 'a land and people alive through the enabling power of words and drama'.[199]

The views of Standing—a poet, a Professor of English, director of the Creative Writing Program and a writer-in-residence at the prestigious Wheaton College in Massachusetts—reflect the international reputation that Wongar/Bozic was able to achieve despite his detractors in Australia.

Another American reviewer, Chuck Moss, went even further, describing *Karan* as a 'languid, horrifying and surreal' portrait. 'The casual genocide, the bored, technological Auschwitz, the peeling patches of skin and gleaming surgical instruments add up to a world monstrously evil in its calculated banality'. Moss described *Karan* as a 'mood piece, another protest against the atomic age'.[200]

Wongar's gloom is heightened in the third novel of the trilogy, *Gabo Djara* (1987), in which heaviness and bleakness abound despite its early whimsy. The central character is a green ant who visits Federal Parliament and other public institutions. The ant discovers on his hazardous forays into the gathering places of human decision-makers that rocks of uranium are to be found in mountains ready to be exploited by the likes of Sir Rock-Pile Ore and Sheikh Dollar. The green ant motif, one of the most potent in Aboriginal folklore, is used throughout by Wongar.[201]

Gabo Djara describes how the *balandas* (Europeans) cannot understand how those bits of rock which they call uranium are so precious to the original inhabitants and how their extraction degrades the environment. To Wongar, the land has become a place where 'plants had forgotten how to grow', and 'against some of the white man's machines, the spirits could be powerless'.[202]

A consensus among Australian reviewers is that if B. Wongar had not been plagued by the continuing doubts about his identity and integrity, his novels would be in the forefront of literature.

A sense of foreboding also permeates *The Ruthless Garden* (1993), the first novel by journalist Margaret Simons, who examines several characters of South Australia's Riverland area coping with the threat to the Murray–Darling basin. Simons, who spent most of her childhood in Adelaide, describes how vast tracts of this 'food bowl for the continent' would be 'laid waste' within a generation unless drastic steps were taken.[203]

Simons, born in London and who migrated to Australia at age eight, dissects how across the millennia this vast country had been carefully tended by the Ngawait Aborigines, many of whom perished, as did their land, with the onset of European settlers who, with increasing vigour and zeal, sought to 'tame' this 'wild' environment. The tone is set with the motto outside a country church: 'God gives enough for man's need; but not for his greed.' The fate of the land proves to be tragic, as are the fates of most of the relationships upon it.

A very human work embracing a range of issues including multiculturalism and personal growth is Lolo Houbein's *Walk a Barefoot Road* (1988). Born in the Netherlands in 1934, Houbein emigrated to Australia in 1958 at age 24. She treats her girlhood experiences of Nazi wartime occupation as a metaphor for the environmental destruction of her adulthood; as with responses to the Holocaust, Houbein traces the responsibility of individual apathy, recalling how when news spread of the Holocaust nobody believed such a thing could happen. Similarly, nobody in Australia realises the tragedy of environmental destruction until it is too late. As in World War II, tragedy which no one foresaw strikes and 'the future began to engulf them'.[204]

Houbein also laments the passing in Australia of 'age-old skills' and the manic zeal for progress at any cost. 'Whether it was a human being, a building or a tree, [it] had to make way for something younger, newer,' she writes.[205]

In the various strands of *Walk a Barefoot Road*, Houbein presents the thesis that the entire ecology (and mankind's collective experiences) is indivisible, and all the parts of the whole must be equally appreciated and understood.

Lolo Houbein's writings have included contributions to the ecologically conscious magazine *Grass Roots*, and since 1966 she has written a range of short stories, essays and poems. True to *Grass Roots*' conservationist philosophies, in *Walk a Barefoot Road* Houbein describes the South Australian highlands community of Eagletop as a place where thrift and a careful allocation of resources prevail. She lyrically examines the dynamics of all the workings of nature, especially in the surrounding forest.

Apart from bucolic musings, there are hard-headed practicalities which could be straight out of the advice pages of *Grass Roots*:

> Once she discovered how much affluent society had to spare or simply threw away, it was hard to stop her enthusiasm for collecting the city's surpluses . . . Piles of cloth samples from a curtain shop, a box of out-of-date seed packets,

> reject T-shirts from a clothing factory, dented tins of paint, and always bags of cabbage-leaves and vegetable refuse from the market . . .[206]

Houbein has intricately outlined her philosophies in her autobiography *Wrong Face in the Mirror* (1988, 1990), in which she describes how she started a branch of an organisation called 'Men of the Trees', which has a plantation program 'to combat soil erosion and climate degradation' and 'to talk to farmers about the need for trees on their land'.[207]

A practising Buddhist with a Jewish background and a world traveller who has lived a varied and unconventional life, Houbein abhors a Great South Land where 'the topsoil is blowing away, crops are diminishing and where some States are already perceiving environmental degradation and a drain on natural resources caused by increasing tourism'.[208]

ALL OF the above novels, despite their vast variety, are suffused with concerns about the ecology. They are observations about the shortcomings of societies which allow our world to be harmed.

Melbourne academic Laurie Clancy said such fiction reflects 'what we all fear and believe, that our environment is decaying and that our world is being destroyed'. Speaking in 1999 after his retirement from the Chair of English at La Trobe University, Clancy said even satirical and humorous novels had a capacity to ignite outrage and deep reflection about ecological concerns.[209]

The role of novels in inspiring and educating Australians about environmental concerns was neatly summed up by *Sunday Age* reviewer Helen Elliott in her discussion of Blanche d'Alpuget's *White Eye*. Elliott said reading this book left her convinced that 'if there is to be a future there must be courageous individuals with a clear moral vision who are prepared to act and live upon their beliefs'.[210]

A large body of Australian fiction—from the 'classics' of Xavier Herbert and Patrick White to the New Wave of Peter Carey to the thrillers of Peter Corris to the sci-fi and futuristic tales of George Turner and others—is there to guide us all into a more enlightened new era.

3

NON-FICTION
Potent touchstones

THE truth about what's happening to the Australian environment is often just as dramatic, as evocative, as insightful—indeed, as strange—as any fiction. For the past thirty years, a broad spectrum of Australian writers have informed, inspired and warned readers about the destruction of our environment. Those who have tackled important environmental issues in non-fiction books are either 'ordinary' citizens with no specialist knowledge about the ecology, or they are ecological specialists who have summoned their resources to communicate with a wider audience.

Australia is fortunate in having many ecological 'experts' whose fine writing skills are accessible to the general public. For instance, research scientist Tim Flannery in *The Future Eaters* has been able to excite and galvanise a wide readership with his descriptions of prehistory and botany; in other hands such a book might not have been read outside the scientific community.

All of the specific issues affecting the Australian ecology have been written about by highly skilled and committed authors. The seemingly mundane subject of poisons, pesticides and toxics has been covered by thoroughly effective communicators like researcher Sharon Beder and two Victorian journalists, Linda Cawcutt and Catherine Watson. The history of British atomic weapons testing in Australia has been written about in three separate books, each with a quite different perspective. And even the theological aspects of environmental philosophy have been raised by major writers such as biologist Charles Birch and broadcaster Paul Collins.

The importance of such books in alerting Australians to environmental problems has been defined by Sharon Beder, who in *Global Spin* and elsewhere has written fervently of the need to counter propaganda by governments and corporations intent on causing environmental destruction.

Non-fiction books aimed at a wide reading public are an effective way to introduce and stimulate environmental ideas, according to researchers like Peter Hay, the co-ordinator of the Centre for Environmental Studies at the University of Tasmania. Hay, who runs a course in 'Literature on the Environment', said non-fiction books are a primary teaching tool and many of the works are a 'major literary form, on a par with poetry'.[1]

Hay noted that 'popular' writers like Tim Flannery had been able to achieve a 'high market penetration', and Flannery's scientific background ensured that a book like *The Future Eaters* had a credibility and influence that far surpassed other environmental histories. Hay said the various books about complex issues like the British bomb testing at Maralinga were valuable not only in describing what happened but also in revealing the mentality which allowed it to happen.

The significance of a widely read writer like Eric Rolls also could not be underestimated, according to Hay, who said few books had achieved the impact of *A Million Wild Acres* and *They All Ran Wild*. Hay said Rolls was able to win over a large and loyal readership and 'tap into the bush mythos' with books that were accessible and beautifully written. He described Rolls as among those idiosyncratic writers who have had 'a great impact on the way Australians think and see themselves'.[2]

Many environmental activists throughout Australia have acknowledged the importance of non-fiction books. Among them is David Haigh, prominent in the struggle against developer Keith Williams' Port Hinchinbrook tourist resort in North Queensland, who said such books make local campaigners realise they are 'part of a bigger world'. Haigh, a senior lecturer in Environmental Law at James Cook University in Townsville and a researcher in world heritage, the president of the Friends of Hinchinbrook Society from 1993 to 1999, said inspirational and informative books helped to connect a remote place like North Queensland to the wider intellectual community.[3] Furthermore, fervent campaigners such as those against the Hinchinbrook resort had a 'self-motivation regardless of the inputs' of books, but widely read and influential studies like *The Future Eaters* strengthened their belief about the 'righteousness of what they are doing', especially on issues such as the loss of natural environment throughout North Queensland. Haigh added that one of the problems for North Queenslanders was their 'remoteness from intellectual places' and this often was a big hindrance to their local campaigns because often it was difficult to get writers, photographers and artists from the southern States to take an interest in what was happening up north.[4]

The growth of non-fiction environmental literature coincided with political developments such as in 1972, when the newly elected Whitlam Government created a separate Environment portfolio, setting a precedent for recognition of green issues. Themes of destruction—and species extermination—predominate in the classic works of the important writers who subsequently emerged.

ONE of the most important writers to explore Australians' capacity for destroying their environment is Tim Flannery, who in *The Future Eaters* (1994) shows how, because of the latest technology and a prevailing culture of exploitation, we are facing doom. Flannery's warnings are dramatically powerful as he asserts that, 'we will, I fear, see a wave of extinctions'. These extinctions will be 'so vast as to dwarf anything that has gone before. For we will have become the exterminator species that broke all the rules. The one that could take not only all the resources of rich lands, but of poor ones as well'.[5]

In one of the most compelling and creative works dealing with the fragile environment, Flannery likens the ecology to a machine which over millions of years has been regularly disrupted, with European settlement 200 years ago a precursor for disasters on a far wider scale. The First Fleet 'once again brought the machine which drives Australian ecology to its knees', and this machine, complete with 'new cogs and gears thrown in', created the inevitable havoc.[6]

Tim Flannery, born in 1956, was brought up in the outer Melbourne suburb of Sandringham. While growing up, he noticed urban decay and wildlife at risk—and even as a youngster he felt the need to find answers. Flannery at first studied English at La Trobe University before deciding to pursue a scientific career, doing a geology course at Monash University and then writing a PhD thesis on kangaroo extinctions in Zoology at the University of Sydney.

His career path would lead him to head the Mammal Section of the Australian Museum and become its chief research scientist, and this work would be at the core of *The Future Eaters*. In the book, Flannery shows how for thousands of years the arrival of man in any new surroundings has led inevitably to the extinction of many species. To Flannery, mankind is the main culprit rather than such factors as climatic conditions, ice ages, continental drift or fires: '. . . species [which became extinct] share little in common ecologically. They are all similar, however, in being either close competitors of humans in their expanding ecological niche, or their prey.'[7]

The Future Eaters includes many poignant descriptions of how species disappeared from the face of the planet. Emotive and heart-rending passages describe how Lord Howe Island, in contrast to neighbouring Pacific Islands, had remained pristine until the time of the First Fleet. Flannery shows how on Lord Howe—'the last South Pacific virgin to be deflowered'—the arrival of Europeans led almost instantly to the extinction of species ranging from white fowl to the most exotic of parakeets.

Flannery describes a 'vast research effort' in the 1990s to come to terms with the spoliation of the Pacific nations and to explore what can be done towards future preservation. Flannery regards this research as 'in some ways the most important being undertaken in our region' and he tells of being confident that there will emerge 'guidelines for the management of the "new" lands in future'. To Flannery, Lord Howe is a window of many wonders:

> Despite the terrible destruction, the island is still an extraordinary place . . . Were it not for Lord Howe we may never have known in our hearts what a remarkable world the first Australasians stumbled upon.[8]

Flannery describes the Australian ecosystem as 'extremely fragile' with all of its components 'entirely dependent upon one another to make the system work'. Flannery warns: 'Once a few key species have been removed, the entire co-evolved structure can collapse.'[9]

One of Flannery's main examples of this is the Great Barrier Reef, where 'the too-rapid removal of parts of the system (such as large fishes) can have disastrous effects'. Flannery joins a host of writers, such as poet Judith Wright, in expressing concern about the Reef's future, especially the way in which phosphates flushed into the system are seeing a monoculture replace wonderful diversity: 'Today, the relatively small commercial and recreational fisheries . . . are having a profound impact as predators, algal grazers and other types of fishes are selectively removed from the environment.'[10]

Flannery examines a host of other threatened regions throughout the Australian continent—stretching from Queensland's rainforests to the heathlands of south-western Western Australia.

Like many of the other writers discussed in this chapter, Tim Flannery fumes at Australians' stupidity and the culture which has made them unfit for sensible stewardship of their land:

> I quiver with rage when I think that it was for racehorses that the Yawarawarka and Dieri people of South Australia were barbarically slaughtered and their cultures destroyed. It was only for this that future Australians were deprived of 23 species of native mammals, all driven to extinction in our dry country. And it was for nothing more than this that the very soil of northern South Australia itself was lost, never to be replaced. The deserted and desolated wastes of this pastoral country, much of it now relinquished by leaseholders, is a pathetic monument to extraordinary folly.[11]

Similarly, Flannery backs up the beliefs of a growing list of Australian thinkers who have bemoaned how European Australians have been unable to learn from the land's original inhabitants, and have not properly tapped into more than 40,000 years of 'co-adaptation' with the environment. 'Without it (Aboriginal culture), we have no precedence; no guide as to how humans can survive long-term in our strange land,' Flannery writes.[12]

Like that other great ecological thinker Eric Rolls, Tim Flannery urges a greater flexibility in people's ideas about what strategies should be adopted to achieve a more sustainable ecology—such as the introduction of flora and fauna absent from this continent for 60,000 years whose presence would achieve dramatic improvements in our ecosystem.

Flannery's specific proposals include the releasing of a colony of seals into Sydney Harbour to coincide with the 2000 Olympics. In various public statements throughout 1997, Flannery said he envisaged the introduction of the seals as part of a wider plan to reintroduce natural wildlife throughout Sydney—including wallabies in parks, reptiles in playgrounds and flora everywhere.

Another controversial suggestion in the same year was that Australian laws be changed to allow native animals to be kept as pets rather than the more traditional—and imported—cats and dogs. Further, he railed against the prevalence of lawns in suburban backyards, saying that these consumed too much precious water, and that they should be replaced with native tussock grass or heathland.[13]

Flannery has been chastised by several prominent critics including Melbourne nature photographer Harry Nankin (see Chapter Four). Nankin, the author of several ecologically aware illustrated books, describes *The Future Eaters* as 'deeply flawed', 'politically naïve' and 'counter-productive' in terms of seeking solutions to profound environmental problems.

Nankin takes issue with how Flannery 'condemns mindless boosterism [for development] yet condones the spread of mining'. Nankin also rejects Flannery's central thesis that Australia's population (and, by extension, immigration) needs to be slashed—Flannery believes the ideal population for Australia is between six million and twelve million—to prevent future ecological damage. By stressing population, according to Nankin, 'Flannery largely avoids the equally urgent but more complex questions of technological, economic and lifestyle change in ameliorating environmental impact'.[14]

But even Nankin concedes that Flannery's major achievement—including his television documentary based on this book—is as a populariser who gives wide publicity to many vital issues which need to be raised.

The Future Eaters has won praise and condemnation for its controversial views, but its writing was a triumph of Flannery's persistence. During its research phase, he became the victim of the same thing that has transformed Australia's ecology—fire. His home in the southern Sydney suburb of Jannali was overwhelmed by bushfires, destroying Flannery's computer, the floppy disks which contained the text of *The Future Eaters*, as well as all of his hard copy including his library. 'It made me realise the true implications of living in such an unpredictable land,' Flannery said of the ordeal.[15]

Despite the heart-breaking loss, Flannery began work afresh from a workstation provided to him by a colleague at the Australian Museum, and fellow scientists rallied to his cause by lending him a range of material with which he could resume his research.

Flannery used his high public profile and the wide recognition of *The Future Eaters* to publicise a range of environmental causes including coming out in staunch opposition to the Port Hinchinbrook development in North Queensland. Flannery joined leading British environmentalists David Attenborough and David Bellamy in attacking the $100 million residential and resort plan next to the World Heritage-listed Hinchinbrook Passage. The environmentalists said old-growth mangroves and a dugong population were at risk because of the development.

Always the inveterate campaigner, Flannery in July 1997 joined in the condemnation of plans to develop Sydney's North Head, where the city's last bandicoot colony was at risk. Flannery told a meeting of 300 people that preserving the long-nosed bandicoots was the 'most important' issue in which he had become involved. 'For me it comes from the heart,' Flannery told the meeting organised by a group called 'The North Head Alliance'. 'The loss of the bandicoots would be a tragedy for our city . . . The loss of the bandicoots is a far more important issue in the long term than any solar powered city.'[16] Flannery came out in support of the local issue at the time that he was appointed chairman of Australian Studies at Harvard University, a post which began in September 1998.

Interviewed before he left for Boston, Flannery hit out at unthinking and unrestrained 'exploiters' who were 'raping' Sydney's natural environment. He publicly called for the setting up of a Property Crimes Commission 'to investigate every major development in Sydney over the past 30 years' because Sydney had 'outgrown itself to the point of self-destruction'. Flannery said major reforms were required because 'we live with a series of antiquated laws and attitudes which allow the theft of the commons to go unpunished'.[17]

In the same month—August 1998—as a three-part documentary series began to air on ABC TV based on *The Future Eaters*, Flannery achieved a celebrity status colourfully described by Mark Whittaker in the *Australian*:

> In the time it's taken for Flannery's book to sell more than 40,000 copies in Australia, he has become one of those public intellectuals whose opinion is sought by journos on subjects as diverse as the meaning of big breasts and penises, possum meat exports, love, reintroducing seals to Sydney Harbour, and the big one—limiting immigration. It's part of his appeal that he just says what he believes, without thought for what others might say or think of him—sometimes taking himself into territory where no multiculturalist dare tread. Stick your head up, somebody's going to come along and kick it. Now, with a television series to stick his head up even higher, you'd think he'd be investing in crash helmets, but Flannery's enthusiasm enables him to keep going unscathed.[18]

ANOTHER scientist who has popularised complex ecological issues is David Smith, who has called for reasoned debate and a less emotional approach towards nature preservation. In his landmark *Continent in Crisis* (1990), this Melbourne zoologist who has taken environmental issues to a wide public in television nature programs, describes Australia's ecological prognosis as 'alarming' but at the same time challenges the conventional wisdom about what needs to be done. He outlines actions to preserve rainforests and coral reefs—both of which have a 'three-dimensionality' and are 'bewilderingly complex'.[19]

Among Smith's long list of concerns are the linked problems of salination, soil erosion and land degradation—all of which have turned vast tracts of Australia into a desert, with fertile farmland being lost at the rate of one per cent a year.

As a solution Smith points to the need for effective management and preservation to undo the destruction wreaked ever since the earliest days of white exploration and colonisation. He laments the deforestation which he says happened in 'a systematic and devastatingly thorough fashion'.[20]

On many issues, this book is more gloom than optimism and presents many unpalatable realities such as how the accepted 'scientific' ways to achieve higher productivity resulted in catastrophe for the Murray–Darling river system:

> Once a source of life-giving water which enabled the city of Adelaide to develop and thrive, the mighty Murray today is barely drinkable, and is laden with salt . . . In some areas the soil salinity is more than 1,000 times higher than normal, and very few plants can tolerate that . . . the damage is extremely difficult to undo . . .[21]

Continent in Crisis is also a celebration of wilderness and tells of how rainforests are a cornucopia of wonders, among whose glories are their capacity to allow us to go 'time travelling':

> Rainforests can tell us about our country's origins and its future. In practical terms tropical rainforests represent a living link with the past and a repository of potentially useful information and materials. In aesthetic terms rainforests pose a challenge to our present view of ourselves and our economically driven society, for if, having acknowledged that rainforests are beautiful and irreplaceable, we remove them for short-term profit, then our society must be judged as shallow and short-sighted.[22]

Smith, born in Melbourne in 1948, reiterates many of the warnings which have emerged in green literature but at the same time stresses that each contention needs a scientific and logical basis—that is, 'well-researched, unemotional management strategies must back up all the hype if there is to be any hope of long-term success in conservation'.[23]

Smith at first considered a career in medicine but his interest in biology took over and he studied zoology and physiology. While graduating in science from the University of Melbourne, he also pursued his 'passion' for film-making and his 'flair for communication' and released several teaching videos while he was a lecturer in zoology. This in turn led to Smith becoming the 'resident zoologist' during 1981 and 1982 on one of the leading late-night variety shows on commercial television, 'The Don Lane Show'; Smith shared with Don Lane's audience films and research on 'strange animals' such as eels and rare lizards, and he saw this as an opportunity to 'talk about all sorts of interesting things' and to 'share with ordinary Australians stories from the hallowed halls of academe'.

Between 1986 and 1988, Smith was the producer and host of ABC TV's 'David Smith's Earthwatch' program after he had been a senior researcher with the ABC's Natural History Unit, where he worked on the 'Nature in Australia' series with John Vandenbeld (see Chapter Four). In the late 1980s he set up his own production company, ImaginAction Films, as well as bringing about the projects which would result in *Continent in Crisis* and *Saving a Continent*.

In *Continent in Crisis*, Smith vividly shows the Great Barrier Reef as one of the most fascinating and intense theatres of conflict that exist in nature. Smith tells of how corals 'go to war':

> The territorial battles between neighbouring corals, although acted out in painstaking slow motion, are fought just as ferociously as are those between predator and prey on the African plains. When chemicals signalling the presence of a potentially threatening neighbouring coral are detected, the coral sends out long sweeper tentacles. These are armed with batteries of vicious stinging cells, and their mission is one of search and destroy . . . As a defence . . . corals adopt many strategies, including armour, camouflage, the incorporation of toxic or unpalatable chemicals into body tissues, or the adoption of a cryptic or concealed way of life . . . all have their equivalents in terrestrial ecosystems.[24]

In the years following publication of *Continent in Crisis*, Smith observed many encouraging signs of changes in attitudes and positive steps towards more sustainable developments, and his growing optimism is reflected in an expanded and updated colour edition of the book, *Saving a Continent* (1994). 'I believe the days of doom and gloom are essentially behind us,' a more confident Smith wrote in the updated book.[25]

Smith—who has also spread his environmentalist message in two novels, *Freeze Frame* (1991) and *Total Containment* (1993), and has been a Visiting Fellow in Zoology at the University of British Columbia and a senior guest scientist at the Swedish National Science Council—urges his readers to adopt more green-friendly consumerist attitudes.

In 1999, Smith saw many of the problems raised in his books as continuing, and he said he agreed with Tim Flannery's thesis that Australia was 'bursting at the seams' with a population of twenty million and that there needed to be a check on unfettered growth. Smith was also concerned at how Western nations were 'using technology to cheat nature' and that the bases of growth were 'fundamentally wrong'. He told of humanity ignoring the 'basic rules by which the planet is governed' and he said a 'new phase in our existence' was needed.[26]

The trends in Australia towards economic rationalism are incompatible with sensible ecological policies, according to Smith. 'Economics is a theory, not a science, and it remains comparatively primitive and should not be the basis for the nation's development,' Smith said.

As an example of the unsuitability of current economic modelling Smith cited the 1989 Exxon Valdez oil spill in Alaska, where an oil slick covered 1770 kilometres of coastline. Despite the overwhelming devastation to the Alaskan ecology, the spill was a boon to the local economy, according to Smith: 'There was an influx of workers in Alaska to clean up the mess, and the locals benefited because hotels were full and there was lots of spending.'

A HISTORIAN who has become an influential voice in forging environmental consciousness is Professor Geoffrey Bolton. His minutely researched and passionate *Spoils and Spoilers* (1981, updated 1992) describes and warns how a range of issues, including uncontrolled urban growth, is causing an unfortunate flow-on effect. He shows, for example, how the influx of people during the gold rush had far more severe ecological consequences than anything that had come previously.

Bolton shows how during and since the gold rush trees have been felled with progressively increasing manic zeal. He describes the serious consequences for climate and environment as last century's colonists—'they hated trees'—treated lumber as a limitless resource and paid no heed to how its felling would affect the land.

Bolton, born in 1931 and educated at Perth State School and the University of Western Australia as well as Balliol College, Oxford, was one of Western Australia's most prominent academics. Among his postings were stints at the Universities of Western Australia and Queensland and the Australian Studies Centre at the University of London. Since his retirement in the late 1990s, Bolton has devoted himself entirely to writing.

Throughout his long writing career, Bolton has tackled a large variety of topics ranging from a history of the Perth street where he was born and raised, to books about the exploration and settlement of Western Australia, to Queensland politics—and, of course, the environment. In his later career, Bolton would be recognised as a major figure giving voice to Aboriginal issues and the role of women, both topics having been generally neglected in earlier histories of Australia. Bolton is widely recognised as a commentator of quiet reason and rationality, being described by writer Deborah Gare as 'a figure of moderation in Australian historiography [rejecting] extremist views of either end of the ideological or political spectrum'. Gare described Bolton as 'one of the nation's greatest writers and visionaries'.[27]

In the largely updated and reworked text of *Spoils and Spoilers*, Bolton describes the overwhelming destruction of forests:

> In one day in September 1873 19 schooners loaded with pine sailed from the Richmond river, and in one week one million feet of pine and cedar were sent away from the district. Each year in the early 1870s the Richmond produced 5.5 million feet of pine and ash and 3.5 million feet of cedar.[28]

On a philosophical level, Bolton echoes the beliefs of Rolls, Flannery, Smith and others in claiming that colonial and pioneering attitudes were incompatible with a careful and caring husbanding of Australia's resources. 'It became a mark of Australia's standing as a democratic society that the poorest man had the right to kill as many animals as he liked without interferences from the laws or from the power of the rich,' Bolton writes. Furthermore, throughout all of Australian history, 'conservationists seemed to fly in the face of [the] ruling ethos'.[29]

Geoffrey Bolton synthesised much of the thinking that went into *Spoils and Spoilers* when he delivered the ABC Boyer Lectures of 1992. In his talk, he spoke of 'the need to advance environmental protection and management . . . and to set an example for neighbouring countries'.[30]

FAR more passionate and far more controversial in their critique of Australian ecological behaviour are the books of William J. Lines, whose *Taming the Great South Land* (1991) is a savage and angry chronicle of the manner in which Europeans, by 'reason and violence', inflicted their hegemony on a new land. Lines describes how under the doctrine of Terra Nullius (not discarded until after more than 200 years of white colonisation) not just the land but the original inhabitants were treated as merely there for exploitation under a mercantilist code.

Lines records how 'men and machines . . . united to clear out all accessible stands of cedar from the entire east coast of Australia'.[31]

He is graphic in his depiction of the totality of the destruction and the distorted visions which generated it: 'The stump, a symbol of nature subdued, became an enduring image in Australian history. The bush, like the wildlife and Aborigines sheltering within, stood in the way, not only of order and light, but of progress. The stump represented victory.'[32]

Taming the Great South Land provides an extended potted history of Australia, including not only political and social factors but also the relationships between different classes of the populace and how they treated the ecology. And since the First Fleet, according to Lines, the way the land and water were treated was very shoddy indeed—the land was blighted by salination, topsoil was regularly lost, there were plagues of imported pests, and the waters became poisoned and covered with algae.

Lines tells how such a total, crazy subjugation of the ecology could be possible only with the appropriate cultural underpinning: a conformist mentality kept the population quiet while a draconian censorship silenced or marginalised the occasional voices of rationality (such as scientist Griffith Taylor and explorer Douglas Mawson in the early part of this century). Lines shows that as the land is oppressed, the voices raised in its defence are suppressed: he relates how throughout the 1970s the Queensland and West Australian governments 'adopted repressive legislation to silence critics of development, increase police powers and prevent public meetings' which might challenge or seek full disclosure of projects potentially harmful to the environment. Similarly, he describes the policy of the Hawke Government (between 1983 and 1991) towards the Australian ecology as 'abetting dismemberment'.[33]

Taming the Great South Land was hailed in 1991 by the *Sun-Herald*'s critic Judith White as 'the most important single work on Australian history' since Robert Hughes' *The Fatal Shore*, which had appeared four years earlier. '[It is] a crucial advance in understanding the source of the environmental crisis we face today [and] in that respect it is a book of international importance,' White wrote.[34]

Lines documents the unfairness of most of the instances of exploitation and describes some of the activities, especially woodchipping, as economic insanity:

> Woodchip royalties were so trivial they did not even cover the costs of regeneration. The States not only gave the forests away [for example, to the Japanese] but, by providing the roads and wharves, actually paid for foreign companies to take them away.[35]

Lines is simultaneously angry, dramatic and outraged at Australians' subservience in the face of their own destruction. The book's epilogue is nevertheless a rousing manifesto for humanity—in all its glory, creativity, originality, imagination and potential for good—to do something about it.

But throughout Australia's bleak history the potential for good was rarely exhibited, and in several subsequent books Lines gives personalised accounts of how individuals were turned into destroyers by the prevailing attitudes.

In *An All-consuming Passion* (1994), Lines examines the life and times of Georgiana Molloy, an Englishwoman thrust into the alien frontier society of Western Australia in the early 19th century. Molloy—a gardener, plant collector and horticulturalist who devoted much of her life to nurturing the flora of the colony—died of a broken heart after years of ill health partly caused by the stresses of challenging the prevailing culture. The story was first told by Alexandra Hasluck in her biography *Portrait with Background: A Life of Georgiana Molloy* (1960), Lines meshes Georgiana Molloy's life with Western Australia's environmental history.

As Lines traces Molloy's life, we meet the people actually responsible for the earliest outrages as delineated in *Taming the Great South Land*. In this case the taming was done during the extensive settlement of the Swan River region in 1829. This settlement, as expected, was an ecological disaster:

> Disembarkation activity exacerbated sandhill erosion and sand became a byword for Fremantle. Blowing sand made living conditions unpleasant . . . [This happened not long after the arriving British ships] discharged cargo onto the open beach. Surf damaged the goods, and waves washed some goods out to sea . . .[36]

These first British settlers and explorers also conducted frenzied and indiscriminate sealing activities whereby 'in less than a year they extirpated all the local seal colonies'. Lines then documents how, having decimated seal life up and down the West Australian coast, they then moved onto the next phase: 'they turned their guns against the swans, ducks, kangaroos, wallabies, and any other creatures that might provide food or good sport . . . Migrants hunted with such avidity and lack of discrimination that observers soon complained of the rapid depletion of wildlife.'[37]

Documenting the voyage of Georgiana Molloy, Lines observes how when her ship, the *Warrior*, arrived at the Swan River, 'the new colonists had reached an already impoverished land'.[38]

In a theme which he would elaborate in later books, Lines dissects how the settlers of Australia were culturally influenced by those back home in Britain, where 'progress' had eliminated the forests:

> The transformation of the English countryside symbolised and epitomised the triumph of man over nature. Progress from forest to field marked the course

> of civilisation. Early nineteenth century Britons viewed forest as synonymous with wildness and danger. Forests were an abomination . . . Only in leaving and destroying forests would people find civility . . . The [WA settlers] thus came from a long tradition whereby to cut down a tree was to strike a blow for progress . . .[39]

Through the eyes of the nurturing and thus often horrified and wronged Georgiana, Lines shows too how the overwhelming religious and philosophical underpinning of almost all settlers was to exploit the environment relentlessly. 'In the new world of Australia they anticipated wringing human plenty from the wealth of nature,' he writes.

These themes are taken up in an even more personal way in *False Economy* (1998), in which Lines details the experiences of his own ancestors during the making of modern Australia, which since Federation has been under the thrall of economic rationalists for whom progress was everything.

False Economy details how in the immediate aftermath of World War II, the national ethos of unfettered 'progress' was totally ingrained and cut across all political and philosophical boundaries and 'all agreed on the need to develop Australia's boundless resources,' Lines writes. 'To the leading men of Australia, no limit to development seemed apparent. The idea that the planet and Australia were finite would have struck them as absurd.'

Such thinking was enshrined immediately after World War I, when in 1918 journalist E.J. Brady defined the unlimited nature of Australia's exploitative potential in the suitably entitled *Australia Unlimited*. Lines traces the irresistible rise of a delusional mentality that sprang from *Australia Unlimited* and the politicians and boosters who enshrined it:

> Gold embossed, gorgeously bound and profusely illustrated, the book was a paean to Australian development. Readers welcomed the message that Australia's untouched resources needed only the energy, pluck, and courage of the pioneer to turn them to profitable use . . .[40]

Long before the Snowy Mountains scheme changed the course of rivers in the 1950s, Australia's topography was converted at will and pleasure to the needs of development—for example, fifteen per cent of the natural flow of the Murray River was diverted in the early years of this century. Lines reveals how the earliest portents of disaster—such as salinisation and land degradation—were ignored by a brainwashed and compliant populace.

Lines began school in Perth at a time when the ethos of grandiose development was in full bloom, when the Snowy scheme and international events like Sputnik and American marvels showed that nothing was unattainable. From his own adolescent experiences and the onrush of history, Lines also believed that the Australian mania for growth was folly.

As Lines was growing up, he saw urban blight spread from his home at Gosnells and the Great Australian Ugliness take hold. Much of this ugliness is described by Lines in the equally personal *A Long Walk in the Australian Bush* (1998), which describes a journey by the author and his American companion Carol Hartland across a 650-kilometre trail through the Bibbulmun region in south-western Western Australia. This track, which shows the scars of the 'progress' which Lines deplores, begins in Perth's outermost suburbs where he grew up and then spreads out across jarrah and karri forests whose destruction Lines further documents.

In a direct progression from *False Economy*, Lines describes the Bibbulmun as a microcosm of a greater madness, leading him to deplore the situation throughout Australia. 'Nowhere else on the planet have so few people pauperised such a large proportion of the world's surface in such a short period of time and for such transitory gain,' he rages.[41]

Throughout the couple's odyssey along the Bibbulmun track, the author gives a history of each plant and piece of soil they encounter. As in all his writings, Lines marvels at what is still somehow left of nature's splendour and he laments most loudly for all that's been lost. He shows how since the earliest settlements in Western Australia, the jarrah and karri forests were seen as having no intrinsic worth or beauty and therefore were only a resource to be exploited.

From the earliest pages of *A Long Walk*, Lines reveals the long history of the utter desecration of the State's forests:

> Previous and present generations had torn out many of the first leaves and passages of [the original forests], mutilating it in many places.
>
> Today, the forests not only *look* different from what the Aborigines knew and the first European settlers found—they *are* different. The annihilation of old growth, invasion by pests and diseases, contamination of surface and underground water sources by toxins and salt, punctuation by roads, clearcuts and mines, a decline in diversity of all types of plants and animals . . . are not merely signs of mistakes in the application of logging methods or management, they are evidence that those methods are inherently destructive and self-limiting.[42]

Along the track are countless examples of the trashing of what should be a glorious region, and Lines largely blames the government instrumentality responsible—the West Australian Department of Conservation and Land Management—which offers no sensible advice to anybody wishing to visit the area and does nothing to ensure an orderly and proper upkeep.

The acronym for this government agency is CALM, but Lines details how in the midst of the supposed tranquillity of the Bibbulmun there is the growing irritation of noise pollution. He echoes the concerns of writers like Colin Thiele (see Chapter Eight) who are appalled at how the ordinary citizen behaves when confronted with a treasure that needs to be treated with tender loving care: 'All the camp-sites we had passed through so far had been strewn with rubbish. Rubbish also lay along the track. In the bush itself we saw many burned-out, rusting car bodies, discarded as thoughtlessly as they had been acquired.'[43]

The several books by William J. Lines are a counterweight to those who accept and are silent about environmental atrocities, Lines became radicalised at an early age and his many books of rage are the result.

A LEGENDARY figure among environmental authors over the past three decades is Eric Rolls, who like Lines and others questions Australians' suitability to tend their environment. In his classic *A Million Wild Acres* (1981), he dispels the myths surrounding the second wave of settlers in the 19th century, describing the arrivals as 'difficult, crude, vigorous, dishonest, selfish, violent' although conceding that often they were also 'capable, adventurous and extraordinarily adaptable'.[44]

What is ostensibly the history of one forest—Rolls' much beloved Pilliga—becomes a microcosm of Australia's wilderness, and the book analyses the shortcomings of those who are its carers. According to Rolls, Australia was developed by rough-hewn people operating without hindrance: 'If these men had remained in Britain they would have had no influence on their times. Society would have restrained them. In Australia they moved outside the law . . . The greater the achievement the greater the violence directed towards that achievement.'[45]

The notion of decency, of acting in a courteous or considerate manner, was foreign to the nation-builders who needed to be brutal to survive, according to Rolls. He does not exclude himself from the broad brush, confessing that 'this book is not written by a gentle man'. Rolls sees himself as a descendant of a 'tormented community' with its share of 'hard gobbets'.[46]

The Pilliga as microcosm and metaphor is stressed by Rolls: 'What happened in the Pilliga forest happened on a lesser scale in most of Australia's forests. It was only by concentrating on one that I could bring the men and animals to life. Each district has its own life.'[47]

Born in 1923 in the north-western New South Wales town of Grenfell and raised in Narrabri, Rolls has been inextricably linked with the Pilliga forest, which he describes so eloquently in his books. He continued to farm at Barradine until the 1990s, when he moved to the city, to the inner Sydney suburb of Glebe.

In *A Million Wild Acres* he describes the stunning variety, habits, colours and preponderance of every type of species in the forest, at the same time debunking some commonly held beliefs such as that Aborigines tended the land much more caringly than would the European settlers. He cites as an example the wastefulness of the way in which Aborigines collected their timber, whereby 'a huge tree was destroyed for a couple of metres of bark . . . Aborigines used bark extensively for canoes, shields and dishes but they left wide strips between each cut'.[48]

Throughout the book Rolls stunningly describes the life cycles within the forest, meticulously examining every creature and its habitat. He calls for all people to become aware of what gives a forest its dynamism, so that everyone can act sensibly and with sufficient knowledge in response to the bewitchingly complex ecosystem. At the same time he notes that the Pilliga 'is a profuse and accelerated world that did not exist in such intensity until modern times'.[49]

Dissecting how an ecosystem can come under risk, Rolls gives details of actions by the Coonabarabran Shire Council between 1973 and 1975, when there were plans for an international airport to be built at the Pilliga forest. Describing such plans as 'greedy and irrational', Rolls slammed the attitudes which allowed such possibilities to emerge:

> The [council's] members once more saw the forest as waste land to be utilised. They slightingly referred to it as 'The Pilliga Scrub'. Orana Regional Council supported them. Narrabri later took up the proposal and pursued its own ends. In 1979 local councillors were still whispering 'The Pilliga Scrub' to politicians uneasily considering new airports.[50]

Rolls questions the desirability of public recreational access and points to the possibility of damage being caused by roads built throughout rainforests: 'Probably permanent roads in any nature reserve are undesirable. In forests they should be restricted to essential firebreaks. Nature reserves should not be regarded as drive-in botanical or zoological gardens.'[51]

Rolls, who in November 1993 at age 70 became a regular columnist on environmental issues for the mass-circulation *Sun-Herald* newspaper in Sydney, says people need to become active to save the forests, whose supposed guardians have shown themselves to be inadequate to the task. In *A Million Wild Acres*, he cites forestry authorities and national parks and wildlife services throughout Australia as lacking the clout or the will to preserve the environment.

A Million Wild Acres has become a much-loved classic, and its widespread appeal led to a greater popular understanding and appreciation of the bush. Although its scientific accuracy has been questioned by experts, it is acknowledged as a touchstone by many creative writers grappling with environmental issues and has been cited by many activists as a source of inspiration.

Fourteen years after the book's publication, in 1995, Rolls described the environment as 'the single most important issue' in the New South Wales election campaign. 'Our land, rivers, seas, farms and fisheries are in such a degraded state that only immediate and comprehensive work can maintain life as we know it,' he wrote in his *Sun-Herald* column.[52]

A similar marathon effort of research went into another Rolls classic, *They All Ran Wild* (first published 1969, updated 1984), for which the author said he scoured more than 7000 newspapers and more than 300 books, journals and documents. The book provides instances throughout Australian history of politicians and public servants treating the environment with cavalier disdain, humbling—and in some cases ruining—any expert or official who gave them facts contrary to their own narrow interests.[53]

Amid the awe-inspiring research and data, *They All Ran Wild* makes poetic and sweeping observations of the amazing sounds and synchronised movement of the birds he sees on the wing. He points out that native species may be outnumbered by imports by a ratio of up to 9-to-1; and he shows how the multitude of creatures which roam Australia are equally at risk from imported species of animals and the greedy, uncaring and unthinking humans who bring them in. Rolls states that the typical Australian is unconcerned about the intrinsic wonder and beauty of creatures.

As for Australians' incapacity to preserve and to recognise the sanctity of indigenous wildlife, Rolls is hyperbolic: 'Many Australians would slaughter their fellows if their skins were worth marketing.'[54]

Rolls carefully explains the peril faced even by Australia's national symbol:

> With shooters, boners, exporters, pet-meat dealers, and governments all hankering for the utmost profit from kangaroos, and with graziers wanting to eradicate them, those interested in wildlife will probably be overridden. Any decision made is more likely to be expedient than wise.[55]

As for solutions, Rolls is frankly pessimistic. He sees 'money, more money than anyone has ever considered' as the only antidote to mass destruction of our wildlife.

In *They All Ran Wild*, Rolls eerily predicts the dangers of bushfires such as those which devastated vast areas of New South Wales in early 1994. He describes fire as 'one of the worst dangers' against which man has to sensibly prepare.

Australians' lack of suitability to be custodians of their land is again powerfully raised by Rolls, who shows how the essential culture needs to change before any improvements are possible:

> One fault of the Australian man-on-the-land is his penchant for hard work. He believes that it is disgraceful to sit down and think. If he is not losing sweat he is not working. If he were found reading a book on soil-conservation at 10 o'clock in the morning, he would feel ashamed. But if someone found him ring-barking the last four trees in one of his cultivation paddocks and called him hard-working, he would feel proud.[56]

They All Ran Wild is a call to arms, seeking commitments—intellectual, activist, financial—and a plea to fight complacency. Rolls tells every one of his readers that those who vandalise the environment must be challenged.

Despite his strong and passionate commitment to seeking a proper environmental balance, Rolls in the early 1990s fell out with the conservation movement. 'I used to belong to all environmental organisations . . . I now refuse to belong to any,' he said in 1993. 'They are continually making fools of themselves because they will not do the work to find out what has happened to areas. A lot of their ideas are quite dangerous . . . locking up some forests will kill them quicker than logging.'[57]

Environmental problems permeate Rolls' writings, including his newspaper columns, and topics include salination, landcare schemes, chemicals and fungicides, the ecological risks from damming and canal housing developments, the spread of bitou bush and noise pollution (especially Sydney Airport's third runway). He also takes in-depth looks at specific issues such as sandmining at North Stradbroke Island, in which he carefully outlines the effects on rare flora and fauna.

Despite the contradictions within him and the flawed nature of some of his writing and research, Eric Rolls is a major and significant figure who has inspired a generation of Australians into caring and thinking about the environment.

MANKIND'S growing incursions on nature are extensively covered by another controversial figure, Percy Trezise, in his *Last Days of Wilderness* (1973), set in the wilds of the Cape York peninsula and which dramatically explains how the region's rainforests will be wiped out within the next 100 years at the present rate of their destruction.

Trezise, who describes journeys into the area which he undertook with experienced fellow trekkers, reveals how rainforests, which in prehistory had abounded, by the early 1970s covered only one per cent of the peninsula. He blames the 'ravages' of hillside burning, the incursions of the timber industry, tourist and building developments and a range of 'other exploiters' for the crisis facing rainforests.[58]

The combustible combination of man and nature has been devastating over the millennia, according to Trezise, who points to another of the great but not universally recognised destroyers of rainforest: cyclones. Trezise shows how in modern times cyclonic winds completed the ravages already begun by the cleared edges of forests (where roads, railways and industries have caused significant damage).

Trezise, born in 1923, is a writer, painter, anthropologist, rock art explorer, bushman, pilot and chronicler of the Cape York peninsula, where he lives in the small hamlet of Laura.

He describes the upsetting of a delicate ecological balance on Cape York, such as introduced pigs depleting the frogs which in turn ate the jellyfish which now threaten the lives of human swimmers. But the greatest environmental havoc has been caused by the introduced cane toad which has brushed aside almost every living thing with which it has come in contact.

As have writers like Rolls, Trezise demythologises many of the popular images of the bush. He singles out popular novelist Ion L. Idriess, who was among writers who 'had been in the bush all right, but saw little and misconstrued much of what they did see'. Trezise accused such writers of sensationalism, claiming they created false perceptions about Aboriginal life, bush ecology and the problems of rural Australia.[59]

Trezise is a maverick North Queenslander and his philosophies and actions are not always consistent with those of the Down South greenies, many of whom he publicly castigates for being out of touch with the real problems and solutions in remote regions like Cape York. In 1980 he allied himself with mining giants like Comalco and MIM while helping to organise a wilderness congress in Cairns in 1980; this congress also had links to the apartheid regime in South Africa and was therefore out of step with progressive elements in the green movement. But at the same time this rugged individualist spoke of his preparedness to use his Cessna aircraft to 'bomb the first oil rig erected on the Great Barrier Reef'.[60]

Trezise has been equally concerned about the over-development of regional centres like Cairns, which he believes has been ruined by the onrush of tourism. Trezise spoke of such places losing their soul and called for a moratorium on growth. 'We've got to decide how many people can visit the Reef and the rainforest,' he warned in 1988. 'We must have limits to growth.' Trezise was described by the *Australian*'s journalist Nicholas Rothwell as being part of a movement whereby 'in the minds of today's north Australians, a new vision of the tropics is being born'.[61]

ANOTHER contrary and controversial view of green political activism is presented in *Fighting Over the Forests* (1990), in which Ian Watson investigates why the confrontations between greenies and loggers became so ugly and why there emerged such a breakdown in communications between those with opposing viewpoints. Watson, a Sydney historian who researched environmental politics for more than a decade, examines several forest-related conflicts, beginning with the struggle over Terania Creek on the New South Wales north coast.

Watson describes the Terania battle as seminal and as 'one of the earliest and most bitter confrontations' which became a forerunner to the 'acrimony' that became synonymous with the forests struggle. Watson delineates the context of how Terania preceded far more intense and wide-ranging conflicts, each of which added to the gulf between powerful and incompatible forces:

> With their siege at the Terania Creek basin, the rainforest conservationists unleashed a new strategy of non-violent direct action, an approach which was to win national publicity for their Tasmanian colleagues during the 1983 campaign to stop the flooding of the Franklin River. Together with the loss of jobs in the timber industry, these direct-action tactics transformed local disputes over land-use management into major political and social conflicts which have persisted to this day.[62]

To get to the core of the dispute and discover the human element, Watson travelled to the Terania district to interview people on both sides of the issue, including conservationists and timber workers. He had also visited university and State libraries where he scoured journals and documents including government reports.

Watson was born in 1954 in the southern New South Wales town of Goulburn. He was raised and spent much of his life in the Blue Mountains, receiving most of his tertiary education in Canberra in the early 1970s. He became a historian with a PhD from the Australian National University, where his thesis was on 'a class analysis of environmental problems'. The New South Wales North Coast forests campaign was the case study for his PhD thesis, and this became the core issue in the book.

His main activity at the time was dealing with unemployment issues in the industrial coastal city of Wollongong. In a 1999 interview, he stressed that then, as now, he felt that environmental and social issues could not and should not be separated, and that the social conflict inherent in such environmental struggles was harmful and counter-productive to all involved in it.

Watson, who had interspersed teaching with his other work, described what it was like to talk to the participants on opposite—and seemingly irreconcilable—sides of the logging issue. 'Some conversations took between two and three hours, and when I spoke to timber workers the suspicions that were there at the beginning turned to trust and openness,' he recalled. 'I asked them for details about such things as the machinery they used and how they did their work. I asked them to "tell me about the conservationists" only at the very end of my interview, by which time they were happy to share their innermost thoughts and concerns.'[63]

Explaining his intricate interviewing process, Watson said working-class people like the timber workers 'appreciate telling you about their life and how they do their work'. He said almost every interview had the workers telling 'long, rich anecdotes' and from this grew 'a strong rapport from people who enjoyed reminiscing'. Despite some initial doubts about talking to someone who was obviously an academic and not of their own background, all of the interviewees, according to Watson, felt good that someone was taking them seriously and cared about their lives and prospects.

Watson said the timing of the interviews proved to be exceptionally fortunate in that 'it was late enough for the passions of the forest battles to have calmed down but it was early enough for everyone to remember exactly what had happened and what their feelings had been at the time'.[64]

The project that became the book stemmed from what Watson saw as the upshot of his own background and experiences: he had been an environmental campaigner, joining Friends of the Earth mainly out of concern about uranium. Parallel to his feelings about the need to protect the environment was a worry about the way environmental politics was heading, that the 'class dimension was being neglected' and that no thought was given by well-to-do environmentalists for the many people who might lose their jobs for the sake of conservation. Watson's own environmentalism grew from his enjoyment of bushwalking and love of wilderness areas but that 'as a socialist, issues like unemployment are equally important to me'.

In *Fighting Over the Forests*, Watson describes the chasm between the opposing sides—'two groups of people whose working lives and cultural worlds seem a universe apart'—and in the book's most valuable contribution shows how each side has its own tradition and folklore which it holds dear and which cannot and should not be ignored by its opponents. Watson says it needs to be recognised that what modern greenies see as vandalism was once believed to be inspirational and heroic. Without a significant change in these value systems, Watson believes, the attitudes of both sides will harden, at horrendous cost to the environment.[65]

To be a timber worker was 'presumed to require great bravery', writes Watson, who explains how this is ingrained in a 'heroic mythology' as exemplified in a popular 1930s film, *Tall Timbers*. Thus when a greenie challenges a timber worker, this is a reflection on the latter's 'masculinity . . . manual work and personal self-esteem'.[66]

When confronted by conservation activists, timber workers saw not only their jobs but also their way of life and culture threatened, and they were prepared to fight back. Watson explains that the Terania protest was a battle to win hearts and minds in the wider community rather than to physically prevent the logging. The campaign was simultaneously effective as a political weapon but less than satisfactory in that the alienation and hostility between the two sides intensified.

In a review of conservationist writings throughout the late 1960s and early 1970s, Watson likens the themes to those of 'the early English nature poets' who had similarly cried out against 'the depredations of early industrial capitalism'. Watson describes the crusading nature of much of the forest activism and points to 'a strong sense of moral indignation'; he quotes individual ecologists who saw their movement as similar to the campaigns to rid the world of slavery, to end child labour and to achieve universal suffrage.[67]

Grass Roots magazine described *Fighting Over the Forests* as a useful study which enhanced knowledge about what direct action entailed. The book was 'enthralling reading for anyone concerned over this, or indeed any other environmental issue'. The *Grass Roots* review added: 'It is to be hoped that it will enable protagonists from both sides to gain some understanding of, and empathy for, the opposing viewpoint.'[68]

In 1999, Watson described the mainstream conservation movement as 'too inward and conservative' and a far cry from a social phenomenon that had once been 'a progressive force to reshape society but which had failed to deliver'. He also criticised the green groups for having an 'evangelical fervour' as well as an inability to compromise. This was compounded, he said, by an arrogance that only the environmentalists had good intentions 'and that the fate of the planet was entirely in their hands'. Watson, who since 1994 has been a researcher at the Australian Centre for Industrial Relations, Research and Training at the University of Sydney, said any fight for the environment lacked its full meaning if it ignored social justice.[69]

Almost a decade after the book's release, Watson regretted that it had failed to 'intervene' in the polarisation between those fighting green battles. 'I had hoped for a developing consciousness whereby all parties could sit down together in one room to discuss solutions acceptable to all, but this never happened,' he said. 'There still seems to be no acceptance of the view that direct-action blockades such as at Terania can antagonise an entire community against conservation.'

THE poet Oodgeroo Noonuccal turned to non-fiction to share her environmental concerns. In *Stradbroke Dreamtime* (1972), Oodgeroo, then still known as Kath Walker, presented a collection of reminiscences and retold Aboriginal folklore, both of which poignantly described and defined the depletion of her beloved island (about sixteen kilometres off-shore and about an hour by barge from Brisbane) and the destruction and indignities inflicted by Europeans on her race.

Oodgeroo, who described herself in 1977 as the best-selling living poet in Australia and second only to C.J. Dennis, writes of how Stradbroke Island looked 'years before the greedy mineral seekers came to scar the landscape and break the back of this lovely island'.[70]

Blaming the 'greedy, thoughtless, stupid, ignorant' people who are guilty of an 'assault' on nature, Oodgeroo looks at the fauna which had evolved across the millennia on this beautiful Queensland island and the tragic fate in store for even the simplest creatures like nautilus shells:

> The island is different now. Civilisation and man's greed have chased away our shy nautilus shells. Motor-cars belch fumes over the land, and the noise of industry drowns out all other sounds of life. Men's machines have cut and maimed and destroyed what used to be.[71]

Stradbroke Dreamtime is simultaneously graphic in its litany of specific cases of destruction yet poetic in its celebration and sacred reflection of the island's glorious and varied flora and fauna. Oodgeroo writes of the place where she spent her formative years and where she returned in 1970 after a career as major literary figure and Aboriginal activist throughout Australia. North Stradbroke—or, 'Moongoolba' as Oodgeroo calls it, denoting its sacredness to her people—would be her permanent home for the 23 years until her death in 1993.

Oodgeroo Noonuccal was born on North Stradbroke in 1920 as Kathleen Ruska, one of seven children, and for most of her life was known under her married name of Kath Walker. In *Stradbroke Dreamtime* she says the devastation on her beloved island will inevitably rebound on mankind, which 'too will suffer . . . his ruthless bulldozers are digging his own grave'. She elaborates with dramatic rage:

> Stradbroke is dying. The birds and animals are going. The trees and flowers are being pushed aside and left to die. Tourists come to soak up the sunshine and bathe in the blue Pacific, scattering as they go their discarded cans and cigarette packs and bottles and even the hulks of cars.[72]

Among the series of short yet extremely sensitive vignettes, the author includes aspects of Aboriginal culture such as tribal laws which forbid the killing of animals other than for food; in one reminiscence, she tells of how her father punished his children for killing a kookaburra. Oodgeroo explains the simple, sensible practicalities of the Aborigines as they harmonise with nature. Among anecdotes from her youth, she describes the Aboriginal harmony with nature and contrasts it with how Europeans despoil unthinkingly:

> We would gather enough crabs to feed the tribe, and no more . . .
> There were always plenty of crabs for the picking then. It is not so now. [White] man's greed and his careless upsetting of the natural balance of life have almost wiped out the mud-crabs of Stradbroke.[73]

The second section of the book presents 'Stories from the Old and New Dreamtime', many of which clearly reveal Aborigines as the wisest guardians of the planet with philosophies, values and totems which respect the environment.

Years later, in an interview at North Stradbroke Island in July 1986 with literary commentator Candida Baker, Oodgeroo elaborated on the devastation she felt about mining: '[The miners have] destroyed and flattened the land, and the island can't live without its sand dunes. We need the dunes to protect the island.'[74]

In her later years Oodgeroo had not mellowed one iota. She described the horrors at Stradbroke, seeing the destruction in a much wider and more disturbing context:

> . . . basically the sandmining companies are the most destructive of almost anything that has come to Australia, expect perhaps the horned and hoofed animals like the sheep and cattle. But the sandminers have destroyed so much. You can't cut down a 300-year-old tree and say it will be replaced. How do you replace it?[75]

Oodgeroo spoke eloquently to Candida Baker about how on Stradbroke the 'co-existence' ordained by nature was shattered by mining and other developments which had caused 'systematic' degradation since 1946.

Discussing the significance of *Stradbroke Dreamtime*, Oodgeroo said in a 1977 interview in *Meanjin* that she hoped readers would become 'more conservation-minded because of these stories', stating she intended her future literary output to highlight contemporary conservation issues.[76]

The quintessence of Oodgeroo's stance towards the future of North Stradbroke, and symbolic of all issues of sovereignty over it, is the continuing attempt by governments and developers to link it by bridge to the mainland. When asked by interviewer Gerry Turcotte if the bridge was any closer to being constructed, she responded with typical feistiness: 'They keep wanting to, but it won't be put in, no way. Eighty-two per cent of the people on Stradbroke—black and white—don't want the bridge put in. It's only the greedy politicians who want the bridge. They want to turn it into a little Manhattan for tired politicians when they retire. A stately home away from home. There's been a big backlash reaction against it from the people on the island. We don't want to be part of the mainland. They'd ruin it if they put a bridge across. She's only 23 miles long . . .'[77]

Throughout the 1980s, Oodgeroo was especially active in spreading the message about 'Straddie' and other issues to the widest possible public, and she granted interviews to reporters from popular publications as varied as those aimed at corporate women and those of the more tabloid variety. In 1987 she told Christine Hogan, editor of the upmarket *Portfolio* magazine, of her 'gloomy' outlook about the future not only of Straddie but also of the world. After being alerted by Oodgeroo, Hogan recognised 'some appalling cases of destruction' on Stradbroke such as the vandalism of sacred sites. 'Within the last few years, the bora ring at Amity Point was bulldozed to make way for a stand of Geraldton Wax trees,' Hogan wrote. 'As an act of cultural vandalism, it cannot be underestimated.' Oodgeroo had told Hogan how such actions affected white history as well as indigenous sensibilities. 'That bora ring was the first seen and recorded by John Oxley,' said Oodgeroo in her appraisal of what the Bjelke-Petersen Government had done to the Queensland heritage.[78]

In the same year, Oodgeroo granted an interview to Susan Hocking of the mass-circulation *Australian Women's Weekly*, to whom she gave some of the most portentous and significant material ever to appear in that publication. 'The nuclear threat takes precedence over everything else, even over land rights,' she told Hocking. 'Let's face it, if we don't stop this nuclear madness, there won't be any land, Aboriginal or otherwise, to fight over.' After informing the *Weekly*'s complacent readership about her life on Straddie and what her people had endured, Oodgeroo added: 'Our kids are as threatened as children anywhere and Australians had better realise that. They have tended to think of nuclear arms the way they regard Aborigines—"If you can't see 'em there's no problem".'[79]

This message was considerably stronger than the fervent plea for restoring indigenous vegetation and wildlife, 'the ecology [of Stradbroke] which was destroyed by sandmining', that she had made nine years earlier in *Woman's Day*.[80]

Kath Walker spread her environmental exhortations even to the readers of the ultra-tabloid *People* magazine, usually known for its nude or scantily clad models and an ambience of sleaze and titillation. In 1981, she told *People* readers that conservation rather than specific black issues had become her primary concern. 'Now I'm involved with the conservation scene, with conserving all people, black and white, and our world we live in,' she told reporter Nic van Oudtshoorn. In connection with an exhibition of her paintings at the Brisbane Community Arts Centre, Walker said her artwork dealt with such issues as whale slaughter and what van Oudtshoorn described as the 'detailed beauty she sees in a clump of mussel shells on the beach'.[81]

When Oodgeroo Noonuccal, who had spent the last years of her life on Straddie, died of cancer in Brisbane aged 72 in September 1993, she was recognised throughout Australia as our foremost Aboriginal writer and most prominent activist for causes that far transcended Stradbroke Island and the lands and seas of her clan.

BIOGRAPHIES of the fighters for the Australian environment are rare, but one exception is Peter Thompson's *Bob Brown of the Franklin River* (1984), which appeared only a year after the High Court ruled that the Federal Government had the power to prevent the Tasmanian Government from building the dam which would have flooded the protected areas of the State's south-west.

In the closest study yet written of how Brown's consciousness grew and how he abandoned a medical career to devote himself to the environmental movement, the biography charts his voluminous writings over a decade and his appearance at hundreds of public meetings across Australia to plead for the Franklin to be saved. As Thompson dramatically describes it, in 1976, as Brown's awareness and anger grew, 'Australia's diminishing wilderness was burning at him like a blowtorch'.[82]

Thus evolved the green consciousness of Brown, who was born in 1944 and grew up far from Tasmania at Trunkey Creek near Bathurst in the central west of New South Wales. This was a conservative area where many of the rural folk held like-minded attitudes to those who would fight him in Tasmania. When studying medicine he, too, shared many of the beliefs of his neighbours and he voted for the Country Party (as the National Party was then known). At the same time, though, he developed a love of bushwalking and nature which would crystallise into his abiding concern for the wilderness. After he became a GP, Brown moved to Tasmania to become a family doctor in the Launceston area.

Thompson sees 1976 as a pivotal year in the making of Bob Brown as Australia's most influential eco-activist. The year began with Brown fasting on the top of Hobart's icy Mount Wellington in protest against the visit to Tasmania of the nuclear warship USS *Enterprise*:

> [This gesture] had not sent the *Enterprise* scurrying in retreat but it did demonstrate something else. Bob Brown had very capable instincts when it came to organising public protest. It was imaginative, peaceful, stirred curiosity and publicity. One person, acting alone, could hardly have achieved more in such a short time.[83]

Also in 1976, Brown went rafting down the Franklin in a glorious experience which would change his life and perceptions. And in August of that year a gathering of concerned citizens at his home in Liffey would expand into the Tasmanian Wilderness Society, which in turn would become the start of the Australian Wilderness Society and the beginning of national green politics. For the next quarter-century Brown would variously lead the fight against the Franklin dam, sit in the Tasmanian Parliament as a Green Independent and then as a Green Senator for Tasmania.

At the same time Brown's study of philosophy had convinced him of the efficacy of Gandhian civil disobedience and non-violent protest in confronting authority. Each of these strands would climax in the dam blockade which in turn would lead to the High Court ruling of 1983.

Biographer Thompson, born in 1952, was active in the Franklin campaigns. His background as a writer and broadcaster included a Master of Public Administration from Harvard University, a Master of Business Administration from the Australian Graduate School of Management and an Arts degree in politics and history. In 1999, he had achieved 25 years as a broadcaster, mostly with ABC Radio as anchorman of the highly regarded 'AM' program. Among his journalistic assignments were covering the fall of the Berlin Wall, the collapse of European and Soviet Communism, the Tienanmen Square massacre and the big breaking stories from all over Australia during his six years at the helm of 'AM'.

In *Bob Brown of the Franklin River*, Thompson discusses the spirituality which imbued Brown in his early odysseys and how it would affect his evolving political consciousness. Referring to Brown's unconventional deeds, Thompson notes that 'few people in the world have had the experience of travel on a pristine river . . . to join a stream not far from its source and be part of its great journey to the sea has a certain life-changing power to it'.[84]

To lead the fight to preserve the glorious Franklin Brown conquered awkwardness, shyness and naivete, to the extent that he could 'move the audience with the remarkable word pictures which flowed when he talked of the river'. Thompson continued: '[Brown] appeared never to weary. It seemed that every time his mind went back to the river his tongue was freed.'[85]

Thompson shows how the moderation and calmness of a self-effacing and modest man allowed him to win hearts and minds to the cause where more strident tactics might have failed. At the same time, as Brown became utterly committed and obsessed with the Franklin cause his exposure to realpolitik widened and he learned how to mix practical needs and idealistic aims. According to Thompson, Brown had arrived at the conclusion that 'a majority of Tasmanians would support the Franklin if they were not threatened by the stridency of the conservation message'.[86]

MANY of the writers of non-fiction books have imbued their ecological themes with a deep spirituality. An additional sense of reverence dominates the works of two thinkers who have directly noted a link between a person's inner self and his or her relationship to the planet. A direct link between spirituality and the ecology is made by Paul Collins, one of Australia's leading religious writers and broadcasters and long-time editor of ABC Radio National's religion programs.

In *God's Earth* (1995), Collins makes a direct call for all religions to be more relevant in coming to terms with the destruction of the planet. Collins, a prominent Catholic thinker, shows how philosophers since antiquarian times have constantly grappled with formulating theories about humanity's place in the environment; he shows how for 2000 years the Church has been extremely destructive, formulating edicts which are inconsistent with preservation of the earth and the cosmos beyond.

Collins, a turbulent priest who in the late 1990s was under investigation by the Vatican for writing a critical book about the papacy, begins his theological discussion with rare bluntness as he describes how all of humanity is facing an unprecedented crisis and that 'never before have human beings exploited, damaged and degraded the earth to the extent that we have'.[87]

Unless the Church's teachings more adequately reflect these catastrophic realities, Collins sees no hope, especially as 'even with our increased ecological knowledge we still largely lack the conviction, willpower and generosity of spirit to do anything about it'. Collins' prognosis is bleak: 'Without regard for the future we have unconsciously rendered the earth less and less inhabitable for both our human future and that of other species . . . many of which we have already driven to extinction.'[88]

Collins, born in the Melbourne suburb of Richmond in 1940, submits that religion—and, indeed, everything else—would be rendered meaningless if a battered environment shattered the quality of our lives. Those seeking to prevent destruction need a stronger religious zeal to achieve their goals; and those with a religious conviction will need to recognise the overwhelming importance of retaining an ecological integrity. Collins believes that any religious thought which ignores ecological issues is not only a contradiction but an irrelevancy.

Without the earth's ecological splendours as we have known them, there can be no religion; indeed, all are meshed in an unbreakable link, Collins writes as he warns of what spiritual life would be like after an environmental holocaust:

> . . . the poetic, mystical core of our spirituality and culture would dry up, for there would be nothing to renew and nurture it. Our human imagination would stultify without the inspiration of natural beauty and ecological diversity, and we would lose the ability to conjure up the new possibilities that actually drive us creatively onward. With the loss of the power of imagination we would descend into a hopeless, mad world in which we would most probably destroy ourselves . . .[89]

Collins takes a historical journey to show how religious thinkers through the ages have interpreted the Bible in very many ways, each taking a completely different interpretation of mankind's role in the environment. Collins cautions against any concept which has mankind at the centre of the creative process, reducing everything else on the planet to a role simply to nourish and enrich mankind. He rejects the 'mechanistic approach' of too many of his fellow theologians who believe that 'nature [is] a vast, inanimate storehouse of useable "resources" that are there to be exploited for the short-term advantage of humankind'.[90]

'It is this attitude that has brought us to the tragic and disastrous environmental impasse that we face today,' writes Collins, who scans the vast literature from Australia and abroad of major theological thinkers who have grappled to find a means for religion to be a positive force to achieve environmental salvation.

God's Earth was published almost ten years after Collins first joined the ABC in 1986, having spent the previous seven years at Harvard Divinity School. After his retirement from the ABC, Collins settled in the township of Michelago, near Cooma in the New South Wales Snowy Mountains country. *Age* reporter Shane Green has described the immediate environmental concerns confronting Collins at Michelago, which 'sits amid the barren, brown hills, over-cleared and over-worked since the area was settled in the 1820s, and lately parched by drought'.[91]

Prominent religious commentator Elaine Lindsay said *God's Earth* dealt with 'an important subject with clarity and passion' and she described it as the first book by an Australian author to synthesise 'the thinking of international eco-theologians into everyday language and encouraging us to think about our environment in terms of their insights and our own spirituality'. Lindsay, co-editor of *Women-Church*, an Australian journal of feminist studies and religion, wrote in the *Australian*: '*God's Earth* will cause members of the Progress at Any Price Party and Christian fundamentalists to froth at the mouth'.[92]

RELIGIOUS writing also directly addresses itself to ecological problems, with one of the foremost figures tackling the big issues being leading Australian biologist Charles Birch. A prolific author, Birch makes panoramic philosophical sweeps across every aspect of global destruction and depletion. Birch is a writer and thinker described by prominent US environmental scientist and population expert Paul Ehrlich as 'the most distinguished ecologist in the world'.

In *On Purpose* (1991), Birch sheets much of the blame for humanity's devastation of the environment on the 'dichotomy' between itself and the rest of nature—and for this he in turn blames much of biblical teaching. Birch is not afraid to take theological dogma to task: 'The book of Genesis refers to man as "made in the image of God". A widespread, yet superficial, interpretation is that no other creature is so made . . .'[93]

Birch, born in Melbourne in 1918, had a long and distinguished career at the University of Sydney, teaching Zoology and Genetics for twelve years until his appointment as Challis Professor of Biology in 1960. He retired in 1984 and continued his writing and activism. Between 1970 and 1990 he was moderator of the World Council of Churches' committee on science, technology and the environment and he has been a staunch advocate of animal rights; in 1993 he publicly criticised Japan for its policies on whaling.

In this book, Birch warns theologians that they can no longer ignore environmental issues; they cannot remain oblivious to the social and political movements which have made the ecology central to many people's concerns. Birch says religion must find a potent response to the technological age, otherwise its message will become part of the 'empty clanging of a funeral bell':

> Modernity and its conception of progress has dead-ended in world wars, the Holocaust, genocide, the exploitation of Third World countries, the increasing pollution of the earth, the disappearance of resources and the terrible spectre of omnicide.[94]

Continuing this theme in *Regaining Compassion* (1993), Birch dramatically tells of humankind's capacity to destroy what could be a paradise and he sees his contemporary generation as 'caught in the middle [facing] a spiritual perplexity without parallel in all history'.[95]

Birch, who is credited with having taught Australia's first tertiary course in animal ecology in the mid-1950s, campaigned throughout his long career to establish ecology as a rigorous science at all tertiary institutions.

He saw the upsurge in learning about environmental issues as being an engine of a growing grassroots awareness which would be the only hope for effective change. 'I still believe that ideas are the most powerful thing in the world,' he said in a 1980 interview in which he stated that politicians could not be trusted to responsibly surmount the environmental crisis. 'We should have more faith that ideas can transform the world. That's the hope for the future.'[96]

In *Regaining Compassion*, Birch defines a 'litany of crimes' against the ecology committed by his contemporaries:

> Topsoil disappears at the rate of one football field each second. Arable land is covered with concrete at the rate of three football fields each minute. Forests disappear at the rate of four football fields each minute. Species disappear at the rate of 100 a day. Add to this the greenhouse effect and the hole in the ozone layer and it becomes obvious that our present treatment of the earth cannot continue forever.[97]

Urging science and religion to learn from each other and feed off each other, Birch describes a crisis which threatens to engulf us all if attitudes do not change. 'The world is dying,' he laments. 'We are on an unsustainable course. . . Our relations with other living creatures [are] that of a despot. We are fast destroying the environment on which all life depends.'[98]

The significance of Charles Birch on worldwide environmental thinking was best encapsulated by the *Sydney Morning Herald*'s feature writer David Leser. 'Birch was always a nomad,' Leser wrote. 'He wandered across boundaries linking biology, ecology and genetics to economics, politics, philosophy and theology. He believed that the world was living a terrible lie based on false thinking.'[99]

A PARTLY religious theme permeates an otherwise very temporal work, Hesba Brinsmead-Hungerford's *I Will Not Say the Day Is Done* (1983), in which the celebrated author of children's literature (see Chapter Eight) gives a very grown-up and frightening account of the many fights to preserve Tasmania's rare wilderness areas.

Brinsmead-Hungerford gives personal portraits of the many ill-fated fighters for the Tasmanian south-west such as Brenda Hean, who perished in a suspicious plane crash at the height of the Lake Pedder controversy. Sharing a passion deep within her soul, Brinsmead-Hungerford describes the many campaigners for the preservation of Lake Pedder:

> All of them [were] particular people; people apart . . . It is the earth itself, that must have its martyrs . . . My story is of a lake that, some claim, had a soul. My story is of the people of the Lake; who were called to its cause; of how it was slain; yet, will always live, as a symbol of the Earth itself. Because of those particular people, those who follow will never say the day is done.[100]

Brinsmead-Hungerford speaks of the great danger in cutting the 'umbilical cord' which she believes binds humans to the earth. She chronicles the ultimately tragic lives of those with an evangelical mission to save Tasmania's irreplaceable natural treasures. 'It is interesting to remember that so many of the Lake People were either artists or folk of deep religious faith,' writes Brinsmead-Hungerford. 'Olegas [Truchanas] was one of the artists, Brenda Hean was for faith.'[101]

She angrily contrasts these noble and devout people with those who sought to indiscriminately develop the south-west. 'The engineers [of the Hydro-Electric Commission] were ignorant of ecology,' she writes, and notes that 'the pendulum is swinging towards evil, by the frightening force of ignorance'. She decries the calibre of Tasmanian politicians and decision-makers and laments how 'it is irony that the earth and its future should rest in the chubby hands of such as Tweedledum and Tweedledee [successive premiers of Tasmania], nursery figures, surely. In the vastness of time and space, we are of course all nursery figures. Yet the planet earth is ours to destroy.'[102]

As for Hean's mysterious disappearance, the author believes that Tasmanians—and indeed the entire nation—cannot rid themselves of the stench of evil until they acknowledge her martyrdom. Ultimately Brinsmead-Hungerford is an optimist and she looks to a 'wiser generation' which will emerge to 'break down the dam' which is flooding Lake Pedder and 'let in the wilderness again'.[103]

NO GREATER or more brazen environmental outrage has been committed in Australia than British atomic weapons testing throughout the 1950s, and a wide variety of non-fiction books has chronicled the extent of the destruction. These books powerfully complement novels (such as Xavier Herbert's *Poor Fellow My Country* and Dorothy Johnston's *Maralinga My Love*), poems, plays and films which have described the horror and stupidity.

The first of the British trials, at the Monte Bello islands off the West

Australian coast, began in 1952 and would continue at the South Australian outpost of Maralinga throughout the next decade, taking the concept of Terra Nullius to its extreme: that is, the inhabitants of these regions, the Aborigines, did not exist and the land they inhabited was deemed worthless.

A growing literature including Robert Milliken's *No Conceivable Injury* (1986) has shown how when Prime Minister Robert Menzies gave almost carte blanche (and secret) permission to the British to conduct their tests, the fate of the Aborigines, their lands and their waters was of no concern to him. Milliken tells of how this was imperialism at its most outrageous—not even the Menzies Cabinet was given information about crucial aspects of the negotiations with London—and genuine environmental considerations did not exist at the time.

Milliken shows how the leading scientists associated with the nuclear tests cared little or nothing about the land or its original inhabitants. Leading Australian scientist Sir Ernest Titterton, for example, admitted that as he prepared for the Monte Bello tests he had never seen an Aborigine and was 'incredibly ignorant' about them. The callous attitude, as extracted from Titterton's papers and testimony to the 1985 McClelland Royal Commission, is dramatically presented by Milliken:

> In discharging his brief to Menzies, [Titterton] had not thought it necessary to carry out an investigation, for himself, of the local sources of population [on the West Australian coast] near the Monte Bellos . . . His university 'would have taken a great deal of umbrage had I suggested that I should become an anthropologist, and start investigating the people living on the north-west corner of Australia'.[104]

The Royal Commission also heard how for the principals in the British nuclear tests 'the only source of information about the Aborigines was the *Encyclopaedia Britannica*', highlighting how Australia's remote lands and their inhabitants were not included in the annals of 'civilisation' and were in every way expendable.[105]

Milliken reveals how the tests—and the attitudes which allowed them—would have disastrous and enduring results on the vast areas where Menzies unleashed the deadly experiments. 'Several sites at Maralinga are still contaminated with plutonium, which remains radioactive for thousands of years and is capable of causing cancer in humans if inhaled or ingested,' he writes.[106]

No Conceivable Injury resulted from Milliken, as a journalist working on the *National Times*, having covered the McClelland Royal Commission. In what would be his first effort at book writing, Milliken had envisaged a 'quick book' when the project was commissioned by Penguin but, as he noted in a 1999 interview, 'it turned into 18 months of solid work'.

Milliken, who was born in Taree in 1946, had an extensive journalistic career behind him when he embarked on the Maralinga book, having received a broad training on tabloids and broadsheets and having covered everything from local rounds in Sydney to the parliamentary press gallery in Canberra. Milliken, an Honours BA in Political Science from the University of NSW, took the classic journalistic route of working as a copy boy at Sydney's *Daily Telegraph* (at the time owned by Sir Frank Packer), joining the *Sydney Morning Herald* as a university graduate and then moving to the Fairfax stablemate, the *National Times*, where he wrote in-depth features.

In 1999, he told of how in writing the book he received valuable support from Commission staff, who provided him with a stream of declassified documents which formed the backbone of the awful revelations. He travelled with the Commission to Maralinga in 1985 and returned to the site two years later to check on the clean-up which had been guaranteed by the Australian Government. Several other journalists and interested observers went on these trips, including Dorothy Johnston who included her experiences in the novel *Maralinga My Love*.[107]

After the release of *No Conceivable Injury* Milliken became a successful and influential freelance writer, and his next book project was *On the Edge*, in which he reported his many environmental concerns about rural Australia, and many of his writings are suffused with a growing range of ecological concerns (see Chapter Four).

No Conceivable Injury was launched in Sydney in 1985 by the Commission's head, Justice James ('Diamond Jim') McClelland, who also alerted the Australian public to several decades of government culpability and the need for adequate compensation to all the victims of the testing.

The book lists the various horrendous injuries to the people and the damage to everything the bomb tests touched, and Milliken shows the unsatisfactory nature and incompleteness of the 'clean-up' which the Australian Government was forced to order in 1967 at the Brumby site near Maralinga. 'The bomb sites are as desolate today, the trees as broken and dead, as the days after the bombs went off,' Milliken furiously concludes in the book.

And speaking more than a decade later, in 1999, Milliken said he was appalled at how Australia was seen as a nuclear-waste dumping ground for the world. 'It's as if we've learned nothing from Maralinga; it's history repeating itself,' Milliken said. 'It's now as it was when the British came in without question. And the way Australia is treated is based on false premises. Our government should realise we are not a nuclear power and that the Outback is not an empty place. People live there, and it's the home of wildlife.'[108]

THIS theme was also evident in an earlier book, *Maralinga (British A-Bomb Australian Legacy)* (1982), a collaboration between a journalist, Adrian Tame, and a physicist, F.P.J. (Francis Patrick Joseph—known as 'Rob') Robotham, thereby combining the expertise of an experienced researcher and a highly respected scientist. The authors show how, three decades after the explosions, radiation levels in the Maralinga area were 'still considered hazardous enough for the Federal Government to warn visitors against staying longer than five hours'. As for the subsequent cover-ups, Tame and Robotham tell how the British attitude, in particular, amounted to 'the hideously inadequate approach to the value of human life'—especially Aboriginal life.[109]

Tame and Robotham present a history of all the scientific discoveries, beginning in the late 19th century, which unlocked the atom. They mock the claims that the Australian tests were 'safe':

> . . . the same Titterton who had spent much of his working life helping design bigger and better bombs was now responsible for the safety of the Australian population following the testing of such bombs. One way, perhaps, of finding out if poachers make the best gamekeepers. The Committee's name (the Atomic Weapons Tests Safety Committee) is also interesting—a safe, atomic weapons test is an intriguing concept.[110]

Robotham, born in Southport, England, in 1935, had spent his early scientific career at the UK Atomic Energy Research Establishment between 1957 and 1966, specialising in radiation safety. In 1966, he came to Australia to join the Atomic Energy Commission establishment at Lucas Heights in Sydney. At that time, the University of Melbourne was looking for the right person to establish its Radiation Protection Service, so Robotham quickly moved from the government to the private sector and remained in that position for the next 25 years.

But Robotham took a very different path to most other scientists at that time by expressing his professional concern about the nuclear industry. He joined the anti-nuclear and peace movements in the early 1970s, being especially vocal against the French testing in the South Pacific. In 1972, he travelled to Tahiti with a delegation led by the Labor Party's prominent peace activist Jim Cairns. Robotham's links to the ALP would strengthen over the next few years, and in 1977 he was the party's unsuccessful candidate for the federal seat of Casey in Melbourne.

'Because of my professional expertise, and because I was no longer working for the government, I was freer to speak out on nuclear issues,' Robotham recalled in 1999.[111]

Robotham said his involvement resulted from his name coming to the notice of 'nuclear veterans', including those who had been at Maralinga. Robotham told of 'a great deal of interest' emerging about Maralinga as early as the 1970s, long before the McClelland Royal Commission. 'Veterans were getting their stories published and there were many sympathetic journalists running with the story, especially in South Australia,' Robotham said. 'There were poignant stories of veterans being interviewed on their deathbeds, of such things as Aborigines walking through Ground Zero.'[112]

Speaking out was an act of defiance and bravery by Robotham, who put his job on the line at the University of Melbourne. He received some 'very difficult letters' from the university's vice-chancellor, Sir David Derham, whom Robotham described as 'a rigid conservative and member of the Melbourne Club'. Robotham was undaunted, and when the Hawke Labor Government was elected in 1983 with a commitment to 'look into' what happened at Maralinga the scientist was there to contribute.

Robotham described the findings of the Australian Ionising Radiation Advisory Council as a 'whitewash'. Its 1983 published report said that stories of illness and injury at Maralinga were 'just that, stories, with no basis in fact'. According to the report, those living near Ground Zero had become blind from such things as measles, not fallout.[113]

Despite the AIRAC report, the Hawke Government's Minister for Resources, Senator Peter Walsh, felt there were a sufficient number of unexplained incidences of cancer in the South Australian desert to begin a chain of top-level scientific inquiries that culminated in the McClelland hearings. In the aftermath of the commission, Robotham was involved in several court cases in which veterans were suing for compensation. 'There has been little compensation,' Robotham said in 1999, 'but one of the positives from the commission is that some good progress has been made in cleaning up the testing site.'[114]

In retirement for some five years and running a bed-and-breakfast in the Victorian spa town of Hepburn Springs, Robotham said in 1999 that he was still involved in radiation matters and retained links with such groups as the Northern Land Council. He said a key aim was to keep the public informed about what happened at Maralinga, especially in light of how the McClelland Commission had to 'beat the British Government about the head to get information'. Nuclear safety is also the key, according to Robotham, who spoke of the need to ensure tight regulations and equally stringent enforcement on all things nuclear.[115]

The journalistic collaborator on this project was another Englishman, Adrian Tame, who was born in Luton in 1944 and worked in Fleet Street and throughout North America before arriving in Australia in 1973. He worked mostly on News Limited newspapers until 1986, when he switched to public relations; after 1994 he set up his own company in central Victoria which allowed him to combine journalism and PR—as well as to write books and carry out the sort of research which made his contribution invaluable in the Maralinga project.

The Maralinga collaboration reflected the breadth of Tame's interests. A later book, *The Matriarch* (1996), was a biography of Kathy Pettingill (matriarch to a notorious criminal family in Victoria) that made it to No. 2 on the best-seller lists. Tame was also involved in films and in the mid-1970s made a documentary about the Hell's Angels called *Thunderground*, and in 1983 an award-winning TV documentary, 'Street Kids', for the Nine network. Tame described his body of work as 'a bit of everything', and his strong interest in environmental issues resulted in the Maralinga involvement.

Among Tame's earliest and most significant stories with a strong environmental concern was for *Australian Penthouse*, an unlikely magazine to run a major article about the links between the Agent Orange defoliant in Vietnam and the sort of pesticides approved and used by councils in Australia. In Tame's mind a further concern would evolve organically—the relationship between the problems faced by Vietnam veterans exposed to Agent Orange and the 'nuclear veterans' seeking just compensation for what happened to them at Maralinga. In a 1999 interview, Tame remembered the period 20 years earlier, as the idea for the Maralinga book first emerged, as one when 'few Australians saw anything other than a massive benefit from the atomic blasts in the South Australian desert'.[116]

The collaboration with Robotham, from which would also emerge a long friendship between writer and scientist, had an inspired beginning. When the idea for a Maralinga book emerged, Tame decided to interview several likely collaborators because 'I am not a technical person, and I needed a scientist with the widest expertise'. Tame had heard of Robotham at the University of Melbourne and at their first meeting Robotham struck him as an 'erudite and learned professor' with whom a degree of correctness and formality would be expected. But just after their first meeting in Robotham's office had begun, the phone rang and Tame was shocked to witness a torrent of foul-mouthed abuse being hurled down the line. The reason for the expletives was that the person at the other end had decided to pike out of a game of five-a-side soccer; so, when Tame advised Robotham he could make up the numbers, the ice between them was instantly and everlastingly broken.

Seventeen years after the release of *Maralinga*, Tame said he derived great pride in the book being used widely as a reference text during the McClelland Royal Commission. But his delight is tempered by his belief that 'there still has not been a satisfactory conclusion' to the McClelland findings.

'Almost half a century after the tests took place, there is still bickering about compensation and the clean-up,' Tame said. Disagreeing with Robotham about the rate and effectiveness of the clean-up, Tame claimed he was 'not at all impressed' by the effort and in his opinion the British Government was continually 'dragging its feet' and still was not spending 'the millions needed' to clean up large areas of South Australia. Tame said the tribes worst affected by the blasts had hardly had any contact with Europeans, therefore warnings in English to stay clear of the area were meaningless. 'Because of such utter lack of forethought, many of the local tribes continued to sleep in Ground Zero,' he said.

Tame described how both he and Robotham, two Englishmen, had felt 'a degree of conscience' about what their nation had done in the name of imperialism to their former colonial subjects. 'I owe a debt to Australia, and my work on the Maralinga book and related projects is one way to repay that debt,' Tame said.[117]

As the 1990s drew to a close, Tame said he was concerned about a range of ecological issues, especially the fate of the bush, the 'decimation' of the forests and the pollution of the oceans around Australia. He described the logging throughout Victoria as 'appalling' and decried unthinking tourist developments such as plans to 'concrete over' national parks throughout Victoria, including Wilsons Promontory.

On a global level, Tame said he was especially 'horrified' by the movement towards genetically modified foods, a trend which he saw as 'something out of Frankenstein', an unspeakable act by 'petty' mankind to interfere in the workings of nature, and related actions like cloning had 'a potential for disaster'.

A THIRD book about British weapons testing, from an international perspective and written by two British television journalists, is *Fields of Thunder* (1985), in which Sue Lloyd-Roberts and Denys Blakeway further describe how the authorities in both London and Canberra saw the Australian environment as just an expendable resource.

While Blakeway examined the situation in Britain and the testing in areas outside Australia, Lloyd-Roberts looked at Maralinga. In what now reads as horrendous irony, the authors quote the Menzies Government's Minister for Supply Howard Beale in the official justification of what would happen at Maralinga: 'The whole project is a striking example of inter-Commonwealth co-operation on the grand scale. England has the bomb and the know-how; we have the open spaces, much technical skill and a great willingness to help the Motherland.'[118]

Blakeway and Lloyd-Roberts see such a stupid and irresponsible mindset as part of an 'incompetence and mismanagement' from which Australians would suffer for 'thousands of years'. The authors portray a scene of widespread radioactive pollution and poisoned landscapes.[119]

As a result of her Maralinga involvement, Lloyd-Roberts travelled to Chernobyl in the Ukraine, site of the world's worst nuclear reactor disaster in 1986. She also witnessed the ecological ravages in the Volga region and the Far East of the former Soviet Union. In June 1999, she had just returned from Kazakhstan, from where she reported on the effects of Soviet nuclear testing. 'Babies are still being born deformed because of the testing,' Lloyd-Roberts said. 'Kazakhstan had to endure non-stop bombing for 20 years, and today it has the highest infant mortality rate in the world.' Of the post-Soviet era, Lloyd-Roberts said that the 'lies are still continuing, and even in the post-glasnost era the Kazakhs remain very ignorant about what happened to them.'[120]

THE poisoning of Australia's lands and waters is central to many books written over the past thirty years, one of the most dramatic being Sharon Beder's *Toxic Fish and Sewer Surfing* (1989), which analyses how sewage has always flowed into the ocean off Sydney and how rivers in the region have become fouled. In this influential and thorough work, Beder reveals how a too-cosy relationship between government, the bureaucracy and business has allowed unfettered pollution, and how any critics within and outside the system have regularly been silenced. Beder exposes conspiracies and scandals of the highest dimensions.

Toxic Fish was launched during a two-day conference called 'Clearwater Revival' on 11 and 12 November 1989 in Sydney. This conference, organised by groups like Friends of the Earth and Turn Back the Tide, was aimed at addressing the urgent problem of ocean pollution and Beder's involvement was in recognition of her already high public profile on the issue. At the conference, Beder—at the time also in a group called the Society for Social Responsibility in Engineering, as well as being the co-ordinator of environmental studies at the University of Sydney—appeared alongside key politicians and representatives from major environmental groups including Greenpeace.

The high-profile conference put into the public spotlight the way in which untreated waste was being dumped into Sydney's oceans and it aimed to put forward more sensible disposal methods.

The extent of ocean pollution was revealed on many fronts throughout 1989, including evidence of fish caught off Sydney being contaminated with organochlorine pesticides and heavy metals. For a time there were bans on fishing near the ocean outfalls and the public became more aware of the extent to which Sydney's waters were dirty and unsafe.

As *Toxic Fish* was being launched, Beder campaigned against toxic industrial waste being allowed into the sewerage system, and in one interview after another she noted how industrial polluters should be made to treat their waste.

Beder had a significant role in a sequence of events which resulted in the New South Wales Government led by Premier Nick Greiner committing $6 billion over twenty years to clean up the mess in all of Sydney's aquatic areas. At the time Beder noted wryly that it took more than 100 years for any action to be taken since the first public discontent emerged in New South Wales about how their waterways were being treated.

In the book, Beder shows that members of the public are denied adequate information and have no form of effective redress against a system of administration that allows widespread pollution. 'The role of the public in pollution control has been severely limited,' writes Beder with angry understatement. 'When the public is denied input into the legislative process, then levels of pollution that are determined to be "acceptable" are based on judgments of public servants in negotiation with industry and government authorities.' Or, to put this another way, operators in New South Wales have virtually had a licence to pollute.[121]

Toxic Fish shows that only a combination of direct action and intensive publicity campaigns by environmentalists have put any sort of restraint on such widespread pollution of waterways. Only a major rally featuring a glittering array of well-known speakers was able to galvanise public anger against the fouling of Sydney's most famous beaches.

According to *Toxic Fish,* careers in the public service have been ruined whenever any experts have sought to expose or highlight the extent of the problem. Beder, whose PhD thesis was on the history of the Sydney sewage system and whose background is in engineering, tells how the political and administrative structure is designed to self-perpetuate the situation, and the pollution is fated to continue. She claims instrumentalities like the State Pollution Control Commission, the Clean Waters Advisory Committee and the Water Board have been totally remiss in their responsibilities, especially with vested interests having far too great a voice in decision-making and regulation. The code of silence which permeates the bureaucracy has made it impossible for reforms to happen or for citizens to be provided with proper information.

Beder claims that Sydney's waters were being destroyed by an iron-clad convention of secrecy whereby any 'criticism of the extended outfalls by engineers, where it existed, was fairly well suppressed'; and that members of the tightly knit and collegiate engineering profession 'were not willing to have their names published' on any critique. Beder contends further that 'it's an unwritten part of the engineering ethos not to criticise works designed by other engineers, because this may reflect badly on the profession'.[122]

The way the engineering and similar professions conduct themselves is at the core of many environmental problems, Beder laments:

> Most sewerage engineers are employed by government departments or organisations and those that aren't are consultants dependent on those same government departments for work. So critics face the possibility of severely limiting their career prospects . . .[123]

In this context, and in light of the governmental and administrative cultures described above, Beder explains how most Water Board publicity campaigns such as Surfline, which monitored the pollution levels at beaches, were a cruel sham. In a partly optimistic appraisal, Beder looks at how people power was one of the rare antidotes to such lack of concern, and she describes the anti-pollution rallies and how 240,000 people attended the Turn Back the Tide Concert on 24 March 1989.

Such constant raising of the public consciousness is essential, Beder asserts, because 'the constitutional structure of decision-making that brought that crisis remains untouched' and those upon whom our environmental wellbeing is entrusted 'have lost touch with community needs and desires'.[124]

Beder proceeded to dissect who 'the players' were in the debates over clean water, and she warned readers to be wary of corporate propaganda:

> engineers . . . look for 'cost-effective' solutions that put less value on environmental amenity than the public do; environmentalists . . . are faced with exposing hazards that you can't see or touch; regulators . . . are pulled in various directions by competing interest groups; politicians . . . do not want to risk unpopularity by putting up taxes and being tough on industrial-waste discharges, or by accommodating industry and failing to protect public health and the environment. Too often, they try to salvage their popularity by covering things up.[125]

THE evils of pesticides—and other nasties such as insecticides, rodenticides, fungicides and fumigants—are starkly documented in *Pesticides: The New Plague* (1984), in which authors Linda Cawcutt and Catherine Watson convey the concerns of Friends of the Earth about this deadly and often ineffective yet powerful industry.

Cawcutt and Watson—both Melbourne-based freelance journalists—list poisons by their brand and generic names, and profile each from the latest available scientific and consumer data. On each poison the book tells the reader what it is, what it does, the effects on health, plus an 'ecology rating'.

An appropriately titled chapter, 'Deadly Games: the politics', shows the difficulty of exposing wrongdoing by Australia's chemical and related industries, backed as they are by the most powerful multinational conglomerates. Such corporations are extremely adept at silencing and neutralising criticism of their actions. 'Scientists and academics who speak out on environmental issues in Australia often risk their jobs and reputations,' write Cawcutt and Watson, echoing Beder. They continue: 'Most scientists prefer to avoid public controversy concerning their own research and teaching. This means that it is difficult to find individuals willing to have their case presented.'[126]

Cawcutt and Watson also describe the cosy reliant relationship between regulators and regulated:

> The whole health bureaucracy has been derelict in its duty to inform the public of the health risks associated with agricultural chemicals. They are more interested in not alarming the public. In fact, they will hold back information from the public . . . People are dying now from earlier exposures to asbestos dust because an elitist health bureaucracy kept quiet about the effects on human health all those years ago.[127]

Another chapter, 'Indecent Exposures: the victims', tells of a six-year-old boy who died from exposure to a commonly used herbicide. Cawcutt and Watson list as among the documented effects of exposure to the poisons: severe headaches, depression, emotional instability, nightmares, insomnia, lethargy, irritability, numbness in hands and feet, liver problems, nausea and vomiting, diarrhoea, constipation, frequency of urination, muscular twitching, tightness in the chest, wheezing, shortness of breath, increased sweating, blurring of vision and dizziness.[128]

The authors see the range of pesticides as interfering with the balance of nature. 'If you spray the aphids in your garden, you'll also kill the ladybirds that eat the aphids,' they warn.[129]

Public awareness about pesticides has increased markedly since the release of this book, which nevertheless presents in its chapter 'Soft Sell' a fascinating collection of display ads over the years for chemicals—an alarming historical curiosity and a standing testament of how a gullible public was hoodwinked. Claims were made two decades ago which would be derided today.

Cawcutt and Watson also urge consumer activism to stop the spread of pesticides and carefully detail effective alternatives. They provide philosophical arguments for consumerist tactics and case studies of how this was effective in other fields.

Reflecting on the book fifteen years after its release, Linda Cawcutt Gordon said its greatest achievement was in providing a layman's guide to the extremely technical and scientific realities of poisons. She confessed that her initial involvement in the book project was simply to accept a writing assignment and that there was 'no thought of a crusade at the outset'.

'We wrote the book like reporters would tackle any subject they were covering,' Gordon said. 'The more I got into it the more radicalised I became. I felt it was a story that had to be told.'[130]

Until Friends of the Earth commissioned Cawcutt and Watson to write the book, they had both had varied careers as journalists, with Watson two years earlier having worked on an oral history of farming practices in Victoria, also commissioned by Friends of the Earth. This book, *Boots and All*, was also released in 1984.

Cawcutt, born in the Melbourne suburb of Richmond in 1954, spent seven years of her childhood in England, and entered journalism after completing her tertiary studies. She worked on suburban newspapers in Melbourne, and briefly on a film magazine—and this early experience, as well as giving her a broadly based training, left her appalled at how a lot of vital information never reached the public.

Despite the volunteers at Friends of the Earth being unable to provide any significant promotion for the book, it was picked up by TV current affairs and reached a wide audience. Gordon said the book was one of Friends of the Earth's best-sellers, and was reprinted after the initial run of 3000.

She believes the book is a 'valuable, brief introduction' to a complex subject, stressing the grassroots nature of the issue and answering the questions 'Can I use a certain product or substance in my garden rather than harmful chemicals?'

Linda Gordon, who after the book's publication got married and now has two children, for the past eight years has worked freelance, covering 'the broadest range of topics'. She said she and her husband were environmentalists; they don't own a car and they grow food organically.[131]

One of her colleagues during her days on the Melbourne suburbans was New Zealand-born Catherine Watson, who joined her on the Friends of the Earth assignments because they provided a chance to hone journalistic skills—and provide some cash. To Watson, the job was just something that 'sounded interesting', and only during the research and writing did it became anything more than that.

Watson shared Cawcutt's disillusionment with throwaway suburbans. 'The Friends of the Earth job was a chance to do something more important and more stimulating.'[132]

THE growing problem of Antarctica's grave risk of exploitation is the theme of several books. Prominent literary critic Stephen Murray-Smith gives a personal appraisal of the momentous problems facing Antarctica in *Sitting on Penguins* (1988, 1991). In 1985–86 Murray-Smith travelled to Antarctica on the MV *Icebird*, and saw mining as a 'rape' of the fragile continent.

The editor of the *Overland* literary magazine from 1954 until his death in 1988 at age 65, and a respected man of letters, Murray-Smith laments almost every human activity in Antarctica, such as the building operations at Casey base creating cement dust which was destroying the supposedly protected rare lichens. He also warns of potential disasters such as would ensue if oil drilling were allowed.[133]

Born in 1922 into the privileged environment of Melbourne's Toorak, Stephen Murray-Smith went to the exclusive Geelong Grammar and was a long-time ecological campaigner, with a special interest in Bass Strait. Throughout his life he embraced many philosophies and at different times was a member of both the Liberal Party and the Communist Party, as well as of the Melbourne Cricket Club. Having been a commando in World War II, he would take this fighting spirit into all his intellectual battles, especially the keen environmentalism of his later years.

Describing how about one-third of the Antarctic coastline was earmarked for oil exploration, Murray-Smith gives an evocative appraisal of how this would affect the rare landscape and the wondrous creatures that dwell there: 'the extraction of the oil which we know is there presents hair-raising difficulties. Just think of the fate of a conventional oil rig with an iceberg the size of a football ground bearing down on it. The rig would be brushed away as we brush away bush flies at a picnic.'[134]

Murray-Smith describes the potential consequences of oil drilling as 'a desecration on the grandest scale that even the human species has devised'. Yet if there is a commitment to save the continent, the cause for rejoicing should be unrestrained:

> If we can keep our grimy fingers off Antarctica, if we can keep it free of our droppings, we not only hand on to the future the world's last great unspoiled wonderland, we not only allow that future the privilege of deciding for itself what Antarctica should mean to it, but we have a chance of redeeming something of our own reputation, of showing that we can rise above our viciousness.[135]

From his own observations during the expedition and from his extensive research, Murray-Smith minutely describes how vital is Antarctica in determining the effects of global warming and what is happening to the ozone layer (of which he gives one of the most accessible explanations and definitions found anywhere). He concludes his historical analysis of the ozone layer as being 'one of the most serious threats to life on earth'.

Such observations are consistent in a man who once said of Australia that 'we have buggered up this continent more quickly than any continent has been buggered before'.[136]

Some of Murray-Smith's observations in *Sitting on Penguins* have been overtaken by events—such as there is no longer any garbage left at Australia's bases—but his central thesis and his concerns regrettably retain their potency and relevance.

Above all, Murray-Smith exhorts creative people like poets, artists and craftspeople to become more involved in the fight to protect this vast and rare place.

THE vast challenges involved in getting the environmental message across are highlighted by Sharon Beder in *Global Spin* (1997), in which the author of *Toxic Fish and Sewer Surfing* dissects how vast international propaganda machines can be activated to counter potent green messages. Beder chronicles how an all-powerful push of overwhelming international forces often drowns out the voices of ecology. She describes how so many aspects of Australian life are under the thrall of international conglomerates. From the ordinary citizen to the most powerful policy-maker, all are extremely vulnerable to the pernicious global spin.

This global spin is usually far more vocal and far more widely disseminated than is the green message, and this makes the collection in this chapter an even greater and glowing achievement for its diverse authors.

The books discussed here are an illustrious and brave counter-weight to what Beder describes as 'corporate activism' worldwide:

> Corporate activism, ignited in the 1970s and rejuvenated in the 1990s, has enabled a corporate agenda to dominate most debates about the state of the environment and what should be done about it. This situation poses grave dangers to the ability of democratic societies to respond to environmental threats.[137]

Beder describes how this corporate offensive, bolstered by countless opulently funded think-tanks, has taken on the forms of bogus and misleading organisations and sham grassroots movements. Often such organisations, backed by corporate money and following corporate agendas, carry names and descriptions which falsely suggest that they are in fact pro-environment.

The more powerful the environmental message, the more shrill and intense becomes the corporate global spin. But Beder strikes one optimistic chord: 'Surveys show that the majority of people in most countries are not only concerned about the environment: they think environmental protection should be regulated by governments and given priority over economic growth.'[138]

Having said this, Beder simultaneously warns of the irresistible power of the global Spin-merchants: 'Yet this widespread public concern is not translating into government action because of the activities of large corporations that are seeking to subvert or manipulate the popular will.'[139]

She calls for a better-informed public with a raised consciousness to challenge and fight for their birthright: 'whilst Australians are amongst the world's most environmentally concerned people, their government's environmental record is one of the worst for OECD countries.'[140]

Knowledge is power, cries Beder, and she urges us all to be aware of the realities of all-powerful international propaganda. The titles discussed in this chapter, against all odds, are a beacon of hope to preserve cherished values. They raise the consciousness of concerned and caring citizens who otherwise would be sacrificed to fulfil the aspirations of the destroyers and the apostles of 'progress'.

Tasmanian writer Peter Grant has noted that 'most Australians remain woefully ignorant about the web of life around them and are better educated about the badger than the bilby'.[141] The books in this chapter seek and almost invariably achieve the creation of a broader knowledge and consciousness.

And Tim Flannery is part of a growing contingent of writers who believe, as he puts it, that they have a responsibility to provide 'intellectual leadership' on pressing issues connected with the environment.

The books discussed are a glowing vanguard of hope and inspiration, proclaiming the need to challenge conventional wisdom—especially that which threatens to wreck our civilisation utterly. As Flannery puts it, changes for the good come not from politicians and decision-makers but from within the hearts of ordinary Australians.

The significance of these books and the authors who wrote them is as fellow critic John McLaren said in his obituary to Stephen Murray-Smith: 'Australia is a better place for his life, and would be still better had we heeded him more.'[142]

4

ILLUSTRATED BOOKS

Shining visions

DESPITE the fact that illustrated books are sometimes dismissed as 'coffee-table books', most of the books discussed in this chapter—despite their glossiness and colourfulness—provide in-depth analysis of almost every environmental issue of the past thirty years and often are invaluable reference works. For example, on the Franklin dam dispute alone there are many dazzlingly attractive books featuring the work of leading wilderness photographers like Peter Dombrovskis and Chris Bell. Other books in the vast genre cover all the regions of Australia—all the way from the Great Barrier Reef to Antarctica, from the deserts of South Australia to the wild islands which ring the antipodean continent, even examining underwater ecologies at risk.

Illustrated books rejoice in the glories of pristine and magnificent places, and carefully categorise the risks and the perils in which many fragile areas find themselves. In many of the books, just when readers thought they had learned about almost every creature and every genus of the wild, there is a stunning, glowing photo of something new to stimulate the imagination.

The importance of illustrated books was recognised by influential activists like Christine Milne as pivotal in raising people's consciousness. Milne, a longtime member of the Tasmanian Green Independents who was largely responsible for stopping the Wesley Vale pulp-and-paper operation at Burnie in the late 1980s, said in 1999 that photographs such as those of the beach at Lake Pedder before its flooding were an 'anchor point' in raising awareness and became 'symbolic of a lake that was lost'. Milne said that, apart from their inspirational quality, such books were a valuable resource. She said they were very important to her as a research tool and she often referred to them when writing political speeches.[1]

Most people don't go to wilderness or other remote areas under threat of environmental damage, therefore it is especially important to present publications of good photographic quality, backed up by informed text. An influential wilderness photographer, Robert Rankin, has described the importance of fine illustrated books as highly significant.

Rankin—a prolific author of illustrated books as well as calendars and bush-walking guides—in 1997 told of the logistical challenges, of the hardships and of the fierce and dedicated commitment required whenever a publication is being put together. The Brisbane-based Rankin has described how when he's on a major bushwalking expedition he gets up at sunrise for days on end, and it may take him up to a week to get into a remote area like the wilds of south-west Tasmania. When such an area has been reached, he may have to endure wild weather, and once even had to crawl along the ground for fear of being blown off a ledge. All this follows at least a month of thinking and logistical preparation.[2]

MANY illustrated books are highly ambitious in concept and stupendous in their execution, and among the most profound achievements is *Australia's Wild Islands* (1997), in which Quentin Chester and Alasdair McGregor postulate on how the thousands of islands which surround the Australian mainland hold many keys to the ecological survival of the continent. From the most over-developed and overly accessible of the tropical resorts of North Queensland to the near-Antarctic and remote conditions of Heard Island, the authors show how these far-flung places are central to our understanding.

Between them, Chester and McGregor brought a range of disciplines to this book and the earlier *The Kimberley: Horizons of Stone* (1992). Chester is a respected chronicler of Australia's remote regions; his book *The Wild Calling* (1998) examines the 'unusual aspects of being out in the wild'. In this book, Chester seeks to explain what happens out in the bush rather than present a 'how to' book about venturing into the wild. It was part of a wider aim to reconcile life in the city and being in the bush—and to appreciate and experience what is there.[3]

Chester was born in 1957 at Blackwood in Adelaide and spent his formative years in the Adelaide Hills. He developed his love of nature while roaming around the hills in areas that were then 'wild and woolly' but are now encircled by Adelaide's urban sprawl. He regularly trekked around the Belair National Park, which he described as one of the oldest in Australia.

While in his 20s, in useful preparation for the journeys he would undertake with McGregor, Chester trekked to Nepal and became proficient in rock-climbing and in leading expeditions. Chester, who had studied English and history at the University of Adelaide, honed his prose skills by writing brochures and catalogues for Paddy Pallin and contributing to climbing magazines. With his brother Jonathan, a well-known adventurer, he produced *The Outdoor Companion*, which contained information useful for the tasks he would undertake in such remote places as Heard Island.

McGregor and Chester set out for the wild islands and the Kimberley because of their growing concern about many environmental issues, chief among them the destruction of habitat and the increasing numbers of endangered animals. Chester added in 1999 that they saw their collaboration as an opportunity to take people to wonderful places they would not be able to see for themselves, something Chester had already undertaken in his articles for *Wild* magazine.

Alasdair McGregor, a Sydney artist, was born in 1954 and studied architecture at the University of New South Wales. In the early 1980s he devoted himself to painting, later expanding his interests to both writing and photography. In his collaborations with Chester, he pursued his devotion towards native fauna and the natural landscape, especially in the far-flung places covered in the books. He participated in many field trips throughout continental Australia and the many islands and was a member of expeditions to Heard Island, Macquarie Island, Antarctica, remote parts of the northern Kimberley in Western Australia and Kakadu in the Northern Territory.

Even though painting and other creative pursuits would supersede architecture, his professional training indirectly resulted in McGregor's concern for the wilderness. He spent a year as an architectural assistant in Hobart, and while there he was touched by the protests against the flooding of Lake Pedder. Also while in Tasmania, he walked throughout the island's remote areas.

'The lessons gleaned from offshore islands are fundamental to the conservation of the continent at large,' the authors proclaim in *Australia's Wild Islands*. These islands—covering the whole spectrum of climates, levels of exploitation and degrees of incursion—contain 'an astonishing diversity of offshore habitats—arguably the greatest of any country in the world'. As such, write Chester and McGregor, the islands are microcosms of ecological ills found on the mainland and contain important clues as to how environmental disasters can be avoided.[4]

The text is generally riveting, based on extensive research and the authors' travels across the vast archipelago, and the photography and artwork are among the most impressive compilations found anywhere. Each photo has clarity and vivid colours—for example, purple cliffs, slivers of a sandy shore and all varieties of blue seen off the West Australian coast near Carnarvon looking north along Bernier Island. The immensity of the dune structures at Red Cliff Point on Bernier are dramatically spread across half a page, showing their natural splendour and the damage caused by introduced goats. The authors' historical research reveals that these goats were a significant problem for about fifty years until their eradication in the 1980s.[5]

Geological splendours abound on all the islands, and their scope and significance are graphically conveyed. Among the glorious pictures is an oil-on-canvas reproduction showing the profuse vegetation on Christmas Island—north-west of Darwin in the Indian Ocean and much closer to Jakarta than to continental Australia—which is described in the text as having 'some of the most spectacular natural phenomena and rarest creatures on earth'.

Such visions and descriptions shed the most glowing and effusive light on how the islands ringing Australia offer among the last remaining windows into how so many living things can flourish. But warnings also abound how the ecologies of the islands are often very fragile. Chester and McGregor warn that many of the islands are extremely vulnerable to worldwide trends like the greenhouse effect plus the consequences of 'this dire age of extinctions'.[6]

To Chester and McGregor, many of the islands are 'life-rafts for plants and animals that are endangered and extinct on the mainland'. They constantly reiterate the key point made at the beginning of their book:

> Australia's parlous record of mammal extinctions would be much worse were it not for the existence of these exiled populations. Some nine mammal species gone from the mainland are still extant on islands. The survival of a further 16 endangered or vulnerable mammals may well depend on the existence of their offshore relatives.[7]

The authors describe as 'among Australia's most treasured islands' a cluster off the West Australian coast—the Houtman Abrolhos Islands off Geraldton, the Bernier and Dorre Islands off Carnarvon, and the Dampier Archipelago off Port Hedland—where there is 'immunity from predators' for several species of mammal endangered elsewhere.[8]

But even in such a largely pristine archipelago there can be no cause for complacency as so many human activities cause irretrievable harm. The authors write of fishing in the Houtman Abrolhos: 'Dragging anchors and some of the million or more pots dropped each year have significantly damaged the coral. Recreation has also taken its toll; many reef areas seem strangely empty of large fish, due to over-enthusiastic amateur spearfishing.'[9]

Chester and McGregor lament that the relevant State authorities in Western Australia have failed to adequately protect the flora and fauna under their jurisdiction on these distant locations, and that 'a proposed Abrolhos National Park covering most of the islands, plus an associated marine reserve, waits to be declared'.[10]

Christmas Island, sumptuously represented in paintings and photos, is recognised by Chester and McGregor as a unique area where, somehow, nature has prevailed against almost every incursion: 'Despite 90 years of forest stripping prior to mining, about 75 per cent of the island's magnificent green cloak remains pristine. In 1980 a national park was declared over more than half of the island. It has been twice enlarged since and further extensions are planned.'[11]

But the authors also concede that huge problems remain, and the disappearance of habitat is threatening bird life.

Similarly, at the Wessel Islands on the western side of the Gulf of Carpenteria, the so far hardly touched region is also at risk. The authors point to the threats posed by the increase in shipping activity, oil exploration and over-fishing. As these threats grow, they call for the need to ensure that this archipelago remains 'a chain of precious refuge in a shifting sea'.[12]

Meanwhile, added to such potential threats are the introduced species, diseases and a range of natural and man-made disasters. Chester and McGregor describe how, in the midst of paradise, visitors must be prepared for 'a sorry melange of plastic bottles and bags, nylon rope, polystyrene, cans and packing straps'. They show how human behaviour is putting added pressure on creatures like the species of wallabies which are extinct on the mainland

and now restricted to just a few Indian Ocean islands. Such creatures and an array of vegetation are being 'increasingly consigned to relatively small parcels of habitat . . . Many of our parks, reserves and so-called wilderness areas are themselves virtual islands, surrounded by expanses of agricultural and urban development'.[13]

Australia's Wild Islands provides full-page and detailed maps of every region discussed, and there are fact files putting all the information into a broader context.

Controversial areas are discussed at length, including Queensland's Hinchinbrook Island and Fraser Island. A glorious full-page photo greets the reader showing what is at stake at Hinchinbrook: water ripples converge on sand ripples, with a backdrop of lush vegetation and the stately Mount Bowen. The authors exult at what they found at Hinchinbrook: 'We had taken refuge in the jaws of a fantastic creature standing at the gates of creation.' While rejoicing in the island's natural splendours and unique features, they warn of the impact of Keith Williams' 2000-bed resort and 234-berth marina at Oyster Point on the North Queensland mainland near Cardwell. Chester and McGregor are horrified at the prospect of such a development, which would contrast markedly with the sensitive, low-key and unobtrusive tourism found at many national parks.[14]

Of equal concern is Fraser Island, whose sands are shown as 'abraded into fantastic forms'. The authors recount Fraser's history of human brutality and resource over-exploitation, and how genocide and disease had decimated the sand island's Aboriginal population. 'By 1948 nearly 70 per cent of the original timber reserves had gone,' they write of the famous tourist hub of south-eastern Queensland. They similarly reveal how as soon as the threat from sandmining had been eliminated, the tourist onslaught became a huge problem.[15]

They are scathing about the impact of the growing hordes of visitors, whom they sarcastically refer to as *homo touristus fraseri*:

> Since the late 1960s, when the annual influx was around 5,000, this pilgrimage has grown to 350,000 strong . . . [Nature was] overtaken by an erratic procession of vehicles that repeatedly scattered crested terns trying to congregate on the sand. For much of the day the spectacle of the Pacific Ocean thumping onto the shore was lost in a blur of Pajeros and Land Cruisers.
>
> The sheer volume of visitors to Fraser poses a substantial threat to the island and its frangible sandscapes. Dune erosion, litter, degraded campsites, lake pollution and track damage—all are plain to see. Effective management of the island is further complicated by legacies from the recent past, including resort development, a network of forestry roads, and makeshift settlements. There is even a private housing estate. But perhaps the most intractable legacies are the competing expectations among the island's visitors.
>
> By late afternoon the traffic on the beach resembled suburban rush-hour as convoys rumbled past, racing to fishing spots and campsites in the north. Fraser may have been spared the physical ravages of sandmining but, according to some onlookers, the soul of the island is being quarried by tourism.[16]

For every gasp of delight by Chester and McGregor about the natural glories of Australia's wild islands there is a cascade of concern—such as the silver, lead and zinc project which is further shattering the tranquillity of the Sir Edward Pellew group of islands on the western edge of the Gulf of Carpenteria. Where dugongs, turtles and an array of water birds still dwell, the scene is set to become as frenetic as that described on Fraser Island.

And then there is the ever-present menace of leaking oil tankers. Chester and McGregor describe how one oil slick caused significant damage at the Archipelago of the Recherche in the Southern Ocean off the West Australian coast where a tanker ran aground in 1991. Seal pups and seabirds were the main victims of this accident, and the authors warn that a tanker disaster usually has catastrophic consequences.

Another painting shown in the book is an oil-on-canvas reproduction of Christmas Bells on Maatsuyker Island off Tasmania. It shows a thick and variegated canopy shielding a panoramic view to a hilly and gentle land-mass beyond yet another stretch of water. Beyond this tranquillity, itself so much at risk, is south-west Tasmania, where giant development plans have often threatened pristine beauty. The fights for this region have been documented in many of the classic illustrated books discussed below.

An earlier book by Chester and McGregor was *The Kimberley: Horizons of Stone* (1992), in which the authors detail 'a failure of will, a failure of imagination' in recognising and saving the splendours of this vast area of Western Australia. They liken the remoteness and uniqueness of the Kimberley to that of the far-flung islands and they describe the 'challenge' to preserve and sustain it.[17]

Describing the 'great wildness' of the Kimberley, Chester and McGregor also warn of how even its remotest areas are being threatened by feral animals and indiscriminate burning. 'Many of the more accessible patches are being trampled, allowing savanna grasses to invade and fire to eat insidiously at the forests,' they warn. They point to the depredations of a variety of creatures including cats, horses, donkeys and cattle.[18]

Chester and McGregor are appalled at the extent of introduced species throughout the Kimberley:

> . . . wild cattle now roam unmolested in some of the roughest countryside imaginable. They can be encountered in the most surprising places. From deep in the canyons of the Cockburn Range to luxuriant rainforest margins on the Mitchell Plateau, feral cattle are unfortunately everywhere.[19]

Furthermore, there have been 'critical depletions' of food and resources available to the more vulnerable creatures. This is sure to have severe ecological consequences:

> Stretched to the limits of survival, the grasp of the most affected species on the ecological life-raft has always been tenuous. In many ways the shift of resources away from the native fauna resembles an increase in aridity, which has meant a decrease in the carrying capacity of the environment. Hence the declines and extinctions have been worst in the arid and semi-arid regions.[20]

Describing the effects of over-grazing and the rampages of feral creatures, the book features evocative photographs of foliage and fauna—including some of the more than two hundred bird species found in the Kimberley. There are unique sights like the Parry Lagoons near Wyndham, vital feeding and breeding grounds for the equally unique water birds.

Again stressing the singularity and vulnerability of the Kimberley environments, the authors point to the absurdity of development plans such as a tourist resort on the Isdell River.[21]

Chester and McGregor describe immense problems such as land degradation and the many commercial activities which have already happened or are in the planning stage for the Kimberley. They warn of the rapid growth in tourism, which at the time of their writing had risen dramatically to more than 250,000 official visitors a year. They consider the Kimberley as one of our most precious laboratories for ecological research and as such should be preserved and appreciated.

ONE of the watershed events in the history of Australia's environmental movement was the saving of the Franklin River and the south-west Tasmanian wilderness in 1983, and illustrated books have captured every aspect of this titanic and glorious struggle. Among the first books to show what was at stake in the pristine region was David Neilson's *South West Tasmania* (1975), which sees that region as a microcosm of global environmental problems. Neilson—a photographer, mountain-climber and trekker—approaches the region with awe, revealing it as a place filled with a strong spirituality.

Added to the glowing photographs by Neilson and his fellow bushwalkers is one from NASA: the earth as seen from the Apollo spacecraft. Then a photo by Neilson shows the wild and remote country surrounded by the glistening waters of Louisa Bay on Tasmania's south coast reflecting the sun's dim glow.

The book was produced during that crucial time between the flooding of Lake Pedder and the fight for the Franklin. It is one of the earliest expositions of what was at stake.

Neilson, born in 1946 in Melbourne, started bushwalking as a student at Monash University in the mid-1960s. He travelled to Tasmania's south-west during holidays with his university bushwalking club while studying for a degree in civil engineering. The plans to flood Lake Pedder triggered environmental concerns which would result in him abandoning his chosen career path and taking up photography. In 1999, Neilson recalled that the Monash bushwalking club was different from the norm in that it became concerned with environmental issues like Lake Pedder. 'Bushwalking clubs—including some in Tasmania—were often not interested in conservation issues,' he said. 'The clubs didn't want to get involved in politics.'[22]

In *South West Tasmania*, Neilson reveals himself to be sadly and presciently attuned to the coming environmental struggle in Tasmania:

> Despite society's growing awareness of the value of wild lands, further threats to the natural condition of the South West are still likely, for unplanned growth and indiscriminate development will be present for some time. One of the many lessons to be learned from the Lake Pedder controversy is that the public must be made more aware of the value of places such as this before they are threatened.[23]

Some of Neilson's writing is a bit wooden yet he reveals the intense secrecy with which Tasmanian government research into dams was carried out, and in which the whole process was rorted and manipulated by the all-powerful Hydro Electric Commission. The story is backed up by powerful black-and-white shots such as Hanging Lake seen through early-morning mists with Mount Bobs in the distance.[24]

Neilson describes the interaction of himself and fellow trekkers with this seldom-trod terrain—and he allows his readers to commune with the colossal significance of it all. The panoramas and vistas of the pictorial essays are profound, but so are the simple photos of vegetation, those indelible signs of nature at its most potent and precious.

South West Tasmania warns of the calamitous consequences of unbridled development. Neilson sees it as a form of mania which must be subdued and a quest begun for a more sustainable solution:

> And so man comes to wild lands armed with his machines. He comes to destroy something of whose value he is largely unaware. He comes because it has been the way for centuries past for man to assert his supremacy over the environment, to wrest a living from his harsh surroundings. Rarely has he been preoccupied with the consequences of his actions. Progress and development are ingrained traditions of our society.[25]

Describing the potential loss of 'a truly priceless wilderness area', Neilson gives specifics of in-progress and pending actions which threaten disaster:

> The Gordon River road and the Scotts Peak road have thrust man's civilising presence deep into the centre of South West Tasmania. The construction camps, quarries, earthworks, dams, water storages, power-stations, and the permanent town of Strathgordon which have followed, have destroyed the natural condition of a significant area.[26]

Neilson proceeds to describe a similar fate awaiting the vast area 'south of the Lyell Highway' and he suggests that there is a profound urgency in dealing with the problem. As Neilson noted in 1999, delays in the book's release meant that Lake Pedder was already flooded by the time it came out, but it helped to raise people's consciousness to the Franklin struggle that lay ahead.

The various wilderness struggles galvanised Neilson, who in 1983 was employed by the Australian Conservation Foundation (ACF) to produce their first wilderness diary. He continued to produce this diary for the ACF along with various posters and small books for the next ten years. In 1986 he also produced the first landscape calendar issues by the Wilderness Society.

After leaving the ACF in 1993, he wanted to get back to doing his own photography and spent the next year taking photos throughout Wilsons

Promontory, the fruits of which became the book *Wilsons Promontory (Coastal wildness)*, which was released in November 1996 and coincided with the announcement of controversial plans to build a hotel within this national park.

Wilsons Promontory was released by Snowgum Press, a publishing house he jointly runs with his partner Karen Alexander, who was the Wilderness Society's Victorian campaign co-ordinator during the Franklin dispute and who later became the ACF's environment manager. As well as publishing other illustrated books like *Patagonia (Images of a wild land)* (1999), Snowgum produces posters and cards which reflect the environmental concerns of Alexander and Neilson.

In 1999, Neilson retained a broad interest in most ecological issues, foremost among them preserving wilderness and natural areas. He was also involved with groups trying to reverse the flooding of Lake Pedder.

THE seminal work on the Franklin and related Tasmanian issues is the mammoth *The South-West Book* (1978, 1983), in which editors Helen Gee and Janet Fenton, with the participation of a great number of writers and photographers including Bob Brown, presented the most definitive document of what would have been lost if the Franklin dam had gone ahead and the South-West had been allowed to be fragmented by development.

Gee and Fenton have accumulated the most comprehensive collection of black-and-white historical photos on the subject. These were gathered from individuals, newspaper archives, government departments and instrumentalities, universities, art galleries and a host of other sources. *The South-West Book* is a widely acclaimed in-depth look at the rich natural heritage which would have been adversely affected by the planned hydro schemes, mining and tourism.

This voluminous book is often overly complex and technical—the sections on archaeology, for example, may challenge even students of this science—but at the same time it is a comprehensive and authoritative pictorial and word study of what would have been lost to humanity had the dam been built. In this vast book the idiosyncrasies of dozens of writers and photographers shine through, each with his or her own niche of expertise or concern.

In one of his earliest literary contributions, one of Tasmania's novice green political activists is described simply as 'Bob Brown of Liffey, a medical practitioner in Launceston'.

Among the book's treasures are black-and-white reproductions of paintings by William Charles Piguenit, described as the first Australian-born professional artist, featuring the King William Range and Frenchmans Cap.

The jagged peaks and inaccessible valleys of this rugged and remote region leap from the pages. Chapter after chapter wanders off the beaten track—among the raucous mutton-birds and to such remote south coast islands as Maatsuyker and De Witt. Among the spectacular colour photos are Chris

Bell's vision of Eldon Bluff and Lake Ewart—a blur of water, mist and mountain-top as blue slips into grey and into brown.[27]

The South-West Book has added poignancy given that at the time of its original publication the Franklin's fate was uncertain, and the book might have been an obituary for the river rather than a celebration.

In 1999, Helen Gee recalled how the compilation of *The South-West Book* had begun 'in my head' during a train trip from Melbourne to Sydney just as the Franklin dispute was beginning and she was 'in desperation' about how a dam would destroy the region. She felt people needed to become educated and aware of the issues and the facts. 'In Tasmania, we who cared for the South-West were all doing what we could to try to prevent a repetition of what had happened to Lake Pedder, and I realised the importance of making people aware of what might be lost, of what was at stake,' Gee said. 'After Pedder, we wanted to prevent this kind of thing ever happening again.'[28]

Gee was a student protester during the Pedder campaign, taking out supplies and participating in vigils at the lake. Born in Launceston in 1950, Gee was raised in the small country town of Westbury, thirty kilometres from Launceston, and had a rural upbringing until age nine. As a youngster she spent a lot of time in the bush, later gaining an intensive knowledge of the south-west. At university, she developed a keen appreciation of the need to educate people about issues and that the biggest problem was an uninformed public.

From the Pedder campaign Gee learned that those fighting for the Franklin needed to demand access to government information and she developed 'a passion to never take "no" for an answer'. The hunt for information, and disseminating that information to the public, would eventually involve 200 people in the project which took three years to complete. All of the participants volunteered their talents and Gee fondly recalled that they all had 'the sense of being part of an important project, of everybody wanting to do what they could, and that time was of the essence in getting the vital information out to the public, unlike at Lake Pedder where the momentum came too late'.

Gee, Fenton and their comrades had to surmount the overwhelming secrecy under which the Tasmanian hydro authority and the Government operated, and they had to unearth from an obstructionist bureaucracy the engineering plans for the proposed dams. They travelled extensively, spoke to people everywhere, ventured frequently into the bush, put on slide shows for schoolchildren, stood on street corners to sell badges—all on a shoestring budget and at often considerable personal sacrifice.[29]

Describing the atmosphere and the intensity of the times, Gee noted that 'people gave years of their lives to the Franklin campaign'. In her own case, she left a job with the Parks and Wildlife Service to devote herself full-time to the book and to campaigning. 'People involved in the campaign were very frightened that the rivers of the south-west would be lost,' she said. 'We all believed that our lives were nothing in comparison.'

The South-West Book was produced in the labour-intensive days before PCs, which meant that all of its text was churned out on a typewriter; if new facts had to be inserted or changes made, this usually meant huge amounts of retyping.

Gee, who has spent the past twenty years on a farm at Buckland about ninety minutes' drive up the east coast from Hobart, spoke glowingly about the Franklin victory but said that ecologically things had gone 'sharply downhill' in Tasmania in the 1990s, with export woodchipping 'through the ceiling' and rainforest still being logged. Gee noted that logging was a more difficult issue than the Franklin dam because it could not focus minds on one specific problem area. 'Logging involves one coupe here, one coupe there—and it's usually up to small communities to tackle local forestry issues,' Gee said. 'In the late 1980s there have been some major wins on environmental issues, but it's getting harder to fight on an increasing number of fronts.'

Gee, her husband Bob Graham and their two teenage children run sheep and cattle on their farm. They lead local campaigns to save forests. Nearby is a woodchip mill which spurs them to manage their own land according to sound environmental principles, retaining 75 per cent native forest cover on the property. At the same time Gee is working on an oral history of the Tasmanian forests campaign for which she has already interviewed more than a hundred people.

Gee's collaborator Janet Fenton was born in 1951 and brought up in the far south-west at Port Davey. There, among the mountains and the button-grass plains, she gained her first environmental awareness about areas that needed to be protected. Gee and Fenton met inevitably at Port Davey as university students. In 1999, Fenton described the collaboration as between two friends with a 'mutual interest in conservation' who felt there was a need for a resource book on the south-west. 'We discovered that information about the south-west was quite dispersed and in scattered archives,' Fenton said. 'We—like many other concerned people—felt such important information needed to be collated.'[30]

In the years leading up to the compilation of *The South-West Book*, Fenton held a number of odd jobs including school librarian, and her expertise in zoology and botany added to the insights of Gee, who provided the environmental and political experience for the book.

Fenton's environmental passion also flowed from her father, Denny King, who had owned a tin mine but was a nature enthusiast, a bird-lover and collector of flora specimens. The environmental credentials of King, who died in 1991, were such that in 1999 his biography was being written by South Australian writer Christobel Mattingley, a keen environmentalist and better known as the author of children's books (see Chapter Eight). Fenton said her father was 'extremely useful to scientists because he was such a skilled observer and able to share rare information about the area'. She was similarly inspired by the surroundings of her upbringing. 'I couldn't help but care about the place, being surrounded only by plants, rivers and animals'. And her conservationist ethic flowed from her parents' activism in such things as pushing for the establishment of foreshore and island reserves in the south-west.[31]

In recent years Fenton, who since 1975 has been living with her husband Geoff and two children at Longley, about fifteen kilometres south of Hobart, has been painting and teaching art, as well as being involved in wildlife issues. She has kept up the lease on her father's place in Port Davey and among her concerns in 1999 was increasing tourism.

On wider environmental issues, Fenton sees agriculture as a big problem, especially on the Australian mainland. She is concerned about the loss of species diversity through forest practices and the 'running of cloven-hoofed animals all over the place'.[32]

The spirit that enabled Helen Gee and Janet Fenton to produce their monument to Tasmanian environmental sanity is summed up by Gee, who stated that 'I have never accepted compromise regarding my freedom to speak out and express myself on such important matters'. Fenton believes the book's major achievement is that it captured the 'passion' of that era, and that it is 'still referred to as a valuable resource and historical document'. The loud and inspiring voices Gee and Fenton unleashed touched the feelings of people throughout Australia and helped preserve the majestic rivers of Tasmania's glorious south-west.

A LATER book, *Wild Rivers* (1983), is a lyrical and pictorial tribute to what would have drowned and disappeared forever had the ancient river regions of Tasmania's south-west been flooded. Written by Bob Brown, with photos by Peter Dombrovskis and drawings by Peter Jackson, it chronicles the deeds of all the exploiters since the Tasmanian colony was founded.

In this large, spectacular, colourful, vivid and dazzling book, Brown records rafting along the Gordon-below-Franklin and tells of many of the characters who had visited this remote and inaccessible region over more than a century. Brown describes the mammoth book as an ode to the 'inexplicable mystery of nature'.[33]

To complement the vibrancy of the photography by Dombrovskis is the rhapsodic enthusiasm of Brown for this magnificent region:

> I saw a rock set like an ancient sculpture, its smooth bronzed surface protected from the flood debris which must have frequently pounded past. Here was a monument to the age of the physical universe. Its solid permanence contrasted with the current sweeping by. I thought of my own transience and that of all life on earth.[34]

Brown acknowledges that the dam and its related hydro schemes represent profound technical and engineering achievement, but he warns that it also 'embodies everything which is wrong with "progress".'[35]

In this early stage of his political activism, Brown also shows how the hydro plans are not only ecological vandalism but economic insanity as well.

Brown shows how the planned Franklin dam, even by the most generous projections, would have less than one-tenth of the capacity of Victoria's Loy Yang thermal power-station.

Between Brown's dramatic observations are the spectacular panoramas shot by Dombrovskis as he captures the broad sweep of the Franklin River above Collingwood Junction. By way of contrast, Dombrovskis also shares an intimate portrait of lichens on a dead eucalypt at Lake Dixon. Waterfalls and early-morning mists are all seen in every conceivable hue.

The majesties depicted in words and photos in *Wild Rivers* were facing extinction from the forces of 'progress' and Brown exhorts his readers to be active to prevent this from happening. 'The onus is ours,' Brown writes:

> The earth will be a better place for the South-West wilderness; for the rhythm of life in its riverside forests; for the mystery and human heritage of its caves; for the peace and gentleness of its wide river reaches; for the thunder of the torrents echoing through its gorges.[36]

Wild Rivers has an overwhelming poignancy in the realisation that everything depicted in the magnificent photographs and everything about which Brown writes so eloquently would have disappeared had the dam been built.

Another activist/politician who chronicled the unique Tasmanian experience is Norm Sanders, who represented the Australian Democrats in both the State and Federal parliaments. Sanders, in a project commissioned by photographer Chris Bell, wrote *A Time to Care* (1980), released at a time when the fight for the Franklin was not yet won. Sanders documents the outrage felt by himself and other concerned citizens at a range of destructive plans including the Franklin dam and the obliteration of much of the State's forests. The book features in all their glory full-page portraits of some of Tasmania's threatened species (such as pencil pines) which were increasingly at risk.

At the time of the book's release, Sanders was well known nationally in his fight to save the Franklin. Later, as a senator, he became increasingly concerned with environmental issues in the national capital, such as planned developments along the Lake Burley Griffin shoreline. And throughout Australia the American-born campaigner voiced his anger on a range of issues such as the US bases on Australian soil, on one occasion posting an eviction notice at Pine Gap.

In 1989 he likened Tasmania to a 'beautiful woman being pack-raped by corporate interests'.[37] In the same year he lamented how his views were unacceptable to many Tasmanians and that physical violence was a constant reality. 'During the Franklin time it was literally impossible for me to walk into a pub anywhere in Tasmania after eight o'clock without somebody trying to take a swing at me,' he said, also describing how he was the target of abusive phone calls.[38]

Amid the photographic splendours of *A Time to Care* are constant instances of Sanders' rage as he notes that 'wantonly drowning [Huon pines] or killing them in 'regeneration burns' is akin to running a bulldozer through the Parthenon'.[39]

Rage is never far away as Sanders rhapsodises about what would be lost should the Franklin be flooded: 'the South-West is corrugated with range after range of profound mountains supporting forests of unique King William pines on their slopes. The rivers roaring through their gorges below are flanked by Huon pines which were seedlings at the time of the Norman Conquest.'[40]

The photography features vistas of a storm on Tasmania's south coast and snow daisies catching the last light on an alpine herb field as Sanders discusses how mining, logging and the hydro schemes were leading Tasmania to catastrophe. 'Hydro-electric development is the most destructive form of human activity short of war,' writes Sanders, who goes on to mourn the flooding of Lake Pedder, where 'valleys which once supported thriving plant and animal communities are now only barren submarine canyons'.[41]

This book also showcases the political and social philosophies—gleaned mainly from American inspirations such as Henry David Thoreau and the green activists of this century—which have guided the singular career of Norm Sanders.

Chris Bell, contributor to *The South-West Book* and the instigator of *A Time to Care*, also produced *Beyond the Reach* (1990), his 'personal statement' about ecological issues, concentrating on the Cradle Mountain–Lake St Clair region. The book features stunning full-page colour photographs of remote areas.

Three separate and equally eloquent strands run through this book: Bell's own philosophies; a careful, itemised history of Tasmanian development; and the author's personal experiences. It is equally a vision of Tasmania's most hidden splendours and a journey into Bell's enraptured soul.

Bell describes Cradle Mountain–Lake St Clair National Park as 'a place to contemplate . . . a place to observe the forces of nature that still govern the land'. To Bell, the experience takes on the highest spiritual significance: 'It provides a starting point for us to acquaint ourselves with wilderness; it elevates our spirits and, hopefully, stirs our consciences to take zealous care of our priceless treasure.'[42]

Bell sees such discoveries and feelings as vital to forging a new consciousness, to overturn the ethos of development which has caused people to 'increasingly lose touch with the natural world', creating a catastrophe whereby 'the wild face of planet earth is disappearing accordingly'. Bell tells how 'country that was wild in recent memory is now ploughed, settled or stocked' and he warns that 'there is not much time to safeguard earth's remaining treasures'.[43]

Chris Bell's admonitions to humans prepared to allow such destruction is as powerful as any work out of Tasmania—or indeed from anywhere else in Australia in the era when a growing number of citizens fought to instil new values of conservation and care:

> It seems incomprehensible that the wilderness—which once fed us and clothed us, and which still comforts and inspires us—has been so abused. For almost all the three million to five million years of human evolution we displayed deep attachment to our wilderness, now we see only its potential material wealth.[44]

Chris Bell arrived in Tasmania in his early 20s as a direct result of the Lake Pedder dispute. 'I got a ticket, and I went straight down,' said Bell as he recalled first learning of Tasmania's environmental plight. 'I discovered pictures of Tasmania which amazed me, as I hadn't known that such alpine areas existed in Australia,' Bell said in 1999.[45]

Born in 1950 at Singleton in the New South Wales Hunter Valley and brought up in various areas of the State (at Oberon in the Central West and at Kyogle near Byron Bay in the north), Bell discovered the wonders of wilderness at Kanangra-Boyd National Park while a youth at Oberon. 'This was raw nature,' Bell recalled. 'It was amazing. It was untrammeled land. There were no human imprints there.'[46]

After he completed his schooling, Bell joined the Navy where as part of the Survey section he 'saw incredible things' off the Papua New Guinea coast such as rare turtles and birds. After learning of Lake Pedder being at risk, he went to Tasmania to try all he could to stop the dam. Bell, who had dabbled in photography and regularly gone bushwalking from the time he arrived in Tasmania, held various jobs, including being a sprinkler fitter, until he embarked on producing books and making a living from his photography.

Bell's commitment to showing the wilderness to the public led to setting up Laurel Press, under whose imprint he published *Beyond the Reach*, which was two years in the making.

In 1999, Bell described Tasmania's ecological struggles as far from over and the world as in deep strife. He said it was a sad thing that the Tasmanian Labor Government in the late 1990s was 'intellectually no different' to the one which embarked on flooding Lake Pedder in 1972.

Another concern for Bell is the global population pressure which has resulted in such things as the forest fires which at the time were raging throughout Indonesia and beyond. In 1999, Bell stated that one of the things about *A Time to Care* which satisfied him the most was that 'several people went to live in Tasmania and fight its environmental battles as a direct result of reading the book'.[47]

MORE superb photographs and yet another perspective on the Tasmanian splendours are found in *Above Me Only Sky* (1981), in which Martin Hawes captures the spirit of the south-west in the years when the Franklin River was at risk. Responsible for the text and most of the photos, Hawes sees the wilderness from the perspective of a longtime bushwalker and mountain-climber.

While in high school Hawes started bushwalking and by the early 1970s had become aware of the environmental campaigns swirling around him. He tells not only of dams and major industrial and resource projects but also of more basic dangers to the wild places resulting largely from the thoughtlessness of humans.

While studying Pure Mathematics and Science at the Australian National University between 1974 and 1976, Hawes commuted from Canberra to the Tasmanian wilderness during his vacations 'to keep my sanity'. On bushwalks he took photographs, and by the late 1970s was doing this on a semi-professional basis. *Above Me Only Sky* is the result of Hawes' collected photos from this period. 'I put on slide shows at university and got good feedback,' Hawes recalled in 1999. 'At first I thought that an audio-visual package would be the way to go, but then decided I did not have enough material. But I felt I had enough material for a book, and this became *Above Me Only Sky*.'[48]

As an avid adventurer, Hawes questions the need for roads across a wilderness area and shows the far-reaching consequences whenever they are built. 'To put in a road is not only to scar the face of wilderness; it is to destroy its very soul,' he submits. He catalogues the folly of such disruptions:

> As far as the South-West environment is concerned, the primary danger is that of fire. Invariably, the introduction of roads into the South-West has led to an increased incidence of fire, thus endangering the alpine, rainforest and native conifer communities, which are not fire-adapted, and disrupting other established ecological patterns.[49]

He sees roads across a wilderness as a symptom of humankind becoming divorced even more from nature. The child of 'Ten Pound Poms', Hawes was born in England in 1956 and arrived with his parents in Tasmania in 1968. He extols the 'intrinsic value' of the pristine environment where 'one can stand entirely apart from the works of mankind and behold nature . . . merely to know that such country exists is a spiritual solace, a refuge for the modern mind in our overcrowded, too-accessible world'.[50]

Describing the south-west as 'one of the few such temperate regions still in existence', Hawes calls on all those who venture there to be prepared for sacrifices and hardships, and to then enjoy the glorious rewards for having done so. 'To go into wilderness is to discover joys that no creations of our own can ever give,' he writes.[51]

Pictorially, Hawes displays the most horrific of contrasts between Truchanas' panorama of Lake Pedder in 1970 and Gee's photo of rubble and wreckage at the Gordon River in 1977. Hawes' words dramatically highlight the vandalism which had been perpetrated:

> The Gordon River runs cold now, its banks covered with slime . . . the river no longer flows with the rains and seasons, but on command from a substation in Hobart. And empty drink cans lie washed up on the rotting shores of the Pedder impoundment.[52]

Hawes' essays are laments exposing the manner in which thoughtless exploitation has destroyed natural magnificence. This applies especially to the waters that run free in the south-west. 'The water that runs in the mountains bears no resemblance to the bland, de-natured substance that comes out of modern taps,' he exclaims.[53]

By the early 1980s, Hawes was spending a lot of time in the bush, having taken a job doing track work for the Tasmanian Parks and Wildlife Service. At the same time he was photographing unprotected areas and he used these photos to publish sets of greeting cards. In the early 1990s he was appointed a track manager, a job in which he wrote a document called the 'Walking Track Management Strategy for the World Heritage Area', as well as setting up a monitoring system for walking track conditions and walker impacts in the World Heritage Area. At this time he became even more concerned about 'the impact of humans in heritage areas . . . and I feared that these areas were becoming loved to death'.[54]

Meanwhile, Hawes continued to write about and photograph the wilderness, and he wrote the foreword to the Dombrovskis calendars for 1999. Also in 1999, his audio-visual package came closer to reality. The photos which formed the basis of *Above Me Only Sky* were collected to become the backdrop of performances in 2000 by the touring Tasmanian avant-garde folk group Alchemy—whose members perform on harp, percussion and various traditional instruments made by Tasmanian craftsmen and who radiate a New Age-style ambience.

ANOTHER natural treasure trove splendidly portrayed in the illustrated book is *Daintree (Where the rainforest meets the reef)* (1985), with text by Cairns-based naturalist Rupert Russell and photography by Leo Meier and Cliff and Dawn Frith. From the opening pages there are lush, magnificent vistas defining what is subtitled 'our last paradise at risk'.

Clifford and Dawn Frith are among Australia's most respected photographers and experts on wildlife and habitat, their work having appeared in innumerable general and specialist magazines throughout the world in all languages. They specialise in what they describe as 'the tropical Australasian region'. Their wide range of interests and hobbies has included ornithological research and they are prolific as photographers, writers and managers of their own publishing enterprise.

Daintree was produced under the auspices of the ACF, whose director Geoff Mosley writes in the foreword: 'It fills one with a sense of awe to know that [the Daintree] is essentially in the state it was in when Captain Cook visited this part of Australia in June 1770.'[55]

The authors describe the region encompassed by the Greater Daintree National Park as 'the most beautiful areas of lowland and upland rainforest, beaches, rivers and mountains in the whole of Australia . . . the beauty and grandeur of the region between the Daintree and Annan rivers has touched the soul of every person'.[56]

Amid such rejoicing comes the warning that logging is destroying the rainforest lushness of areas like Mt Carbine and the Mt Windsor

Tablelands. Chapters are devoted to such issues as land use (with a full-page photo by Meier of vast rolling tracts denuded of trees).

The book also examines the deleterious effects of a range of activities—tin mining, grazing, the expansion of sugar plantations (which require burning, liming, fertilisers or insecticides) and subdivisions. Discussing the last, Russell notes that any settlement close to the region requires the greatest of care and planning, and an attitude consistent with effective conservation: 'It is well to be discontented with the quality of life in our cities, but to seek escape by carrying city-born stereotypes to inflict upon another landscape will do little to bless the refugee and much to blight the landscape.'[57]

In 1987, Russell warned that a rare breed of white-furred possums were under threat from loggers in a patch of rainforest about sixty-five kilometres north of Cairns. In *Daintree*, despite the dire warnings of environmental destruction, Russell retains an optimism that commonsense will somehow prevail to preserve the rainforest's beauty.

Similar celebrations—and identical warning signals—are at the heart of another major project by the Friths, *Cape York Peninsula (A natural history)* (1995), in which that region's biodiversity is shown in all its profusion and splendour.

The English-born couple have become synonymous with North Queensland where they have extensively studied and described the biology and behaviour of birds of paradise, bower birds and other local species. In *Cape York Peninsula*, the Friths present hidden vegetation, rare fauna and Aboriginal rock art unknown to the average person. As Alasdair McGregor and Quentin Chester did in the Kimberley and around their wild islands, the Friths have expanded our knowledge of Cape York.

In a sumptuous book of more than 200 pages, the Friths show every facet of Cape York—especially the rainforests which now cover only five per cent of it—and warn that much of the area may gradually disappear or change irretrievably. The Friths glowingly describe what is there now, then warn of the threats to the rainforest:

> To the stranger, the tropical rainforest at night is an awesome, somewhat frightening place, full of strange sounds, mysterious activity and bizarre creatures. But the animals living there are all too busily intent on foraging and impressing mates to pay attention to human visitors. For humans are a trivial part of their world—that is until they modify or destroy their habitat.
>
> As in most remote areas, human development and settlement are catching up with the land. Sadly, no landscape with high wilderness qualities remains completely pristine or without severe management problems forever. Some landscapes on the peninsula have depreciated considerably in recent years and are still threatened by further decline.[58]

In a chapter entitled 'Depreciation', the Friths discuss how over the past fifty years there have been constant pressures on the Peninsula: 'Clearing of extensive tracts of land by pastoralists, over-grazing and/or over-burning have resulted in severe erosion of some landscapes . . . The Peninsula has also suffered invasions from pigs, horses, weeds and cane toads.'[59]

The Friths describe Cape York, despite its remoteness and inaccessibility to most people, as nevertheless fragile and vulnerable to rapid change. They list among the greatest threats 'extensive clearing, introduction of exotic flora and fauna, over-grazing and unmanaged fire regimes'.[60]

Cape York Peninsula is an in-depth and invaluable guide for anybody wanting to travel to this region, as well as a comprehensive reference work and educational tool.

A LAMENT for how development is ruining much of Queensland is at the core of *The Noosa Story* (1979), in which popular novelist Nancy Cato tells of and presents a photographic panorama of a rearranged ecology, spec building, suburban blight and ugliness superimposed on once-beautiful places. Where there had been a pristine landscape and quiet villages surrounded by natural bush, Sunshine Coast towns like Tewantin, Noosaville and Noosa Heads were becoming 'a monstrous conurbation', Cato writes.[61]

Once in Noosa there had been a 'wide, safe beach shaded with casuarinas and banksias in which rainbow lorikeets squabbled', but senseless development and official folly had ruined all this utterly.[62]

Cato, born in Adelaide in 1917 but in her latter years known as a definitive Queensland writer, sees the Sunshine Coast as a microcosm of the insatiable lust of developers and tourist operators to 'improve' the rare and beautiful areas of Queensland. At Noosa Beach, solitude and majesty have been replaced by the detritus of an industrialised mess-making society:

> Today, monstrous 'vacuum cleaners' suck up water and sand and spew it out in dirty heaps, extracting rutile. Plastic bags, beer cans, milk cartons, old tyres, rubber thongs and tangled nylon lines are mixed with the seaweed at high-tide mark.[63]

The book features pictorial collages based on historical records and private collections, contrasting the loving care with which Aborigines tended the area for generations and how the Europeans behaved, indiscriminately chopping down any trees in their path and despoiling the splendid vistas. Family snaps down through the generations show Noosa's main thoroughfare, Hastings Street, as it looked 30 years earlier, in the 1940s—quiet and empty—and then how it was after a new road was hacked through the nearby pristine bush. Another photo shows how Noosa once enjoyed bucolic splendour, such as a child playing with a living koala rather than a stuffed teddy bear, in a setting of luxuriant bushland.

Nancy Cato, who in 1967 moved with her husband Eldred Norman from South Australia to Brisbane and then to the Sunshine Coast, traces, often with tedious detail, the many excesses of thoughtless town planning, civic negligence and a laissez-faire attitude to 'progress' which allowed the 'rape' of Noosa:

> Development followed the usual pattern: first the explorers and geologists; then the exploiters, tearing everything valuable out of the soil, denuding the hillsides of forests, destroying wildlife, and 'dispersing' the Aborigines, the land's original inhabitants, when they objected to losing their hunting grounds. Then came the farmer-settlers, clearing with cross-cut saw and fire, bringing in cattle and dairy herds to munch the wiry grasses and muddy the creek banks. Koalas, once prolific in the gum forests and often seen about Tewantin and Noosa Heads, lost their habitat. Those that survived were shot out in the great slaughter of 1927, when an open season was declared on their 'valuable' skins. These harmless, infinitely attractive small animals are now rare in the local bush.[64]

Koalas were not the only wildlife either obliterated or dispersed as habitat was removed. Many rare Australian species are among the scattered victims:

> . . . the residents on Laguna Hill woke to find a smoking ruin spread out in the estuary. It was as if a selective atom bomb had struck. Not one living thing was left, except for a large, solitary pine which must have escaped through some oversight. It soon died of exposure.[65]

Quality colour photos are also included among the historical memorabilia, usually belonging to caring locals. There is an aerial shot of the picturesque natural waterways around Tewantin and Noosa Heads in the 1960s—before the worst of the developments. Among the before-and-after portraits, there is Noosa beach with hardly any tourist buildings and below it as it was after the notorious rock wall had been built, forcing those on the beach to huddle even closer to each other.

Cato, who believes she and her husband had conservationist leanings long before this became fashionable, documents the occasional success in the burst of environmental activism since the early 1970s, but she notes that whenever vigilance falters the results are usually catastrophic: 'the Noosa Inlet was lost without even a struggle. Dredging and filling, on a vast scale, altered the river's mouth irreversibly, creating problems that are only just becoming evident, and destroying the wildlife in that area.'[66]

The Noosa Story is a work of the greatest gravity which simmers with outrage at the utter stupidity of those developers and public officials who took a splendid beach contoured by the curves of nature, had it straightened by engineers who placed a modular line of rocks and then dumped artificial sand behind it.

A fervid campaigner, Cato constantly speaks out against anything she believes is an undesirable development, such as a 'horrifying' plan in 1998 to build a Hilton hotel and other structures on Noosa Hill as well as another ambitious project for Settlers Cove which would have impinged upon the 'green backdrop' to Hastings Street. 'There should be no buildings above the treeline,' Cato declared. 'Who wants a great hotel on the green hill?'[67]

Cato has also campaigned against developments close to her home, such as plans to import varieties of trees to replace the existing native flora. 'I like trees that birds love and everyone knows birds hate palm trees,' she said in 1990. The complete environmentalist, Cato became a vocal opponent of motor sport because of the pollution it caused. She took this step even though her late husband had been a prominent racer and car designer (as the inventor of the famous Norman V12).[68]

THE dangers facing Victoria are chronicled by Harry Nankin in *Victoria's Alps* (1983), which shows how Australia's European settlers were unfit to husband what they tended—and how things have not improved too much since then. Writing under the auspices of the ACF, Nankin describes how, according to one of his chapter titles, the area under review was seen only as a 'resource to exploit'.

Listing activities such as grazing, mining (starting with the discovery of gold in 1851), forestry, ski resorts, water utilisation and vehicular access to the High Country, Nankin angrily sheets home the blame:

> European man has traditionally been, and remains, psychically alienated from his adopted environment . . . the inevitable consequence of this perspective on the High Country has been and continues to be that of ever-growing despoliation of the natural environment in a variety of ways and from a range of activities.[69]

Superb colour plates show the panoramas, vegetation and wildlife which are all imperilled in the High Country, and Nankin calls for protection 'soon . . . before it is too late. Before the chainsaw, the bovine hoof and the bulldozer render it a heritage irredeemable, a heritage lost'.[70]

Both black-and-white and colour vistas of the many peaks, such as Mt Buggery, are shown to be threatened by grazing and 'once-only' logging. There are also thorough descriptions of the area—including categories like verdure and firmament—and the creatures within it.

The High Country extends beyond Victoria to New South Wales and the Australian Capital Territory and Nankin shows it to contain 'an assemblage of landforms and living things unlike any other in Australia or the world'.

Nankin pleads for this area to be part of a more extensive national park. To him it is 'a land of treeless highland plateaux and eucalypt-filled valleys . . . the only extensive alpine and subalpine environments on the Australian mainland . . . an area where nature (despite some developmental tamperings by humanity) remains pre-eminent'.[71]

Discussing the creatures of this terrain, Nankin describes the animal life as 'diverse and prolific as its vegetative cover'. He describes 38 species of native mammals, 183 species of birds (which Nankin says represents about one-quarter of Australia's indigenous avifauna), 30 reptiles, 15 species of frogs and toads, 14 kinds of native fish and 'untold varieties' of invertebrates.[72]

Nankin reveals the need for greater public concern as politicians generally are unable to meet their responsibilities on environmental matters.

Born in Melbourne in 1953, Harry Nankin's first exposure to wilderness came in his middle teens when he came into contact with a youth environmental group called 'INSPECT', attending meetings and going to camps where he 'got involved and had my consciousness raised' to ecological issues. At the same time his involvement with another youth group led to camping out in the Lerderderg Gorge near Bacchus Marsh west of Melbourne—which Nankin in 1999 described as 'forest and low plateau country done over by goldmining in the 19th century'—and he became 'enthralled by it all as I had an aesthetic and spiritual experience'.[73]

At the University of Melbourne, where Nankin studied Town and Regional Planning, he joined the bushwalking club and learned more about wilderness and nature. 'I became interested in changing the world environmentally,' he said and told of how he turned to photography and art, having already shown some flair in the latter. Within two years he became a fully fledged wilderness photographer and went to Tasmania to visit Peter Dombrovskis, who became his mentor.

Among the things Dombrovskis taught Nankin was how to market wilderness calendars and become professional in all of his pursuits. The difficulties of combining creative effort with a steady income were highlighted almost immediately as a planned book of photographs of New Zealand's fjordlands proved impractical and far too expensive to produce. Despite working as a wine waiter and having various other odd jobs in New Zealand, 'this book didn't get finished owing to lack of money'.

Back in Australia, Nankin saw an ad in the ACF's newsletter calling for someone to produce a book about the High Country so 'even though I knew little about this area or what I was in for, I said "why not?"' He was able to convince the then head of the ACF, Geoff Mosley, whom he had interviewed for his Town Planning thesis, about the viability of the project. For the next eighteen months he worked on researching, writing about and photographing the areas which would form the basis of *Victoria's Alps*. Although Nankin is disappointed that the original grand pictorial plans for the book did not eventuate, he is encouraged that it ended up as a textbook.[74]

In 1987, Nankin released *Range Upon Range: The Australian Alps*, for which he provided an introductory essay of his conservationist philosophies. The book featured his photographs of alpine areas accompanied by the verses of leading Australian poets such as Douglas Stewart, David Campbell and Chris Wallace-Crabbe.

MORE natural wonders can be found in *Nature of Australia* (1988), in which writer, film-maker and television producer John Vandenbeld collected highest quality pictures from the best still photographers of nature as well as ABC TV cameramen who show the vastness and variety of the continent. Apart from these impressive visuals, the book features a series of essays by Vandenbeld, who tells how the landscape, especially the different types of forests, is a combination of old and new—for example, eucalypts side by side with wooded margins in the New England National Park of New South Wales.

Vandenbeld, a pioneer of the ABC's natural history unit who died of renal cancer in 1992 aged 52, describes a fragile relationship increasingly at risk and shows the wonders Australians have lost or may not enjoy for much longer.

'Southern beech forest . . . still exists in Australia,' writes Vandenbeld. 'To walk through these cool, damp and almost silent forests is to walk back into Gondwana.'[75]

A shimmering, lustrous, dusky full-page photo of broken ice (by David Parer) highlights the importance of preserving Antarctica for the sake of life systems closer to the Australian mainland: 'When the summer sun returns to the Antarctic . . . there is a burst of plant and animal growth, generating a wealth of nutrients carried north by deep currents to sustain the marvellous diversity of marine life in Australia's southern seas.'[76]

Ecological awareness permeates each lavishly illustrated page as Vandenbeld stresses, for example, the importance of climate and explains the formation of deserts—nature's cruel and enduring form of erosion, a cautionary tale of the totality of land degradation. Vandenbeld envisaged this project as a valuable contribution to the Australian Bicentenary, and when the TV program of the same name won a Logie award his acceptance speech referred to many of the environmental problems Australia was then facing.

This is the work of a long-time journalist who began his career at the *Geelong Advertiser* and the Melbourne *Herald* newspapers and then joined the infant TV industry, first writing the weekend news bulletins on Melbourne's Channel Seven and then joining the ABC, where he was part of the Melbourne link for the ground-breaking current affairs program 'This Day Tonight'.

Vandenbeld's environmental concerns are evident in his description—heightened by a graphic, stark photo by Reg Morrison—of a denuded section of Queenstown, Tasmania, stripped bare by the Mt Lyell mining operations. 'Much of the hill country was cleared to supply the mine,' Vandenbeld writes, 'and the sulphurous fumes from its workings killed what remained'. Vandenbeld's anger rises in the all-encompassing description:

> There is a piece of blighted land in Western Tasmania where nothing lives. The earth is as dead as the moon, killed by a mine that took the trees and spewed them back as sulphurous fumes. The scoured hills stand bare and pale as a bone . . . Not far away, loggers are bulldozing a road into one of the island's last remnants of rainforest . . .[77]

Vandenbeld was born in the Dutch town of Haarlem in 1940 and emigrated to Australia with his family from the Netherlands in 1953. He felt keenly the inappropriateness of colonialism, which he blamed for many of Australia's environmental ills. He rails against the 'fixed concepts' of Australia's earliest settlers who 'looked upon [their new land] as a second-class environment at whose demolition the world would shed no tears'.[78]

His widow, Margaret Vandenbeld, in a 1999 interview told of how he had been passionate about the environment and was utterly committed to the *Nature of Australia* project, both the book and the TV series, because 'nobody had done this sort of thing before'. Margaret Vandenbeld said her husband had been ahead of his time in wanting to make people aware of issues like salinity, about which there had been little discussion in the 1980s. 'Not a lot was known then about such things as salinity, the greenhouse effect, the hole in the ozone layer, and people didn't know if these things were true . . . and problems like the El Nino effect were not discussed,' she said.[79]

She said *Nature of Australia* was four years in the making, and 'an enormous amount of research' went into making the TV programs (a joint venture between the ABC and BBC) and writing the book. She fondly recalled how John Vandenbeld started a typically long day at the crack of dawn, accumulated boxes of material and produced copious scripts on an old Remington typewriter in an era when PCs had not yet arrived—even for journalists.

According to Margaret, John was heartened that the book became a recognised reference work for those interested in the Australian ecology, especially as he had set out to 'produce something different . . . something for the future for his young family'.[80]

In the *Nature of Australia* book, Vandenbeld covers the spectrum of ecological concerns, including day-to-day problems caused by rural attitudes such as conservation meaning no more to many farmers than 'trying to stop soil blowing away', to the more far-reaching issues of the widespread contamination risks from uranium mining.[81]

This book's final chapter is a clarion call for care, compassion and commonsense which, if unheeded, will result in us all sliding into doom.

SERIOUS and in-depth analyses of endangered species have been the subject of illustrated books over the past quarter century. Among them is *Australia's Endangered Species* (1990), in which general editor Michael Kennedy describes an environment under grave threat. 'It is critical that every person in an essentially affluent Australia realise that something is wrong,' Kennedy writes. Describing the increasing number of species at risk, he warns that this is but a prelude to 'an environmental degradation that will ultimately threaten us'.[82]

This handsome and far-ranging book, which has a foreword by well-known ABC television newsreader Richard Morecroft, warns that Australian efforts so far at saving species at risk have been found sadly wanting. The consequences of mankind's depletion of this continent's rarest fauna are starkly presented: 'Unless we shape up today, the myriad Australian species represented in this book may well be gone tomorrow. Once they have gone, it is forever, never to appear on this earth again.'[83]

Born in England in 1951 Michael Kennedy played as a child in London's commons and parks where he developed a love of nature and its creatures. Since 1978 Kennedy has been a professional campaigner on environmental issues for various organisations, with a special emphasis on threatened species. Since arriving in Australia as a 19-year-old seeking adventure, his own experiences and his research drawn from publications such as *The Ecologist* alerted him to the issues of whaling, the trade in wildlife, and the need for convention treaties like those applying in Antarctica. Between 1986 and 1987 he worked on the staff of the Hawke Government's Minister for the Environment, Barry Cohen, and for the next two years was a consultant for the National Parks and Wildlife Service. Between 1989 and 1994 he worked with the World Wildlife Fund, posts which prepared him for the dual roles of conservation and animal welfare which he undertook as the director of the Australian office of Humane Society International. His co-director is his wife, Verna Simpson, a former owner of *Simply Living* magazine (see Chapter Five), who was managing editor of the endangered species book.

Kennedy uses his broad range of skills (from computers to running campaigns and teaching others) to alert Australians that 'there are more species in trouble than we thought'. In 1999 he said the list of endangered species had grown in recent years and described the easing of their plight as a 'slow process'.[84]

In *Australia's Endangered Species*, Kennedy stresses the disastrous consequences if biological diversity is not maintained, and he links the simultaneous disasters of species depletion, disappearing forests and air and water pollution. To illustrate the extent of the disasters, Kennedy has collected photographs from a variety of sources, including a vivid quarter-page depiction of mining at Kakadu. And across one-third of a page, grey-streaked among the blue/grey/green foliage, is the Eastern barred bandicoot, described as 'highly endangered on the mainland'.[85]

There are copious illustrations, by leading wilderness photographers like Leo Meier, of all the various categories—all dedicated to the message that Australia's record in species preservation 'is not good . . . in fact, it is disastrous': 'Australia has seen more mammal species go extinct in the last 200 years than any other continent or country . . . Already 18 (10 marsupials and 8 rodents) unique species, found nowhere else in the world, have been lost forever.'[86]

To Kennedy, the very essence of human survival hinges on how we treat all other forms of life on earth. In seeking to show how the consequences of the current disappearance of life-forms could become devastating, he has collected one of the most comprehensive listings of endangered species, as well as maps showing their habitats and in-depth appendices of useful information.

Kennedy is among those who believe that illustrated books not only have a mission to influence people's hearts and minds about species extinction but also have a responsibility to educate.

AMONG the very serious specific issues canvassed in illustrated books is mining at Kakadu, which is discussed extensively in a handsome volume, *Kakadu Country* (1988), with text and photographs by Peter Jarver. By four-wheel-drive and by boat, Jarver went on several expeditions to areas whose marvels he conveys in full-page colour photos of every aspect of one of Australia's greatest national parks.

There are images and descriptions of pristine glories: paperbark trees reflected in the pure waters of Nourlangie Creek are a dynamic mesh of yellow and green; sunrise at the Koolpin Creek rapids is a swirling interlapping of blues; saltwater crocodile (in the process of eating a barramundi) and lustrous varieties of birdlife.

Jarver has been hailed as presenting the Northern Territory with a vividness similar to that with which Peter Dombrovskis has presented Tasmania. The quintessence of the book's Kakadu message is explained in the foreword by the then director of the Australian Conservation Foundation, Philip Toyne, who stated on behalf of the ACF that 'there is no place for miners and their activities within national parks'. Warning about the various threats to Kakadu from miners and developers, Toyne refers to the 'huge responsibility we have to guarantee our descendants the right to experience' the magnificent sights shown and described by Jarver.[87]

Born in Adelaide in 1953, Jarver first went to Darwin while taking a break from electrical engineering studies at the University of Adelaide. He stayed for a year, but the experience was so fulfilling that he went back and stayed for 17 years, first borrowing a friend's camera to photograph the natural wonders.

'Thunderstorms had a huge hold on me,' Jarver recalled in 1999 as he described the dramatic cycle of wet and dry in Darwin, which in the early 1970s was 'just a speck on the Top End' and was a 'genuinely remote place—not a Clayton's wilderness as it is now'.[88]

Jarver, who availed himself of the plentiful and varied work available in Darwin at the time, fell in love with his new surroundings. At Kakadu, he marvelled at the waterfalls and the weather cycle and enjoyed the sense of true isolation. 'The bush developed a sway over me, and I developed a personal relationship with the country,' Jarver said. He was bedazzled by the rich and very diverse ecosystem of the Top End, where there was a profusion of different plants.

At the same time Jarver became passionate about photography, and despite having had no formal training developed the skills necessary to record for posterity the wonders that he was experiencing.

Kakadu Country was the culmination of about eight years of photographing the region—everything from 'skyscapes' to deep within the bush. Because of his feeling of urgency in the late 1970s about the fate of Kakadu, which was being pressured on all sides by a development frenzy, he decided to put a book together. 'I had photographed the region before there was the herd mentality and the influx of tourism,' he said. 'The book also resulted from my natural curiosity to see what was there. I was like most people in not knowing what was there.'

Jarver's book shows Kakadu as it was before it was promoted as a major tourist destination and sprung from his 'burning desire to photograph the area before it was spoiled'. To Jarver, it was also a 'journey of self-discovery, to find out who I was through my photography'. Jarver was driven by a fierce conservationist ethic and was appalled at what might happen to Kakadu. He was inspired by international organisations like Greenpeace which were taking a keen interest in developments in the Northern Territory. Despite the backlash by many Australians against plans to develop the Top End, Jarver witnessed the onrush of tourism and the opening of the Ranger uranium mine and the building of the township of Jabiru which would support the mine. 'I took part in the anti-Ranger protests and at the same time feared that Kakadu would be overrun with tourists,' Jarver recalled in 1999.[89]

By the mid-1980s, as major publishers throughout Australia either rejected Jarver's book as not commercially viable or offered him only 'pathetic' money, he decided to release it himself and set up his own Thunderhead Publishing. The publishing venture was a reflection of Jarver's flair for marketing as Thunderhead had started as a retail photo processing facility in a shopping arcade in Darwin's CBD which did good business with interstate visitors. Publication of *Kakadu Country* entailed raising about $40,000 through family and friends which in 1999 Jarver remembered as a 'scary undertaking, putting everything on the line' to release the book. After finding the right designer and printer, Jarver then worked out a distribution strategy whereby Thunderhead would handle the Northern Territory side of things and a freelancer in Sydney would handle things from there. Jarver recalled with satisfaction that straight after its release in 1988 the book 'took off like wildfire', selling its initial print run of 10,000 and then two years later releasing a further 15,000.

In 1993, Jarver moved to North Queensland to embark on new photographic challenges, setting up a new publishing house, Wildscape Australia, with its headquarters at Kuranda near Cairns. Jarver described Wildscape as being a viable enterprise, mixing local tourist-related material with his more ambitious illustrated books.

But in 1999 much of Jarver's heart and soul remained amid the spectacular skies and the rare ecosystems of the Top End. He was concerned about the Jabiluka uranium project becoming 'another Ranger—a mine in the middle of a national park', something that Jarver abhors because 'you can't separate a mine from its surrounding ecology'. Jarver believes the Jabiluka mine will pollute the broad flood plain which surrounds it. He also opposes the Jabiluka development because he rejects the nuclear fuel cycle and all its connotations. 'Australia has the best solar cell technology in the world and we have huge amounts of sunshine, so why hasn't the Government put money there?' he asked. Further, he believes Australia should turn its back on a commodities-based economy because 'technology is the way to go'.

Jarver, who said he approved of controlled and selective logging on a sustainable basis, said 'historians will view us as criminally negligent' for such environmental follies as the clear-felling of Australian forests. 'I can't comprehend how in a supposedly enlightened age we can allow clear-felling for woodchips,' said Jarver, who in his expanding publishing venture in North Queensland hopes to take such profound ecological messages to as wide a readership as he can.[90]

ANOTHER major issue embraced in the illustrated book genre is the plight of Australia's farmers, whose struggles with natural disasters and an economic recession are documented in *On the Edge* (1992), featuring the words of Robert Milliken and the graphic black-and-white photographs of Lorrie Graham. Milliken, whose other major book *No Conceivable Injury* (1986) detailed the British nuclear tests of the 1950s (see Chapter Three), this time examines another ecological disaster: the longer and more enduring catastrophes of land degradation, salinity and algal blooms which accompanied the rural recession beginning in the late 1980s.

If Maralinga was senseless destruction, so too was the rural crisis which, as Milliken painstakingly describes, led to suicides and shattered lives. A way of life Australians had known for generations was being increasingly snuffed out by powers outside the control of the average person on the land: Milliken examines the role both of local banks and the economic policy of distant, foreign entities.

Milliken's consciousness about rural issues stems from his upbringing on the mid-north coast of New South Wales, where he was born in Taree in 1946. He was a son of the same soil of the Manning River bounded by Wingham and Gloucester that nurtured Les Murray and Ian Moffitt a generation earlier. His earliest education was in this region, before going to high school at Scots College in Sydney and then to the University of NSW.

Environmental consciousness arrived fairly late for Milliken, who grew up in a region where the timber industry was pivotal. It was not until the 1970s that he began to 'rethink all of this'. In a 1999 interview, Milliken told of how he had come to realise that Australia was a fragile rather than a rugged environment and it needed to be treated with respect'.[91]

Also, Milliken told of how in his youth he had taken the coastline and waterways for granted, and it was only after he had embarked on journalism that he discovered all the bad developments along the coastal fringe and how sewerage outlets were inadequate to prevent a ruination of many of the waterways. The nearest coastal resort to his boyhood home was Forster, where in the 1990s there would be an oyster scare, and nearby Wallis Lake was very fragile and over-developed.

In *On the Edge*, as Milliken dissects the disasters, Lorrie Graham's portraiture shows in stark relief the faces of despair, stolidity, hopelessness and fear. The close-ups of lined faces and the scarred backdrops of ravaged soil are vivid and compelling statements.

Milliken examines a cross-section of the causes of the rural ecological disasters—from cotton-growing in Wee Waa to over-use of irrigation in the Murray–Darling Basin. We meet the range of people involved in LandCare schemes, the brave attempts which began to bear fruit in the early 1990s. In words and pictures we see those most affected by the rural catastrophes—how they look, how they feel, and how they see the future.

The authors often allow the victims—and those who have triumphed or who intend to prevail—to speak for themselves and tell of their lives. For example, there is Terry Simpson of the property 'Winjallok' near St Arnaud, Victoria, whose LandCare scheme rejuvenated a spread which had been 'left without the cover of trees . . . The water carried off with it soil particles and nutrients, depleting the soil structure'. Milliken minutely describes the scope of the problem:

> Here was a classic case, repeated on other farms, of the historical clearing of native vegetation causing degradation in places far away from where the clearing originally took place . . . [the land] ceased to capture much of the region's abundant rainfall . . . the effect was like a drought every year.[92]

Milliken then describes how resolute farmers like Terry Simpson were able to turn things around with ecological best practice: by astute planting methods revitalising up to two-thirds of his property and lifting its carrying capacity more than seven-fold.

On the Edge resulted from photographer Lorrie Graham seeking a collaborator for a book about farming which had been commissioned by Simon & Schuster. At the time Milliken, who had become a freelance writer after the release of his landmark *No Conceivable Injury* in 1986, was doing a lot of travelling while covering Australia for the *Independent* newspaper in London. During these assignments, Milliken went to western New South Wales, where he saw at first-hand the rabbit plague of the early 1990s.

Issues of urban blight also concerned Milliken, and during the late 1990s he was part of a protest movement against the high-rise developments along Sydney's East Circular Quay—especially the residential block known as 'The Toaster' which hovered menacingly alongside the landmark Opera House—and he joined about 400 prominent Australians (including poet Judith Wright and novelist Tim Winton) in seeking to have the Bennelong Point peninsula restored to its former charm.

Milliken had been disturbed by Sydney's urban environment generally and trends at East Circular Quay specifically for more than fifteen years as he discovered that planning throughout Sydney had been 'handed over by authorities to developers and buildings had been torn down willy-nilly'. To Milliken, East Circular Quay was just the latest example of Sydney's great natural advantages being squandered. He noted that after a frenzy of demolitions Sydney had lost all its old theatres as well as many lustrous landmarks.[93]

ANOTHER thoughtful and comprehensive illustrated book showing the extent of global destruction is *Earth First* (1987), in which film-maker Jeni Kendell (see Chapter Six) and Eddie Buivids graphically illustrate many of this nation's wonders under threat. Many of the major environmental issues since the mid-1970s are brought to life in this book. There are misty panoramas of western Tasmania in a detailed chapter on the fight against the Franklin dam, including an in-depth biography of Bob Brown which personalises all phases of this pivotal issue. There are various first-person accounts of the Franklin blockade, including by Ian Cohen, who in 1995 would become a Greens MP in the NSW Upper House.

Visible in full focus is the earlier battle over Terania Creek, with the authors examining issues which would be amplified in Ian Watson's 1990 study *Fighting Over the Forests* (see Chapter Three). Like Watson, Kendell and Buivids analyse the obdurate hostility towards green activists by proud timber workers who saw protesters as 'upstarts, stirrers and ignorant visionaries'. Kendell and Buivids explain the misunderstanding and mutual suspicions which would plague any attempt at a mutually satisfactory resolution to logging and related issues:

> The conservationists, on the other hand, believed that the foresters lacked ethical and moral guidelines because forest policy was dictated by economic considerations alone, that the elder generation of foresters was trained only to see forests as a 'vertical collection of logs with undergrowth in the way'.[94]

The deepest divisions would prove insurmountable, with intense hostilities never far from the surface even after the Terania protesters had forced the Wran Government to stop logging, Kendell and Buivids show:

> . . . deep rifts had been created in the local community. The whole region polaraised over the issue. Long-time residents and traditional rural people resented the protesters—they had threatened their heritage, their livelihood, their government and had won![95]

Black-and-white and glorious colour photographs dramatise the forests conflict, showing lush and abundant greenery surrounding the Selva Falls in the border region straddling New South Wales and Queensland. This area had been thriving and indestructible until the arrival of Europeans and now its future could be guaranteed only by the recent phenomenon of national park protection. Photos also show the glories of the Nightcap National Park near Nimbin.

Apart from these visual majesties, Kendell and Buivids discuss the causes of global destruction and the guilt of international organisations such as the World Bank.

AMONG a range of charming 'little' illustrated books is *A Walk in the Rainforest* (1985), in which Dawn and Clifford Frith examine what they describe as 'the world's richest ecosystem'. The book features David Stacey's exquisite line drawings, beginning with a pale yellow robin, a green tree frog and a forest panorama. Dramatic full-page illustrations in the pocket-sized volume depict the stunning variety of rainforests and the life within them.

This was one of the earliest projects by the Friths, whose wide range of interests has encompassed such diverse activities as being scientific and technical advisers to film producers and crews including the BBC Natural History Unit in the UK, the ABC Natural History Unit, Film Australia, Television New Zealand and a host of private film-makers. They have also lectured and given public talks at universities, at the esteemed Royal Society of London, the Zoological Society of London, and various specialist societies and wildlife and conservation bodies. They have led ornithological tours for visitors to North Queensland from all over the world.

Their concern for birdlife brought them together, both having been members of a Royal Society expedition to the Aldabra Atoll in the Indian Ocean in 1971. After much travelling and extensive research, the Friths settled in North Queensland, whose ecology they have described to a wide readership.

With such a range of expertise and a growing concern for the tropical ecology, the Friths note in *A Walk in the Rainforest* that only half of all the rainforests which existed at the time of the First Fleet continued to exist by the end of the 20th century. This 'natural heritage' as described by the Friths now covers only one-quarter of one per cent of the land mass but 'houses well over 50 per cent of all [Australia's] plant and animal life'.[96]

Warning that rainforests were 'still continuing to disappear or be degraded', the Friths dissect the causes of the destruction, glory in the 'beauty, fragility and intricacy' of our most stunning heritage, and agonise at the horrible ways it is put at risk.[97]

The littlest illustrated book of them all is *Poetry in Pictures (The Great Barrier Reef)* (1985), in which poet Mark O'Connor (see Chapter Seven) and photographer Neville Coleman combine their talents to provide potent images of this unique region.

Many of the subjects in O'Connor's verse are captured alongside the text by one of the nation's foremost underwater photographers. The descriptive verse and creative photos also show the reef's life above the ground—the bird life, shimmering scenes at dusk, forest life and the effects of fire. And, of course, there are the corals themselves in a splash and explosion of contrasting colours.

THE significance of illustrated books in awakening and kindling the nation's consciousness towards ecological issues has been stressed by environmentally aware photographers and commentators. It is vital to make as many people aware of what the wilderness is, says Robert Rankin, who sees his own work as an extension of his great interest in exploring wild areas. Rankin regards the advantages of exposing the wilderness to as wide a readership as possible as outweighing any intrusion by cameras and crew.

'Am I being in any way destructive in advertising the wild places?' he asked in 1997. 'I don't think that on balance I am, because if I didn't take my photographs I can't whip up enthusiasm for protest against the increasing encroachments on the wilderness.'[98]

Rankin, who has made numerous science documentaries with the ABC, said his aim was to extend information about the wilderness, not to withhold it—and above all he aimed at accuracy. He said illustrated books had played a role in pushing green issues into mainstream thinking and he noted that as a result 'all political parties now address these issues'.

At the same time, Rankin noted that it was very difficult to maintain the rage about environmental destruction. He said an intensity had built up between the campaigns for Lake Pedder and the Franklin but after that a war-weariness had set in with 'much of the passion gone'—and illustrated books were one way to keep informing the public who could then vividly see for themselves what was to be lost.

Rankin wistfully noted in 1997 that a major pictorial work was needed to depict what was at stake in the Port Hinchinbrook tourist development in North Queensland because, unlike at the Franklin River where an entire landscape was conspicuously at risk, at the Hinchinbrook Channel it is perceived that just the dugongs are endangered and the issues are usually presented in a simplistic fashion.

The significance of spreading the environmental messages through means such as illustrated books is seen in the political realities of Tasmania, as described by Christine Milne. In 1998 she failed to be re-elected to State Parliament as the major parties, in an effort to eliminate 'minor' parties such as the Green Independents, lifted the quota requirement for winning a seat from 12.5 to 16.5 per cent of the vote. Three of Tasmania's four Green Independents thereby lost their seats, and spreading the green message outside of the mainstream political system assumed greater significance.

The last word on the importance of illustrated books is best left to the feisty Nancy Cato, who in 1990 defined the importance of fighting for what was worth preserving: 'Once people stop fighting for the environment, the country's finished, the world's finished, we're all finished.'[99]

5

MAGAZINES

Home-grown gloss

MAGAZINES have a special place in the hearts and minds of their readers. Not quite as enduring as a book, they nonetheless have much more of a sense of permanence about them than a newspaper. As everyone knows, magazines can hang around a home for years, gathering in piles, too precious to throw out. Special-interest magazines create a sense of community and connectedness among their readers. In the environmental movement, magazines such as *Grass Roots* and *Earth Garden* cultivate a feeling of 'family' among their readers, and consciously set out to bring like-minded people together. In these magazines, people who thought they were the only ones who rejected the nine-to-five grind of consumerism and stress find another world of kindred souls.

Magazines also have the capacity to bring the world to the reader, and in Australia we are fortunate to have a fair number of glossy publications that bring the visual splendours of this land to light. *Wild* and *Geo*, in particular, take readers to remote places through their stunning photography and informed, insightful text. These magazines also bring the bad news: that the beautiful places of our continent are under threat.

Magazines have helped to raise the consciousness of their readers to these threats, and made them more aware of what's at stake in the battle for the environment. Both *Wild* and *Wildlife Australia* were founded by environmentalists, who saw their role as educational and who had a strong spiritual motive for wanting to preserve the beauty of the land.

Many smaller publications have cropped up over the years and every environmental organisation has its own magazine. All have helped to inform and inspire their readers—and unite them in a common cause.

AS ITS name suggests, *Grass Roots* is no-frills, basic, simple and heartfelt. From its origins in self-sufficiency, frugal home-making and crafts, it has over the years come to represent the conscience of the green movement. *Grass Roots* is simultaneously a resource, a kaleidoscope of varied reading, of down-home and home-spun philosophies, and practical advice on almost everything under the sun to help readers become self-sufficient—all with an eye towards an environment becoming unsustainable. It involves reader participation at a level rarely found in publications, and this enables *Grass Roots* to share a wealth of expertise and experiences.

There is nothing glossy or elegant about *Grass Roots*—for instance issue No. 4 has lopsided type and a generally cluttered layout.

In every issue there is a range of articles on environmentally sensitive topics like the threats posed by toxic sprays, the benefits of solar energy, composting, more ecologically friendly building design, and the use of seaweed to rejuvenate overworked agricultural land.

Above all *Grass Roots* is a sharing of personal experiences and many articles are poignant vignettes of suffering and hardship (including descriptions of the Depression years) and how resources can be shared and conserved. Its correspondents have ranged from Wild Cattle Island in Queensland, Harcourt in Victoria, Bridgewater in South Australia, Glen Forrest in Western Australia and Blackmans Bay in Tasmania.

Grass Roots first appeared in 1973 as a 'craft and lifestyle magazine'. Issue No. 1 (April–June) described itself as 'produced by a co-operative formed by people who believe the world is improved by sharing' and dedicated to alternative lifestyles. From this first issue *Grass Roots* was in the vanguard of environmental awareness, its concerns ranging from chemical poisons to the destructiveness of big business and to a decaying ecology in general.

Although seeking to attract a wide and mainstream audience, from its earliest days *Grass Roots* appealed mainly to those already committed to self-sufficiency and other environmentally sound practices. The magazine's gospel was a thorough rejection of the rampant consumerist society and the corporate mentality which feeds it. The magazine's founders and joint editors, David and Meg Miller, call for a radical rethink of our lives: 'You don't have to bow to the dragging monotony of set hours, set jobs, set transport and set wages. Throw your clocks away—the time for change has come.'[1]

The self-sufficiency advocated in the first issue extends to information about how to organically grow one's own vegetables, bake one's own bread and make one's own clothes—but there is also plenty of help and feedback for those who wish to stick to more mainstream urban living. Many of the topics in issue No. 1 encapsulate the magazine's direction over the next two decades: 'Getting back to the soil', 'Spinning with a spindle' and 'Building with natural materials'. In the last-mentioned, contributor Steven Power gets to the core of *Grass Roots'* assessment of society's ills: 'Today we just ring up or drive to the nearest shop to satisfy our needs. We live in a "turned-on" world. If you can't get it today it can be delivered in a week. We don't get the chance to develop the ability to turn everyday "waste" into useful articles . . .'[2]

There are regular historical and in-depth insights on ecological destruction, such as that the Australian continent had been free of weeds until the First Fleet arrived and 'all plants had a use in the intricate balance of the ecology'.[3]

The second issue (September 1973) introduced 'Feedback from Farout', later to become 'Feedback/Link-up', which across ten to fourteen pages is a forum for every conceivable topic. Among the earliest collection of correspondence were descriptions of the establishment of communes, a celebration of herbs, a humorous and warm-hearted look at goat breeding, the best way to store vegetables, and tips on baking bread and candle-making.

From its earliest days, *Grass Roots* sought to encourage all its readers to share and generate a feeling of inclusion. The people featured on the covers—front and back usually—are the subjects of the articles within. They'll be the new back-to-the-land family, or the fellow who is building his own home out of recycled materials or someone showing off her organically grown fresh veggies. As one 'Feedback' letter put it: 'I realised that I'm not some kind of nut or freak [by being an environmentalist or questioning the corporate culture], that my way of life is shared by others.'[4]

Many articles, even on the most obscure topic, are more thoroughly researched than is usually found in mainstream news magazines. For example, a study of the herbs marjoram and oregano traces their history all the way back to the Greek legends; an article about chamomile is full of anecdotes, history and folklore including that during the American Revolutionary War, chamomile was called 'the Whigs' plant' because 'like the Whigs, it thrived on being trampled upon'; and an article about parsley is possibly the most in-depth and informative analysis of its kind available anywhere.[5]

Regular features down the years in *Grass Roots* include 'Down Home on the Farm', in which a range of contributors (often Meg or David) chronicle the delights—and hardships—of the bucolic life. Almost every issue has a page of book reviews, which invariably are taut and informative reports on publications dealing with all categories related to the environment and self-sufficient living; there is a kids' page with a wide variety of topics and things to do.

Although predominantly concerned with down-home matters, *Grass Roots* also tackles major global and national issues, including concern about the ozone layer, the destruction of forests, overgrazing, the enormous damage caused by seismic surveys, the growing need for ethical investments, and warnings regarding the bona fides of supposedly 'earth friendly' products. Most of the display advertisements are consistent with *Grass Roots* philosophies, including for such products as solar heating.

Issue No. 16 introduced 'Gumnut Gossip', in which Meg Miller detailed the behind-the-scenes activities in bringing out the magazine. This extraordinary sharing of the problems faced by the editorial staff, their feelings and the issues which concerned them would become a regular feature unique in its honesty and self-criticism. Often poignant, the column would evolve not only into a description of all the problems and issues involved in bringing out each *Grass Roots*, but also discussions about the evolution of the *Grass Roots* ethos. By issue No. 26, Meg tells not only about how the magazine was produced but also about what she sees as 'steps to a better world'. She seeks the involvement of every reader to make the world a better place: 'If we are to survive these troubled times, it seems imperative to simplify our needs, to minimise our consumer dependence by increasing our self-sufficiency to link up with and support our like-minded friends.'[6]

Meg Miller shares her innermost feelings of how the magazine is just one aspect of a total and thoroughly fulfilling commitment:

> *Grass Roots* seems like a flower—we've been through the bud-like stage these last few years and now it is opening out to show its centre. As we here in Victoria watch the economic fibre of the State break down . . . the essence of *Grass Roots* strikes a chord.

Even the magazine itself has a grassroots flavour in the manner it was produced in its early years, and 'Gumnut Gossip' describes each trial and tribulation. In issue No. 34, for example, Meg Miller explains why it was late. 'Drought,' Miller explains. 'Our [printing] equipment has become the victim of dust that is being whipped across the nearby paddocks . . . the typesetter, our pride and joy, is increasingly resorting to producing indecipherable hieroglyphics; on occasion, it has given up the ghost and refused to operate at all'.[7]

Over the years, several articles have raised issues which had not yet reached the public consciousness. Among these was a warning about genetically engineered food, which health expert Maurice Finkel also saw as a microcosm of the wider decay of the planet.[8]

By the 60th issue, in April 1987, Megg Miller (as she was by then spelling her name) tells of how *Grass Roots* had attained mainstream respectability and was no longer just a symbol of the counter-culture. She described *Grass Roots* as 'commercially viable' throughout 'mainstream society':

> A lot of ground has been covered [over 60 issues], in our personal lives and philosophies, in those of our readers and in society generally. *Grass Roots* started at a time when health foods, conservation, alternative energy and schooling and even handicrafts were associated with eccentric, radical thinking people who if they couldn't be dismissed as cranks represented a threat to conventional society. Much of what we take for granted now was unavailable then: wholemeal products, fruit juices, muesli, yogurt, compost, non-chemical insecticides, solar energy.[9]

At the same time, Megg Miller pointed out at various times that *Grass Roots* did not intend to idealise or romanticise the 'alternative' lifestyles—indeed, the magazine constantly stresses the hardships and pitfalls, and relates the problems faced by the *Grass Roots* correspondents, some of whom are quite savage about the pretensions of idealists and zealots.

In Issue No. 132, April/May 1999, *Grass Roots* was celebrating the end of its 25th year of publication, and in this issue we see the winning formula brought out once again. *Grass Roots* has changed very little in that time; it is the same mix of inspiration and practical tips for survival on the land—and a window on this lifestyle for armchair farmers.

There is a look at a big-picture environmental issue too in the two connected stories about tree planting, one by frequent contributor Jose Robinson of Wild Cattle Island in Queensland, who tells the story of how she helped reforest a bit of fire-damaged bush on her property. This is an empowering story about how Robinson brought a forest back to life, lovingly tending the small plants until they thrived. Robinson makes readers feel that every little bit helps, that every individual can make a difference to the earth.[10] The other 'tree' story is more scientific, the latest instalment of a continuing account about how the writer, Bev Lynch of Mt Barber in Western Australia, is attempting to create a working forest.[11]

The main article in issue No. 132 tells the story of a back-to-the-land family, the Hansons, who are also featured on its cover. This story sums up so much of what *Grass Roots* is about, and the Hansons themselves are very much a typical *Grass Roots* family. The story is full of inspiration—a 'we did it and so can you' type of pep-talk—but it is also full of sound advice about the pitfalls and the practical considerations that have to be faced in such a transition. The Hansons, a couple with three sons, were a city family, living in Sydney but dreaming of the bush. They bought a three-hectare block of land near Pomona, on Queensland's Sunshine Coast, and over three years turned it into 'a burgeoning food forest'. The story tells the truth about how hard this was—and how much planning and saving went into it—and never idealises even the 'finished product'. It is still hard to live away from the city sometimes; there are still snakes and spiders on the land and the work can still be backbreaking. But the Hansons believe that it is all worthwhile.[12]

This is very much the *Grass Roots* experience, both for those actually doing it and for those sharing their adventures from the sidelines. *Grass Roots* readers everywhere can identify with their struggles and triumphs—and wish them well.

A PUBLICATION with a similar approach, philosophies and emphases is *Earth Garden*, launched in 1972 by Keith and Irene Smith to promote 'natural' lifestyles and what they described as 'practical ideas and alternatives to the nine-to-five day'. Keith, a journalist, and Irene, a teacher, started *Earth Garden* during their honeymoon and pooled their expertise to produce what they described as 'Australia's first self-sufficiency magazine, promoting practical ecology, recycling and a sustainable lifestyle'.[13]

Like *Grass Roots*, the early issues of *Earth Garden*, which was produced out of the Smiths' home in the Sydney suburb of Epping, were a potpourri of environmental and lifestyle concerns with topics as diverse as Fraser Island at risk, a Sierra Club bulletin, and a strong anti-smoking stance which defined tobacco as a significant pollutant. *Earth Garden* also featured profiles of ecologically committed personalities not necessarily known as such throughout the wider community—for example, prominent artist Clifton Pugh is revealed as a protector of bushlands and native animals.[14]

Earth Garden's wide range of concerns was evident throughout the mid-1970s, with issue No. 13 (November 1975) given over to the bush: 'Essentially we feel Australians still have to learn to live with the bush, how to appreciate it and use it without the destruction that has gone on now for nearly 200 years.'[15]

The next issue, No. 14 (February 1976), warned about rainforest under threat at Terania Creek in northern New South Wales, and issue No. 15 (July 1976) concentrated on the benefits of herbs. Other special issues were devoted to energy alternatives, Aborigines and women.[16]

In late 1978, *Earth Garden* went into colour but retained a down-home touch as Australians from all walks of life, spread out across the continent, shared their experiences and philosophies.[17]

Concentrating on one specific vital concern continued into the early 1980s, with issue No. 32 (November 1981–January 1982) dedicated to trees. This issue collected a welter of personal experiences with exhortations to readers to plant trees, and contributor Charles Massy detailed the monumental problems of land degradation:

> Eucalypt dieback is a problem that is potentially one of the biggest natural, aesthetic and economic disasters to strike rural Australia and especially the tablelands in eastern Australia.
>
> In fact the traditional rural image of Australia of scattered gum trees and the odd wooded area in paddocks may soon be a thing of the past. But dieback is an issue in which *Earth Garden* readers could play an important role.[18]

Just like the 'Feedback' sections in *Grass Roots*, *Earth Garden* sought to instil a feeling of involvement among all its readers, to whom columns were opened up and opportunities presented to air their views and concerns. In fact, there is quite an overlap, with many of the *Grass Roots* names popping up in *Earth Garden*. Jose Robinson, of Wild Cattle Island in Queensland, is a regular in *Grass Roots* and she's had a few pieces in *Earth Garden* as well, including one with a photo of this woman we feel we know so well.[19] And as with *Grass Roots*, *Earth Garden*'s cover models are the subjects of the articles within its pages.

In early 1987, Irene and Keith Smith departed, with Camille and Alan Thomas taking on the dual roles of publisher and editor, producing the magazine from Moreland in Victoria. *Earth Garden* went from quarterly to bimonthly and for the first time, in issue No. 53 in February 1987, included advertising.

Keith and Irene Smith then began to concentrate on writing books such as *The Hard Times Handbook* (1984), *The New Hard Times Handbook* (1999) and *The Australian Self-Sufficiency Handbook (a survival guide for the 21st century)* (1992), the last co-edited by Alan Thomas. 'We're not fanatics about the environment but we try to do our bit,' Irene Smith said in a 1999 interview in which she detailed their lives and philosophies. 'We had solar hot water at Epping, and whenever we stay anywhere, the first thing we do is turn down the hot-water heater. Around the house, I avoid the use of any poisons or pollutants: I've always used sodium bicarbonate to clean the bathroom and kitchen.'[20]

Camille and Alan Thomas retained a keen environmental consciousness for *Earth Garden* and by mid-1988 the magazine was calling for a total reassessment of industrialised society. The letters pages and many articles warned of pending developments likely to harm the ecology and of problems such as dams and potential soil erosion. The Thomases were in the forefront of fighting against the Wesley Vale pulp mill in Tasmania and called for a new attitude towards recycling paper.

By the mid 1990s, Alan T. Gray had taken over as editor, but the mix of big picture and small picture stories had not altered. There were still stories about national and international issues—uranium, woodchipping, energy conservation—alongside tips on organic gardening, DIY crafts and self-sufficiency. In the December 1996–February 1997 issue, Gray himself wrote a piece entitled 'Woodchips, shares and senators' in which he discussed the activities of the then new senator Bob Brown in exposing possible conflicts of interest in the Federal upper house.

In the same piece, Gray expounded on the possible dangers inherent in the new industrial-relations legislation, which was prohibiting so-called secondary boycotts. Gray pointed out that this legislation would prevent green groups—or human-rights groups or any other interested party—from getting involved with industrial action, under penalty of huge fines. Gray wrote that this new law would be a threat to democracy and a serious impediment to protest of any kind.[21]

Earth Garden had clearly moved a long way from handy hints about composting and candle-making. A couple of issues later, Gray again wrote about woodchipping in a story about the blockade and the arrests of protesters at Goolengook, in the forests of Victoria's East Gippsland. Gray's description of the forest is both emotionally powerful and evocative of what he calls 'the government's war on nature':

> This is the Somme-like imagery which confronts you . . . Trees here are up to 600 years old; tree ferns are eight metres tall. The mountain grey gums reach 70 metres into the sky, and this is the home of endangered species like the sooty owl, the tiger quoll and the powerful owl.[22]

But Gray can explain the issues clearly and unemotionally too:

> Many people worry about timber workers' jobs. They don't realise that only 620 people are employed in all Australia's woodchip mills, that the native forest timber industry has shed 10,000 jobs in the past 10 years through mechanisation, and that far more jobs are being created in the forest-friendly plantation timber industry, which now provides 65 per cent of all our building timber.

Gray also takes his *Earth Garden* readers to the blockade at the Jabiluka uranium mine, where he and his wife went and were duly arrested. He writes about life on the blockade, both the inconveniences and difficulties and the fun and camaraderie, and expounds upon the issues at stake—such as the overall wrong-headedness of nuclear energy, the stubborn refusals to acknowledge Aboriginal rights to the land in question, and the high-handed bully-boy tactics used against the protesters. He points out that support for the mine is bipartisan, despite some half-hearted objections over the years from Labor, so that grassroots action is the only hope that sanity will prevail.[23]

Along with all this political activism and involvement, *Earth Garden* has never lost sight of its mandate, to provide advice and support for those trying to find a kinder, gentler path through the world. And often, simple stories of how others are faring on this road are the most inspiring.

Gray is the author, with Anne Hall, of 'Forest Friendly Building Timbers' released in 1998, a booklet to inform consumers about which building timbers are best to buy from the standpoint of conserving native forests. The booklet created a huge ruckus after it was published, and the timber industry sought to have it banned, a stoush which Gray describes in the pages of *Earth Garden*. In a story entitled 'The book the woodchippers wanted to ban . . . now it's a best seller', he writes of the pressure from the National Association of Forest Industries, especially as applied to the BBC Hardware retail chain which was persuaded to no longer stock it. Gray writes about the motives for writing the booklet:

> From now on, using this book, anyone who buys any timber can use their consumer power to become a forest campaigner. Rather than simply saying that forest destruction is bad, we're giving people positive suggestions for good alternative timber products. If consumers and councils take up the message of this book, we can work around the 'political logjam' we face in Canberra with the two old political parties.[24]

This philosophy is very much the heart and soul of *Earth Garden*. The magazine is not just saying that consumerism and rampant industrialisation are bad for human beings. It is providing a sane and sensible alternative to this lifestyle, and lending a hand to those who want to make changes, and to gain some real control over their lives.

AT THE other end of the magazine spectrum to *Grass Roots* and *Earth Garden* are several glossy publications. The glossiest of them all is *Simply Living*, founded in 1975 and which, for a generation, through several changes of ownership and emphasis, was a leading voice for conservation and alternative lifestyles. In its first two years *Simply Living*'s ecological credentials became well established, and it editorialised loudly against such things as the danger of nuclear facilities on Australian soil and the culture of government secrecy to keep information about such installations under strict control.

In a wide-ranging discussion in 1975, joint founding editor David Stewart points to worldwide research showing many illnesses directly linked to the nuclear industry.[25]

As a token of *Simply Living*'s broadly based concerns, in 1977 (Vol. 1 No. 4) it featured a celebration of whales, a warm-hearted appreciation of one of the planet's most intelligent and perceptive creatures. The issue also had articles on crafts, tai-chi, crystals, flower remedies and philosophies connected to the then prevailing counter-culture.[26]

The next issue (No. 5, 1977) continued the counter-culture emphasis with a call for the legalisation of marijuana. While strongly advocating a change in approach to pot, *Simply Living* at the same time documented the great harmfulness of tobacco smoking and called for a greater community recognition of non-smokers' rights. Such concerns were raised alongside more ecologically orthodox topics like the Antarctica treaty and better forms of land use. Simultaneously there were calls for a new public consciousness of ecological matters, such as the growing rate of animal extinctions.

Four years later, in 1981 (issue No. 17), *Simply Living* editorialised on how it saw its key role as a protector of the environment. 'We hope that *Simply Living* will continue to serve as a constant reminder of the Eden that is our planet and the desert it will become if we do not immediately alter our attitudes,' wrote associate editor Rosa Shiels. 'The earth needs to be nurtured as we nurture our children because the survival of the two is inextricably interwoven.'[27]

From about issue No. 18 in 1982, *Simply Living* for a time became cluttered, departmentalised and trivial, with an unclear distinction between ads, graphics and text—although there were occasional in-depth looks at important environmental issues such as tree planting programs.

In the following year, 1983, in what became Vol. 2 No. 1, *Simply Living* changed course again: Verna Simpson and Richard Jones took over as publishers with Robert Swan as editor. The new team also gave the magazine a deep philosophical underpinning, as part of a green movement which 'other community groups can no longer dismiss as a fringe, alternative or too idealistic'. *Simply Living* now sought to become a truly 'popular voice' and widened its concerns to include animal welfare.[28]

This transition issue included a plethora of concerns ranging from a suggested product boycott of French goods over nuclear testing in the South Pacific, many 'Green Pages' for networking, an in-depth interview with Bob Brown just after the Franklin River was saved (including stunning photographs of the Lower Franklin panorama, showing a rafter amid the rapids, and some of the area's glorious wildlife). True to its earlier concerns, this issue also included articles on Zen and astrology. Two issues later, as *Simply Living* expanded to a blockbusting 130 pages, the publishing team renamed itself Otter Publications and dealt with such issues as Pine Gap, pesticides, Antarctica and an even more extensive range of green networking.

Despite many noble aims, several irritants emerged during this phase of *Simply Living*. The year of publication is seen nowhere in the magazine; the book reviews and the book club news give no indication if a title is Australian, and an overall sloppiness is evident, as in the several attempts to correctly spell the name of the magazine's justice correspondent Jocelynne Scutt. Nevertheless, during this time *Simply Living* retained many high-profile contributions such as Les Murray's poetry and Tim Winton's fiction. About this time, Verna Simpson became managing editor, with Pip Wilson as editor.

By the mid-1980s, with the growing green awareness throughout the community, more and more departments were introduced throughout the magazine, including fourteen 'Green' pages with an 'Overview' which dealt with such issues as the politics of rainforests. Various interest groups submitted their own sections, such as the Fund for Animals, Amnesty International, the Total Environment Centre (which warned of expressways being built through national parks), People for Nuclear Disarmament, Animal Liberation (which railed against the live sheep export trade), Greenpeace (concentrating on saving whales and dolphins), Scientists Against Nuclear Arms (writing of the need for a code of ethics within their profession), the Wilderness Society (which detailed its campaign for a western Tasmanian national park), the Doctors Reform Society (detailing its services for the Aboriginal community and activities at its women's centres), the United Nations Association (describing the best ways to mount the most effective peace campaigns) and BUGA UP (on the fight against tobacco companies).

In 1987, Richard Jones wrote an article, 'Is the earth dying?', which reported with alarm on the ozone hole over Antarctica and the consequences for Australia:

> Even if the entire world were to stop using chlorofluorocarbons today (which is of course impossible, bearing in mind the enormous influence that multinational corporations have on governments) there is no way we can stop the millions of tonnes of CFCs already in the atmosphere from drifting upwards into the stratosphere.
>
> We will have to learn to adapt to a world suffering from much higher doses of ultraviolet radiation, warmer temperatures and higher sea levels—whether we like it or not.[29]

By the late 1980s, *Simply Living* joined other publications in recognising land degradation as 'the most fundamental environmental issue of our time'. Contributor Stephen Brouwer estimated that 'one-third of the earth's land surface is now threatened by perpetual desert' and he listed land degradation 'hot spots'. In another article, contributors Jolyon Burnett and Simon Leake described the problems caused by land degradation as 'so fundamental to all of us that it clearly can no longer be left in the hands of politicians and farmers alone'.[30]

At the start of the 1990s, logging and sandmining were targetted as major concerns, especially the fight to save Fraser Island.[31]

The high-quality production of *Simply Living* remained one of its greatest assets, and a full-page photograph (Vol. 6 No. 2) of a frill-necked lizard against a backdrop of lush greenery raised readers' consciousness of the glories of nature. During this period *Simply Living* continued much good work, including a study of decaying river systems, especially the Hawkesbury–Nepean in the Greater Sydney area, with expert comments on what needed to be done to reverse the trend.[32] Then, after issue No. 84 in 1996, *Simply Living* simply disappeared.

TARGETTING a similar audience—but widening its appeal to young, outdoor adventurers—is *Wild* (originally *Australian Wild*), which has environmentalist credentials well beyond surface appearances. *Wild* at first glance is of interest mainly to fit and daring types like mountain climbers, whitewater rafters, bushwalkers, skiers, and canoeists—but it has a pro-environment zeal well beyond what seems to be its mandate.

Wild's guiding light is Chris Baxter, who founded the glossy quarterly in 1981. Baxter has remained its managing editor and publisher for more than eighteen years and more than seventy issues. Early issues concentrated on wilderness, which remains *Wild*'s chief area of interest. Gradually, though, almost every ecological issue facing the nation has become the concern of *Wild*.

'*Wild* will recognise its duty to support the responsible preservation of distressingly scarce wild places,' pledged Baxter in his inaugural issue, and for almost two decades he would not waver from his keenness to spread the ecological message. 'It will recognise also the urgent need to ensure that Australia's unique national heritage is preserved intact for future generations.'[33]

With a keen appreciation of his ruggedly individualistic readership, Baxter similarly pledged that the magazine would avoid being 'obsessed with the politics of wilderness' and instead would be a forum for providing broadly based information, ideas and philosophies.

Beginning at 60 pages and expanding to an average 120 pages and beyond, *Wild* has featured top-quality photography; each issue contains in-depth features and regular departments, many with an emphasis on environmental matters. The ecological theme was evident even in the advertising pages, with the first issue including a full-page ad from the Tasmanian Wilderness Society warning about the Franklin River being at risk from flooding. The Franklin debate was also extensively canvassed in the editorial pages. The ad pages—which increasingly also included classifieds—featured hiking gear, clothing, maps, travel agents, walk organisers and so on.

Many issues also featured a '*Wild* Interview', which alternated between outdoor adventurers and environmental activists; the first issue provided a forum for the director of the Australian Conservation Foundation, Geoff Mosley, while issue No. 2 featured a celebrated rock climber. A later issue featured Bob Brown. The two strands of '*Wild* Interview' coincided in the Summer 1982 issue, with the subject being leading environmentalist Milo Dunphy, who discussed with great enthusiasm his bushwalking activities and feats of adventure, all of which combined to forge his ecological views.[34]

As the fight for the Franklin intensifed nationally, so did *Wild*'s concern—and its coverage of the controversy. In its Autumn 1982 issue, several feature articles highlighted what would be lost if the Tasmanian Government's dam proposals proceeded. In 'The Franklin: Threatened river', Bruce Heckinger describes rafting its wild waters, with spectacular photographs by Peter Dombrovskis (see Chapter Four). The words and illustrations of the article show the Franklin as a place of unique and unspoilt majesty, up to that point untouched by human destructiveness. Heckinger, on the spot, articulates the anger of the nation's environmental movement: 'The Franklin is still threatened by those who believe that its natural power should be harnessed. I vowed to investigate. The more I learned, the more absurd, unfair and obscene it all became.'[35]

For several years throughout the early 1980s, *Wild* examined environmental destruction throughout Australia, and its concerns spread well beyond the Franklin. In late 1982, Baxter lamented in an editorial the destruction being endured by the Howqua–Jameson Rivers area of north-eastern Victoria:

> . . . once magnificent, unspoiled river valleys have been ruined by major logging roads, visible for miles as unsightly scars snaking around the steep hillsides, opening up a wilderness to their destructive legacy.
>
> The logging appears to have been widespread and of a particularly callous nature. Not only have huge areas of magnificent natural forests been devastated but the usual trail of havoc that follows logging is more than usually conspicuous: fallen timber, tangled coppice growth, rutted side tracks, loading ramps, dug-outs and erosion-prone areas denuded of topsoil and vegetation.
>
> In the wake of this mayhem have come hordes of four-wheel-drive vehicles and trail bikes. Drivers of the former often seem to have the environmental sensitivity of a runaway tank—and leave similar scars.[36]

In a recurring theme, Baxter in his editorial accuses uncaring authorities, who regularly were oblivious to flagrant law-breaking, for many such problems.

Over the years *Wild* introduced departments including book reviews, readers' letters and '*Wild* Information'—the last-mentioned evolving into one of the most comprehensive round-ups of contemporary ecological concerns in Australia and beyond. Many of the snippets in '*Wild* Information' are by experts following a particular issue and are thus able to interpret and define each government decision. For example, in mid-1989 '*Wild* Information' gave early warning of the planned woodchip mill at Wesley Vale in Tasmania—a controversy which would galvanise the green movement nationally and reinvigorate green politics in the State. In the same issue were warnings of a planned development at Mt Kosciuszko; '*Wild* Information' examined a proposal being considered by a New South Wales Government ministerial committee and warned how expanded tourism would adversely affect a pristine alpine region.[37]

Later, 'Green Pages' would be yet another department which, State by State, would report on ecological matters of concern. These various departments, from issue to issue, would keep readers abreast of Federal and State politics and would assess the likely performance and examine the environmental credentials of each incoming government. In Autumn 1994, the 'Green Pages' reported on the devastating Sydney bushfires of the previous summer, as well as giving warnings about developer Keith Williams' plans for Hinchinbrook Island, complete with an action plan for concerned residents throughout North Queensland.[38]

'Green Pages' puts a strong focus on wilderness issues. In the Summer issue of 1997, for instance, there is a piece about the presence of snowmobiles in Victoria's Alpine National Park in contravention of regulations; there is a round-up of new national parks, in which it criticises the fact that horses have not been excluded from such fragile places as the Bicentennial Trail and that the 'vocal four-wheel-drive lobby' has had too many concessions made to it. It alerts readers to a threat of more mining in South-East Queensland's Glasshouse Mountains and informs them of a logging moratorium in Victoria's Wongongarra wilderness. *Wild* hands out some company brickbats as well, here singling out Boral and Amcor for their woodchipping activities. The 'Green Pages' end with an 'Action Box' with names and addresses of people to write to, products to boycott and suggestions of actions to take to fight against some of the threats reported.[39] One small criticism of the 'Action Box' is that *Wild* does not suggest alternative products to those it is boycotting; such information would be helpful.

In this issue, *Wild* reports that it has become a founding member of the Conservation Alliance, one of a handful of companies which have pledged to donate a portion of their annual sales revenue to 'carefully selected conservation projects'. The Australian Conservation Foundation's high-profile president, rock star Peter Garrett (see Chapter Seven), had been appointed trustee of this organisation, which was to be devoted largely to education projects as well as preservation.[40]

Although *Wild* is political and passionately concerned about environmental matters, the magazine is by no means all preaching. What it does best is to take adventurers—and would-be adventurers—to some of the most beautiful places in Australia and the wider world and just show and tell how wonderful they are.

The Summer 1997 issue is typical of the way *Wild* operates in that it is a nice mix of inspiration and information, all put together in a very readable, thought-provoking and attractive package. In this issue, for instance, there is a double article about Tasmania's Overland Track. The first part is a reprint of a piece first published in *Walkabout* magazine in 1952 and tells about a journey taken back then. It is a truly elegant piece of prose, written by John Bechervaise, a poet, a former teacher at Geelong Grammar School, a published author and an explorer of Antarctica. The second part, a breakout by Cathie Plowman, is an update of what the track is like today.

Bechervaise writes of a journey undertaken in winter, which he took then because even in 1952 the summers brought crowds to Cradle Mountain. Experienced outdoorsmen today might blanch at his paltry list of equipment for the trek, but his descriptions of the region and its wildlife—including seeing dozens of platypuses frolicking together—are stunning, and truly reveal for the reader the sights and smells and the very feel of the place as it was then. An example:

> We awoke incredulously to heavy snow. A new fantastically lovely world held

> bowed branches and gentle hummocks. From the sombre forest peered out old, bearded faces contemplating us in the great silence, black against the sky. Somewhere in the upper air the wind had veered; now instead of the grey, depressing rain, nature provided her most fragile handiwork—feather-fine dry crystals—to transform the jagged land. We were content to linger and all day the silence grew more intense as the little streams ceased flowing and froze beneath the snow.

The second part is the modern story, by Cathie Plowman. It is full of crisp, detailed information about what to expect, where to stay, and what to take to trek in this wonderful area. It has helpful graphics, too, including a 'Walk at a Glance' panel (a regular feature within *Wild* articles). There is useful information such as location of the walk, grade, degree of difficulty, type of scenery to be expected, and a 'special points' section, which warns of such things as the need for a campfire permit and the cost of park entry.[41]

Wild also has a mission to get as many people as possible out into the bush to have them see for themselves how wonderful it is and learn to appreciate it. Each issue features an 'Outdoor Skills' story, which covers such basic concerns as how to keep warm and dry when camping out, how to eat, what to wear, how to treat blisters, as well as reminders about the importance of maps and basic forest safety. And at least one of the suggested walks will be of 'moderate' difficulty, suitable for any reasonably fit person to attempt.

Each issue also features a thrilling adventure or two. In Summer 1997, there is an account of a kayak trip, island-hopping across the Bass Strait, perhaps the most treacherous stretch of water in the world. The writer, John Wilde, includes some moments of humour—such as when some yachties, themselves pretty rugged adventurers, tried to rescue him, and there are some lovely descriptions of wild islands, pristine beaches and jagged cliff-faces—and, of course, of the raging splendour of the Bass Strait. For this book's authors, who in 1999 had stood on the southern shore of Phillip Island, clinging to a fence to stay upright in the fierce wind and looking on the incredibly fearsome strait, this was white-knuckled reading material.[42]

But the adventure stories aren't just hard-man accounts of derring-do. They are informative, entertaining, and speak with authority and concern about the environment and also about political and human rights issues. In a story about the Annapurna circuit in Nepal, author Gregor Jordan writes of the intricate relations this nation has with its powerful neighbour China. Jordan explains the geology of the region, introduces the wildlife and laments the damage being done to its fragile landscape.

Wild is an adventure magazine—and can be read at that level—but it is much more than that. Its pages are filled with spiritual reflection about what wilderness means to us, about our response to it and our duties to conserve it, about the sacredness and glory of it, and about the meaning of beauty and solitude. In a piece about a trip to the Victorian Alps, Stephen Curtain writes what is almost a meditation about wilderness. In it he ponders: 'Had I come for a touch of soul-searching or to look for a bit of wilderness? Or both?'[43]

Regular contributor Quentin Chester, too, writes a lovely piece about encounters with wild animals and the almost spiritual significance this has for him. He writes that despite having spent so much of his life in the bush, his meetings with animals had been quite rare, and that the animals had always been seen as either a nuisance (they chewed on valuable equipment) or a threat (as falcons can be to rock-climbers) or as cute interlopers. But he writes how his perspective gradually turned around as he began to visit some of the truly wild places of this country. In what can be read as a bit of a sneak preview of his book *Australia's Wild Islands* (see Chapter Four), Chester writes about coming to a lonely spot in the Bass Strait, Albatross Island:

> It was impossible to see the island simply as an inspiring spectacle or some convenient backdrop for my experience. Whatever I might have wanted to think or feel, this island belonged to the birds, it was *their* home. At best, I was a privileged onlooker.[44]

Chester writes that wildlife had come to have an 'emphatic, humbling presence' for him, and that had given him a better understanding of humanity's place in the natural world.

In his editorials, Baxter often reflects on such matters. In the Winter 1997 issue, he writes, under the heading 'The *Wild* person, a reflection on the value of wilderness':

> I know I have not experienced anywhere else the sense of peace, of belonging, of awe and humility or the sense of perspective I find in the bush. These feelings have in no way diminished with age or with my experiences of wild places . . . I now know that I can't live without it and I find it difficult to see how *anyone* can live without it.[45]

In recent years, *Wild* has continued its battle against the forces of darkness, as often represented by government and big business. Baxter's editorials have taken on the Federal Government over such matters as logging and the Kyoto Agreement on greenhouse gases[46] and the Victorian Government over the treatment of national parks. In the Autumn 1997 issue, he lambasted the State Government over plans to develop Wilsons Promontory. The Government was considering such plans as a hotel, roads, and even a walkway to be cut through virgin forest. Baxter called on readers to speak out against the despoliation of this wilderness, and alerted them to what he considered the tricks and stratagems of government negotiators.[47]

Again in the Spring 1998 issue Baxter writes:

> At the depths of the Great Depression of the early 1930s Australians turned to bushwalking in numbers not seen again to this day. But it is sad that the natural world has never been more at risk than at present. At a time when our need for it is greater than ever we sit helplessly by, allowing it to be torn down and built over for short-term economic gain.[48]

Throughout the late 1990s, *Wild* has taken its readers to nearly every corner of Australia and to many places beyond Australia, including Ethiopia's Simien Mountains.[49] It has told us about wilderness under threat, such as Victoria's Wongungarra region, and it has continued to present such issues as logging, tourism, development, as well as Jabiluka and the whole uranium debate.[50] And it has written profiles of such outstanding conservationists as Bob Brown and Charles McCubbin.[51]

Wild is ever-vigilant to threats to the bush—and every issue confronts readers with its beauty, its fragility and its magnificence.

ANOTHER prominent and enduring glossy is *Geo*, founded in 1979 and which, although paying obvious homage to the American *National Geographic*, sought to capture the uniqueness of Australia and its Outback characters and characteristics. Its first issue also promised to reflect concerns for the environment with managing editor Cheryl Barnier warning of the potential destruction posed by developers in Sydney's historic The Rocks district.[52]

Under Barnier's helmsmanship, *Geo* did not deal directly with ecological issues, but its text and photos nevertheless exulted in descriptions of land, water, flora and fauna. In issue No. 2, Barnier warned: 'As technology allows us to reclaim and use many of our "useless" swamp areas, we are inadvertently destroying the environment in which hundreds of beautiful water birds live and breed.'[53]

Issue No. 2 also described the fragile state of Queensland's rainforests:

> When man pushes roads through rainforests, their fringes become prone to landslips, as torrential rain scours out soil . . .
>
> Queensland's rainforests are shrinking a little each year due to fire and bulldozing, although new growth is extending slowly in some undisturbed areas.[54]

This issue also details the delights awaiting tourists to such national treasures as the Great Barrier Reef, but warns of the need to control the number and the behaviour of the visitors. Similarly, *Geo* warns of the disastrous consequences if drilling were allowed on the Reef.

At this early stage *Geo* included very few advertisements, but among them were such large corporate entities as Shell and Qantas.

By issue No. 4 Barnier was replaced as boss by Alfredo Roces, who had the diminished title of 'editor'. Roces sought to broaden the intellectual horizons of the magazine, which henceforth would pledge itself to a more passionate and extensive study of ecological matters. In his first editorial, Roces outlines the directions that he hoped a far more serious and cerebral publication would take: '*Geo* offers more than nature study; just as passionately it seeks to chronicle Australia's culture, its people, rich history and current problems,' Roces declares grandly.[55]

Signalling a shift to what he perceived would become a far more intellectually challenging magazine, Roces continues: 'In future, *Geo* will look into today's Aboriginal community in Sydney, for example, and delve as well into Islam in South-East Asia.' And, in a particularly savage gibe at the editorial policies of his predecessor, he noted that 'in brief, while *Geo* delights in offering light pictorial confections, we are equally aware of the responsibility to provide food for thought'.

Above all, Roces pledges that the magazine would have an environmentalist conscience and he shared with his readers a longstanding passion:

> The basic philosophy concerning delicately balanced symbiotic inter-relationships that ecology preaches applies equally to men and nations. *Geo* will investigate the rich tapestry of life as it is lived, pell-mell, about us, from the deadly serious to the uproariously flippant, from the breathtakingly beautiful to the painfully harsh.

In his florid exultation of *Geo*'s ecological values—'the preciousness of each ticking second, of every microcosm of the universe, of cultural legacies handed down by past lives thousands of years ago'—Roces says the magazine is a 'celebration of life':

> The common denominator is a faith—that by learning with an open mind, by studying cause and effect, by mentally putting ourselves inside the cells of another living being, by a reverence for life in its diverse forms—we may save the whales and all endangered wildlife, nurture a valued heritage, and treat the natural and human environment as precious.

Several articles in this issue reflect Roces' concerns, such as warnings about hydro-electric developments and the threat to wildlife. Indeed, the following year (1980) began with a superb, comprehensive appreciation of whales. At the same time *Geo* was taken over by the magazine division of the Rigby publishing group, and this reinforced its ecological commitment. Later in the year the magazine's concerns widened and book reviews appeared.

By its third year (1981), *Geo* was such a viable concern that it attracted full-page ads from major corporate sponsors such as Visa Card and Travelodge, but Roces assured his readers that editorial policies would remain sacrosanct. The first issue of 1981 (Vol. 3 No. 1) highlights such contemporary environmental issues as the threat to Hunter Valley vineyards from the Kurri Kurri aluminium smelter and the growing exploitation of Western Australia's Pilbara region.

Roces remained at the helm—and *Geo* continued in this consistent vein—until late 1991, when the publication was taken over by Weldon Young Productions, which installed Michael Hohensee as the new editor. The announcement of Hohensee's arrival was curiously accompanied by the inaccurate description of Roces as 'founding editor' (Cheryl Barnier's role).

Hohensee's first issue largely continued in the Roces tradition and included an article, 'Beautiful but vanishing flora' (text: John Benson; photography: Esther Beaton), which showed how human interference was destroying plant life. This in-depth essay included a full-page photo showing naturalist Wynn Jones and a dead dwarf pine, described as 'a victim of polluted waterways, at Wentworth Falls (in the New South Wales Blue Mountains)'.[56]

In its 14th year (1992), *Geo* was again taken over, this time by Kenvale Holdings, which opted for an even more commercially oriented product—including a $20,000 photographic competition, advertised in the 'From the

Editor' column. Despite this—or possibly because of the expanded resources such commercialism entailed—the magazine attracted high-profile contributors like Tim Flannery (see Chapter Three), who wrote of the blue-eyed cuscus.

By 1994, *Geo* went from quarterly to bimonthly, partly as a result of its ballooning corporate support. As a consequence, in the following year (1995), *Geo*'s advertorial pages often were indistinguishable from articles, and the number of promotions multiplied. Many of the short articles in sections such as '*Geo* Talkabout' were often contributed by a variety of organisations such as the CSIRO, resulting in an inevitable upsurge in political correctness and a diminution of individuality.

The copy and general presentation of *Geo* are nearly always terrific. The text is generally informative and interesting and the photographs are always superb. In fact, reading through several years of the publication for this book took days longer than scheduled because there were so many articles demanding to be read carefully.

In the January–February issue of 1996 the magazine shows a nice mix of entertainment, hard science and gorgeous photographs. There is a very good piece, for instance, about the work of zoologist Jim Mitchell, whose mission in life is to rid North Queensland of feral pigs. The story is an informative, unsentimental and environmentally aware look at the problems that stem from an introduced species taking over an area—from soil erosion to stream pollution to competition with native wildlife.[57]

There is a piece on the emperor penguin in Antarctica—*Geo* writes a lot about penguins of one sort or another, mostly to focus on Antarctica, its climate changes and its huge importance to Australia and indeed the world. There is also a beautiful photo essay about the underwater life that has sprung up on and around an oil rig off the coast of Thailand. The story makes no value judgment about oil rigs or the intrusiveness of mankind's endless building and changing; it's more a tribute to life—to the sea's ability to soften and transmute everything that comes in contact with it. Contributor Michael Aw writes:

> The frame of steel beams was encrusted with orange sponges. White, green and red coral trees crisscrossed it in every direction, giving it the appearance of a three-dimensional labyrinth in virtual reality. Looking skyward from 30 metres below, the perspective was reminiscent of the Eiffel Tower . . .[58]

There are still various outdoor exploits for armchair adventurers to appreciate—again in the January–February 1996 issue, a canoe trip down a crocodile-infested river in Western Australia. *Geo* also sees as part of its mandate the writing about different cultures around Asia and the Pacific, with an emphasis on the offbeat or even the downright strange. In this issue, the long-necked women of Myanmar are featured in a story that is interesting despite glossing over the big-picture political issues in the former Burma. The article manages to avoid the hand-wringing hypocrisy that often happens when representatives of the developed world visit poor villagers.

Over the next few issues there were several very informative articles with an environmental message, such as 'Himalayan hemp' by Calvin Michaels. In this piece he tells the story of villagers in the Himalayan region who depend on their hemp crop as their life blood, using it for virtually all of life's necessities, including food in a pinch. Michaels tells the story of hemp in the West, including that US President George Washington allegedly smoked it to relieve pain from his ivory dentures. He makes the serious point, though, that hemp is a clean, green alternative to cotton and is being recognised as such around the world.

> Cotton is the single most polluting crop farmed by mankind today, requiring pesticides and chemical fertilisers to help it grow and chlorine and bleach to clean it. Yet hemp is a weed which can flourish in almost any climate with little external input.[59]

In drug-paranoid America, hemp is imported rather than grown, although major retailers such as The Gap and Neiman-Marcus are selling clothing and other goods made from hemp fibre.

At the beginning of 1997, *Geo*'s 'Comment' piece was entitled 'A cry for the wilderness' and pointed to the main story, a feature about the threatened natural environment of Hong Kong. The story, by Edward Stokes, an Australian who is a longtime resident of that city, tells readers about the wondrous 'green lungs' of Hong Kong, large tracts of wilderness which are little known to foreigners accustomed to thinking of Hong Kong as one great monument to commerce and development. He recounts that within sight of the great skyscrapers are steep and deeply forested mountains, wild islands that are scarcely inhabited and on which it is possible to walk all day without encountering a soul. He tells of the native animals and bird life, which include many species not found anywhere else in the world. But because this land is so close to the heart of the city, it is all under threat.

With the arrival of Peter Stirling in the September–October issue of 1997, first as publisher, then combining that role with editor, *Geo* once again committed itself to an environmentalist stance. In the first 'Comment' piece of his tenure, a change of style is noted. Previous 'Comment' pieces were just pointers to a particular story in the magazine and usually by the journalist who wrote it, but the first Stirling 'Comment', although it is unsigned, works more like a standard editorial, outlining the policy and attitude of the magazine in general and stating a theme for the particular issue. Headed 'Reversing decades of despoliation', it is about Australia's shabby history in caring for its natural environment.[60] It is for the most part a plug for the National Heritage Trust, a government initiative aimed at redressing the wrongs of the past, and as such it is a bit of government boosting. There is no suggestion, for instance, that the $30 million earmarked for national heritage projects might be a drop in the bucket. But it does show an understanding of the problem and a real commitment to improving our treatment of the land. This theme is taken up again and again in later 'Comments', which from then on were signed by Stirling.

In the November–December 1997 issue there are several environmental stories—one about saving endangered species, one about the ongoing battle for the Great Barrier Reef and other coral reefs under threat. The *Geo* 'Focus' story, a regular feature, is about eco-tourism. In that issue, the 'Talkabout pages' were expanded for what has become a regular *Geo* feature, 'This Australia', subtitled 'a round-up of environmental news, events and issues'. This feature continues to be interesting and informative, but it leans rather heavily on government and corporate handouts—and thus tends to be a compendium of good news. In the March–April 1999 issue, for instance, there is a short piece praising BHP for its financing of a University of Queensland study of the bridled nailtail wallaby, a rare animal though one place it is found is near BHP's Gregory coal mine. It also lauds the French car maker Peugeot for investing $18 million to plant trees in Brazil. Now these are worthwhile efforts and they should be recognised, but perhaps with a little less naïve enthusiasm.

In fact, if *Geo* has a fault it is this enthusiasm and an emphasis on good news. For example, in the August 1998 issue, there is an extraordinary piece about whales which, in its optimism, unfortunately misses the mark completely. The story, 'The return of the whale', celebrates the ever-increasing numbers of whales that can be seen in Australian waters now that hunting them has stopped.[61]

There is some fascinating information—about whale songs varying from place to place, about how young whales play with dolphins—all sorts of interesting tidbits, including the best places to watch whales and when. But the overall emphasis of the story is wrong—it ignores the fact that at the time of its writing whales were still under threat, Japan and Norway continued to hunt in defiance of international bans, and immense financial and diplomatic pressure was being exerted internationally—especially by Japan and Russia—to have restrictions on whaling relaxed further.

Stories such as this can do absolute harm to the cause of whales by suggesting that everything is OK now, that whales are back with a vengeance and no longer have to be protected. The story is a sharp and sad contrast with one done a few years earlier about dolphins, which was a truly thoughtful commentary about the human–dolphin interaction at places such as Monkey Mia, where the dolphins come in to be fed by hand. The story evaluated the benefits to humans (which it decided were minimal without supportive education as well) and the risks to dolphins and decided that the whole enterprise was ethically questionable.[62]

But *Geo* is basically a fine magazine and, in March 1999, it 'returned to its roots' and once again became a quarterly publication after five years as a bimonthly. What it does best—and has always done best—is to show us the wonders of our natural world through its glorious photographs and lively and informative articles.

ALMOST a travesty by comparison is *Australian Geographic*, whose role nevertheless cannot be ignored or diminished. In 1996, *Australian Geographic* celebrated its tenth year and a subscriber base of more than 200,000. In its first decade, the magazine largely revolved around its founder and publisher, millionaire entrepreneur Dick Smith.

From the first issue, which introduced a glossy appearance with superb colour photography and design, its environmental concerns were often drowned out by its commercial boosterism. Unlike most other magazines with an environmental thrust, *Australian Geographic* has always accepted major commercial advertising, with most issues boasting full-page colour ads for not only outdoors-related products like adventure and wilderness equipment, but also multinational companies like food chains, vehicle giants and oil companies. The ads also had to reflect *Australian Geographic*'s attitudes (many of them quite enlightened), whereby *Australian Geographic* declined to accept any promotions for cigarettes, alcohol or other products it deemed unhealthy.

The first issue, whose cover featured what would become the magazine's trademark platypus, was more than 120 pages and sold for $6.50 (later it would be for subscribers-only). The issue contained a pretentious listing of 'Our Aims', one of which was 'offering encouragement and financial support for research into Australia's flora, fauna and the environment'—but this fairly environmentally friendly goal was only one of nine listed objectives.[63]

In issue No. 2 was more self-congratulation, boosterism and mutual admiration—and a constant reminder that the magazine and its generous patron were sponsors of a range of expeditions and enterprises. Again there was stunning photography often accompanied by banal and simplistic text. The articles cover a range of interests such as birdwatching, life in the Pilbara and the revegetation of Queenstown in Tasmania, yet most miss the chance to forcefully analyse the relevant environmental issues.

One article in this issue, 'Sewerage: after the flush', is an exception and shows *Australian Geographic* at its best: the story is in-depth (14 pages), it combines personal experiences with voluminous information and illustrations, and deals with all aspects of how Sydney's waste is treated. The article, by *Australian Geographic* regular Paul Mann, is rambling and disjointed, but nevertheless powerfully reveals the important experiments at the Glenfield Veterinary Science Research Centre which found that digested sludge used as a fertiliser might enhance crops—a major change of mindset compared with simply pouring the waste out to sea. The article also alerts readers to the need to much more carefully dispose of cooking oils and detergents.[64]

The second issue also introduced book reviews of up-to-date releases (in this case Bob Brown's *Lake Pedder*), but in terms of quality or insight they pale beside the efforts of magazines like *Grass Roots* or *Bogong*.

By its eighth issue, at the end of 1987, the magazine claimed more than 100,000 subscribers. It remained a concoction of contradictions: on the one hand, seeking to tap into the environmental movement, on the other being insensitive to it—such as devoting its 'Tribute' pages to Len Beadell, who surveyed the South Australian wilderness areas where the Woomera rocket range would be built and where the Maralinga atomic blasts would take place.[65]

Although capable of lauding someone like Beadell as a pioneer, *Australian Geographic* also displayed attitudes far more enlightened and imaginative than the norm. For example, in an 'Opinion' piece at the start of 1988, a progressive economist, Ted Trainer of the University of New South Wales' School of Education, asks 'Is growth the solution—or the problem?' This is brave and thought-provoking writing which went against the manic zeal of boosterism which infected not just the pages of *Australian Geographic* but of much of the nation at the start of the frenzied Bicentennial year.

At a time when gaudiness and excess were the norm, Trainer calls for restraint and more frugal and humble values and for people to question their rampant materialism. 'Let's hope that we in the affluent countries have the selflessness to realise that we cannot continue to grow materialistically,' Trainer exhorts.

Calling for a 'zero-growth economy', Trainer urges readers to change their lifestyles to embrace 'personal growth, developing skills, restoring ecosystems and especially building a friendly and co-operative society'. He pursues the theme vigorously: 'a sustainable alternative society would have to be far less affluent and resource expensive, more simple, more self-sufficient and more co-operative . . . Although we would be producing and consuming much less, the quality of life would be far higher than it is now.'[66]

Meanwhile, in its tenth issue, in April–June 1988, *Australian Geographic* got into the Bicentennial swing well and truly by announcing the magazine had become subscription-only, with over 100,000 on the mailing list and aiming for 200,000.

Increasingly the magazine contained in-depth articles on weighty issues like uranium mining and whaling, but each was very carefully written so as not to offend any company associated with the activity being discussed, especially as such companies usually provided co-operation and information to the writer. Where other magazines may be guilty of political correctness on conservation issues, *Australian Geographic* veered towards an unshakeable corporate correctness.

Indeed, in the magazine's various opinion pages, *Australian Geographic* gave voice to one or other of its favourite hobby-horses, such as what it saw as the need to curb Australia's population growth. *Australian Geographic*'s fuzzy and simplistic views on extremely complex issues were evident in such articles as 'Why small is beautiful', by Rodney Spencer, a member of Australians Against Further Immigration, a political movement of questionable aims.[67]

By its fortieth issue, October–December 1995, *Australian Geographic* claimed 208,000 subscribers and announced it had been sold to the Fairfax media group (which in turn on-sold it in 1998). The editorials were still repeating the platitudes and simplicities of the first issue.

There are times when *Australian Geographic* seems to delight in portraying deliberately contrary notions to those of environmentalists, such as in issue No. 41, in which the 'Opinion' column comes out against World Heritage listings. Its author, a hired spokesman for the pastoral industry, calls for 'commonsense' to prevail, as though groups concerned about Australia's special places were silly. He dismisses the concerns of 'urban conservation groups' and declares that World Heritage listing is 'a blunt political instrument to get green votes'. And he laments the fate of 'impoverished' timber towns—as if conservation were solely to blame for the collapse of that precarious industry.[68]

Another such opinion piece, in issue No. 44, calls for the keeping of Australian wildlife as pets. The author, Dr Paul Hopwood, a senior lecturer in veterinary anatomy at the University of Sydney, makes some valid points, first among them that it is better to be a captive animal than an extinct one. But he fails to address the main issue—that the custom of keeping wild animals as pets soon leads to traffic in those animals and grave disruption to the species and to its surrounding environment. And wherever the practice has been allowed, surrender rates of native wildlife to humane societies are astronomical. Something that looks cute and cuddly as a young creature grows up to be a wild animal—one with needs that can't be met in a suburban backyard or a cage.[69]

There is often a 'Boy's Own adventure' feeling to the stories in *Australian Geographic*. An example of this can be found in issue No. 42 in a story called 'Battalions, boots and bulldust, the Army's new profile in the Top End'. This is a long story—it runs over 30 pages—and it's all about the Army's manoeuvres in the north. There are photo spreads of tanks bashing through the scrub, of more tanks churning up mud, of still more tanks fanning out through wide-open spaces, kicking up masses of dust and frightening every living thing for miles around—you can hear the roar of them just by looking at the pictures. The gee-whiz tone of the piece is highlighted by the breakout artwork, showing all the neat stuff each soldier is carrying in his pack. Now, we acknowledge that Australia has to be defended, and the Army has its job to do, but there are scenes here to freeze the heart of any lover of the bush, and nothing to impress anyone except perhaps a ten-year-old boy. It's especially sad that a nod is given to environmentalism—one of the things the Army says it's defending us from is 'diseased stock' that could harm our farm animals or even our wildlife if it were brought in by illegal immigrants.[70]

Even *Australian Geographic*'s science stories often exude a boyish enthusiasm rather than hard information or thoughtful commentary. In issue No. 54 (April–June 1999), Brian Boyle writes about the work of the Anglo–Australian Observatory at Siding Spring Mountain, on the edge of the Warrumbungle National Park in northern New South Wales. In a story reminiscent of the American *National Geographic*'s long-term love affair with space, particularly the works of NASA, the writer is content to merely extol man's reach for the stars without really telling us very much at all.[71]

In that same issue, Ken Eastwood writes about Victoria's Grampian mountains, which are a very popular national park within an easy drive of Melbourne. It is a story which both delights and troubles the heart of any concerned reader. The photo spreads are glorious panoramic displays that show the tremendous natural beauty of the region, its varied wildlife, flowers, craggy rocks and deep forests. It's a story that seems to have been put together with real love of the Grampians—and of nature generally. And there is some good and interesting information about the local history, including a frank recounting of the fate of its Aboriginal inhabitants. But the story is disappointing too. For instance, it mentions that the park has 800,000 visitors a year, but there is not a word about the perils of overuse. It is just the opposite, in fact, when after a discussion about timbering, only recently stopped in the region, the writer says:

> But during one afternoon visit to Mafeking in the southeast of the park, we saw how well the bush can recover. Late in 1900, 10,000 people, hungry for gold, descended on Mafeking and totally denuded the area. Nearly 100 years later, striding through thick stands of silver banksia, blackwood and brown stringybark while whitebrowed scrubwrens jumped among the bracken, I wouldn't have known.

And he quotes local resident Janet Witham, who is not a scientist of any sort, saying: 'I find that in the Grampians, unless something is constantly used or looked after, the bush takes over very quickly.'[72]

The message here is that care need not be taken because the bush pretty much looks after itself, and that '100 years' in recovery time—if indeed it has truly recovered—is glanced over very lightly.

In the Grampians story, one can see the best and the worst of *Australian Geographic*—and the best is so good that you wish the worst weren't really there. The photographs—and indeed the youthful enthusiasm—can act as a spark to give some people an interest in the outdoors, the types who wouldn't think about it much otherwise. And some of the articles are good enough, and inspiring enough, to set readers off in search of more knowledge or make them want to get out into the bush themselves. But the glossing over of problems, the 'she'll be right' attitude, is downright harmful because it suggests that everything is just fine, that nobody needs to change their ideas or their behaviour, that science will solve everything, that everyone can win. This complacency and kowtowing to commercial interests may well undo any good the magazine hopes to do.

REGIONAL publications with environmental themes have also emerged over the past thirty years, attracting a committed readership beyond their geographical boundaries. Among the best-known and most influential is *Bogong*, launched in Canberra in 1980, seven years after *Grass Roots*' first appearance. The early issues of *Bogong* appeared about once every six weeks, highlighting in its fourteen pages local environmental issues within the Australian Capital Territory and including a few other articles relating to national and international matters. This ambitious magazine, edited by Laurie Shane, was published by the Canberra and South-East Region Environment Centre, whose members contributed most of the articles.

Calling for 'greater public access to information about the environment', the first issue established a framework accessible to a wide cross-section of Canberra residents. From the start, the magazine had an 'alternative' feel to it, being cheaply produced on rough paper, with just a few black-and-white photos and no ads to speak of. As a reflection of its inclusiveness and social compassion, *Bogong* had a separate subscription rate for those without a wage, and for some of its life made a point of stipulating on the masthead against any racist or sexist language appearing in submitted articles. It also explained on its title page the choice of its name, telling readers that the Bogong moth was 'a special feasting food for the Aborigines in this area and is an important cultural link with the land'.

Topics covered in the first *Bogong* include the 'vanishing' rainforests of New South Wales, the dams debate in Tasmania (at a time when the flooding of the Franklin was a bleak possibility), the toxic threat of 2,4,5-T, a call for a nuclear-free Pacific (which would be equally potent 15 years later when French testing resumed), and how the patenting of seed varieties might have grave environmental consequences.[73]

Bogong was able to harness the expertise of Canberra residents who had arrived from all over Australia and who had brought their environmental concerns with them. The Tasmanian article was by Penny Le Couteur, a member of the Tasmanian Wilderness Society and president of the Canberra Bushwalking Club.

Like *Grass Roots*, *Bogong* is an outlet for networking and this began with the first issue, which included calendar entries for the Tasmanian Wilderness Society, the Canberra Bushwalking Club, Friends of the Earth, a just established Science for People group, Canberra Peacemakers and Action for Public Transport. The issue includes 'Environmental News' from near and far, and articles of primarily local interest like a report (by Ian Fraser, project officer at the Environment Centre and later *Bogong*'s editor) on the Goulburn National Parks Symposium which pitted 'greenies' against 'the rest'.

Issue No. 2 (May–June 1980) deals with some of the pressing environmental concerns of the time—such as the Kakadu National Park Plan of Management, Kosciuszko as a biosphere reserve, and a critique of the visit to Australia by Susan George, the internationally known writer who exposed how world debt was helping to destroy poorer nations—as well as local matters and some topics off the mainstream of green thinking.

In an article 'On the smoker as environmentalist', an unnamed reader suggests that tobacco and green consciousness are incompatible. *Bogong* would continue the theme of cigarettes as pollutants, also pointing out that tobacco cultivation destroys inordinate amounts of land.[74]

Among local issues is a report on how the Brindabella ranges surrounding Canberra contain a diverse range of ecosystems which need to be protected under New South Wales conservation and national park laws. Another article extols the pleasures of riding a bicycle around the ACT.[75]

The mix of international and parish-pump topics continues into the third issue with the range encompassing the Maralinga tests, 2,4,5-T in Vietnam, steps towards a nuclear-free Pacific, bushwalking and conservation, and the birds of Canberra's Mt Ainslie.

Maralinga is discussed by Ian Fraser, who describes the developments resulting from the McClelland Royal Commission into the British atomic testing as 'tortuous, multi-stranded and unfinished'. Although the article, 'Echoes from the thunder field', is a rather convoluted piece, it lists the litany of false assurances by Federal ministers since the 1950s—as well as by the Australian and British scientists who insisted all was safe. Fraser focuses on the revelation in Federal Parliament by Labor backbencher Tom Uren in 1978 of the deaths and dangers which could be linked to the atomic tests, and he assesses the most recent documents to emerge and the path-breaking reports by investigative journalist Brian Toohey.[76]

In the early 1980s, *Bogong* was attracting high-profile contributors such as Canberra poet and novelist Bob Brissenden (see Chapters Two and Seven) and the more far-flung Judith Wright (see Chapter Seven), who describes her wilderness experiences. A kids' page also evolved, while the range of concerns widened even more—including the ineffectiveness of Freedom of Information legislation and how tobacco smoking was becoming an epidemic in Australia—with nuclear issues and logging still among the core concerns.

By mid-1981, *Bogong* was among the first publications to reveal the pro-development views of the then much-loved naturalist Harry Butler, who until then had a cult status through his ABC-TV programs on Australian nature and wildlife. Butler would later become an apologist for the damming schemes in south-west Tasmania. In the same issue (Vol. 2 No. 2), *Bogong* presented a poignant description of how plantation forests could in no way be likened to rainforests, whose clearing destroyed habitats forever. In such articles *Bogong* also provided extensive referencing and suggestions for additional reading.

The next issue (September–October 1981) marked a series of changes, including the replacement of Laurie Shane by Ian Fraser, who had been a regular contributor. At the same time the magazine, which had grown to twenty pages, was now under an editorial board from which all decisions emerged. *Bogong* was also developing a persona and image with a homely touch centred on the fun illustrations of Helen Fitzgerald.

The first issue under the editorial collective featured extracts from the transcript of the Terania Creek logging inquiry (which had begun on 21 December 1979), headed by Justice Simon Isaacs. *Bogong* portrayed the inquiry as a farce, with witnesses badgered and experts not able to fully give evidence and opinions, often being restricted to 'yes' or 'no' answers.

By early 1983, *Bogong* was reflecting the political ferment of the time—especially with the Tasmanian dams issue reaching its climax with the election of the Hawke Government. *Bogong* took its readers to the corridors of power, to life on the barricades, and even presented the experience of going to jail for an environmental cause. An article by Skyda Hopkins, 'The blockade', is an on-the-spot description of the Franklin protesters in Hobart's Risdon Jail. The next issue features a commentary, by Tom Faunce, an associate of Justice Lionel Murphy, of the High Court decision which saved the Franklin from flooding.[77]

Partly as a result of its editorial-by-committee, *Bogong* by the late 1980s had lost much of its personal, friendly touch as it aired its internecine disputes in public, revealing the divergent views within the Canberra Environment Centre—especially on whether the centre should accept Bicentennial funding from a Government whose policies and philosophies were under fire from environmentalists.

Despite such discordant notes, *Bogong* continued to raise a kaleidoscope of concerns and each issue gave increased space to book and video reviews which directed readers to the definitive literature on important topics.

By the early 1990s, *Bogong* was still taking tough and uncompromising stands on the issues. It's interesting, for instance, to compare its 'Green Shorts' items with similar stories in glossier magazines. *Bogong* writes about government and corporate initiatives, but it is much more a watchdog than a publicist. In the Autumn 1991 issue, such matters as the environmental impact of tobacco, Australia's greenhouse gas record and the Federal Government's threat to drop endangered species legislation are discussed.[78]

There is always in *Bogong* an emphasis on the care of the bush and it is vigilant about any threats to its integrity, whether they come from invasive weeds, urban sprawl or government interference. In the early 1990s, *Bogong* ran numerous articles about feral plants—how to identify them and what can be done about them.[79]

The magazine has always been interested in the nuclear industry. In the Autumn issue of 1998, Moira Rowland wrote a piece entitled 'Nuclear power: the answer to greenhouse?' The writer, described as 'a disenchanted economist who has been worried about nuclear energy since 1945', perhaps not surprisingly decides that the answer is no. She backs up this stance with a lot of hard information about the risk–reward ratio of nuclear power—and does so in a witty and entertaining fashion. For instance, she says that the political process can help, and suggests we vote for 'the least appalling candidate'. She also mentions Maralinga and Chernobyl in the same breath as comparable disasters.[80]

In that issue, hints were aired, once again, of all not being well at *Bogong*. The issue ran a proper editorial, the first in years, by its new editor Stephanie Haygarth, and the editorial committee was replaced by a 'working group', although many of the names remained the same. In her editorial, Haygarth told readers that the magazine was under siege from economic rationalists and was about to become a victim of its insufficient bottom line. She listed the magazine's achievements in influencing the opinions of decision-makers on such issues as logging and plantation forestry, and asked all those 'lurkers in libraries' to subscribe to the magazine—and to get a friend to subscribe too.

The articles remained hard-hitting. Besides the nuclear story, there was a piece about Australia's dubious deal at the Kyoto conference on climate change. In it, prominent environmental economist, political adviser and author Clive Hamilton says that Australia 'committed fraud on the international community at Kyoto' and 'the world's negotiators have agreed to pay the worst polluter in the world (Australia) to continue to pollute at record levels'.[81]

The next issue, Winter 1998, a rare instance of government boosterism appeared in *Bogong*, a piece called 'Working for the environment'. In it, Graeme Gibson, a publicist for Green Corps, a federal initiative to mobilise unemployed people to work for the environment, touts the virtues of the program, lightly glossing over its many problems. There is also a related piece called 'Earth workers' which praises volunteerism and community action in cleaning up the environment. This is all very well—it is good to work together to clean things up and work for change—but the shift in tone from establishment critic to booster is ominous.

In that issue, the masthead reveals that someone had been put in charge of advertising for the first time, Ruth Pfanner, and in the next, Spring 1998, we see that she has been hard at work. There is a veritable explosion of advertising over the next few months—ads for gardening centres, for organic butchers, for eco-tours, for various guesthouses and mountain retreats and other kindred soul types of businesses. In this issue, there was also a survey of readers as the editor and her assistants showed they were trying to increase subscriptions by giving the people what they want.

The editorial in the Autumn 1999 issue reveals that the staff upheavals have continued. Various people are farewelled and others welcomed (with pictures and little biographical sketches) and it is all portrayed as very amicable, but the strain is beginning to show. It is clear that there are problems with funding—the editor speaks of grant applications—and *Bogong* is not entirely comfortable with having to pull its weight in the free market. *Bogong* suddenly seems dull, as though it is struggling to find a new voice. There is an article, for instance, about the work of Earth Sanctuaries, about putting dollar values on wildlife so that it can be proven that it is more economically sensible to keep kangaroos alive for tourists to look at than to kill them for burgers. This should be interesting—the subject is interesting—but the piece is written by an accountant so it isn't.[82]

Another piece airs the intellectually dubious issue of population control and another is a bit of whimsy about how one can survive happily without a car, but that too is dull and the points raised are all rather, um, pedestrian.[83]

This is not to suggest that *Bogong* has completely lost the plot. There have been several fine articles over the past year or so that continue in its hard-hitting, tough, no-nonsense tradition. In the Spring issue of 1998, there was a good piece about frogs and how they are disappearing from the Australian Alps, as they are from many places in the world. This piece, by two knowledgeable scientists involved with the problem, is *Bogong* at its best.[84]

On the whole, *Bogong* has been a wonderful, individualist voice for the environment over its nearly twenty years of life. It is to be hoped that it manages to find its feet and survive and thrive in the new marketplace reality without discarding its soul.

AN ENVIRONMENTAL organisation which projected itself onto a far broader national stage was the Wildlife Preservation Society of Queensland, whose leading lights—poet Judith Wright (the society's president) and naturalist Vincent Serventy (magazine editor)—were the guiding forces in the establishment in 1963 of *Wildlife in Australia*, a quarterly which broadened the 'scientific' and 'educational' interests of the environmentally conscious group.

The magazine's leadership was an impressive mix, with Wright, one of Australia's foremost poets, being one of the most respected conservation activists, and Serventy, a long-time president of the society and a prolific author, being a campaigner on such issues as land degradation and the destruction of native mammals.

After ten years of publication, *Wildlife in Australia* described its mandate as seeking to 'increase and disseminate knowledge of Australia's natural history, and . . . foster greater understanding of the urgent problems of conservation'. All articles in these early issues were solicited, and writers were unpaid.

The magazine's editorials usually were motherhood statements often sprinkled with classical quotations. The editorials were a mirror of their times, especially those during the Whitlam era and its aftermath. For example, *Wildlife in Australia* commented at the end of 1976 in an analysis of the Fraser Government's performance:

> . . . [1976] has not been a good year for the environment and our previous rapid forward progress has slowed to a crawl.
>
> Australian government action has been pruned on the plea of financial expediency and we can accept such a reason, although with reservations.
>
> Whether present financial expediency has become future environmental policy will be seen clearly in 1977 and it is our task to make sure that apathy on our part will not strengthen apathy at government level.[85]

Three months later, another editorial would again lament the diminution of environmental consciousness but recognised the Fraser Government's achievement in stopping sandmining on Fraser Island—an action described as a 'dramatic and heart-warming victory' which would 'encourage all of us'.[86]

Such a fighting spirit would solidify in the bigger struggle over the next few years to save the Franklin River, and other editorials warned of the mining and export of uranium.

Wildlife in Australia would continue to expand its range of concerns, and early in the next decade its editorials called for greater government funding to deal with such problems as the rampant disease of dieback. The extent of the dieback plague was dramatically explained in an editorial of March 1981: '[Dieback] is now spreading more rapidly and is possibly now starting to be responsible for more deaths of native trees than the axe, the chainsaw or tree-killing hormones'.[87]

In its twentieth year, in March 1983 (Vol. 20 No. 1), *Wildlife in Australia* adopted a new and larger format. Vincent Serventy retired, and the magazine was run by an editorial board with Don Henry (later to become the head of the Australian Conservation Foundation) as chairman and individual issues featuring guest editors. It sought a glossier and more professional approach. The more sophisticated coverage and greater awareness of specific ecological issues also resulted in a name change, to *Wildlife Australia*, to coincide with its twenty-first birthday, in June 1983.

In the summer of 1996, *Wildlife Australia* got a new lease on life when it joined forces with the Queensland Museum, becoming one of its regular productions. This move gave the magazine a comparatively huge cash injection which allowed it to go to full colour, with larger photos and more of them. And suddenly, it had a large pool of expert contributors at the ready. This new-found wealth allowed it a certain largesse, and it immediately instituted an annual photo competition, with total prize money of $3000. (A side benefit of this competition was that there were suddenly a lot of very good photographs pouring into the magazine's offices—photographs which were used in many subsequent issues.)

From that date, there were no more 'editorials' or even 'editors' with the magazine taking its direction from the board of the Wildlife Preservation Society of Queensland. Each magazine was prefaced by a message 'From the Director', meaning the director of the society, and the messages were nearly always about the magazine, rather than about issues—and often still begging for money.

An exception to this was the 'From the Director' column in Autumn of 1997 in which director Pat Comben pleads for commonsense to prevail about the lyssavirus and the equine morbillivirus, believed to be carried by some types of bats. These diseases were very much in the news at the time, and often reported in a sensational manner. Comben's comments point to an article in the magazine by three scientists from the Queensland Museum, all with many years of experience in researching and handling bats. The article is interesting and informative, and makes the point that more human beings die from bee stings and snake and spider bites each year than from anything to do with bats, and they express the strong hope that people will not start maligning—or harming—flying foxes, or bats in general, because of this disease outbreak.[88]

Despite many years of effort to widen the base of its appeal, most of *Wildlife Australia*'s readers seem to be from Queensland, judging from the letters to the editor, questions sent in to various experts—and even the entrants in the photo competitions. The magazine has contained articles about sights and sounds around Australia, including its regular feature, 'Nature Watch', a State by State roundup of what's happening in the natural world, what flowers are blooming, what birds are migrating where. It's a nice, interesting feature that is usually not very controversial.

In recent years, the magazine's focus has largely been on animals, as its name would suggest. There is a lot of 'backyard' stuff—how to help injured wildlife, what birds are coming to your neighbourhood. These are not 'environmental' stories per se, but they help to inculcate a respect and a love for the natural world. There are regular features, such as 'Wildlife Tales', in which readers write in with stories and anecdotes about their experiences with wildlife. One reader, for example, tells of watching a spider build its web, another tells about her visit to a town with a large frog population.[89]

Wildlife Australia gives the impression of a magazine that is trying hard to win hearts and minds under difficult circumstances. There is constant coming and going on its staff—Pat Comben was soon replaced in the top job by Adrian Caneris (in Winter 1998). The magazine takes regular surveys to see what people want to read; they institute new columns and new features all the time to try to create interest, but the quality of the magazine remains patchy. Some of the stories are just not very interesting, and some don't seem to have much point (such as the piece about building a bat sanctuary which bats haven't yet moved into).[90]

Many articles are poorly written, and some of them are of questionable accuracy, despite the long list of scientists involved with the magazine. For instance, Adrian Caneris and Jan Oliver have written an upbeat piece about cane toads which contains the first good news about that pest found anywhere. The article has all the usual gloom and doom—where it came from, how fast it's spreading—but it also tells us that crows and currawongs, and other creatures, have learned how to eat the cane toad without dying. (Simply turn it over and get at its soft underbelly—that's the secret.) Now this may have happened (some crows may be eating cane toads) but the story suggests that nature has triumphed and that cane toads won't be a problem much longer. This is untrue—or at least it's very much out of step with what everyone else is saying—and creates a false sense of security and hope. It also seems to express a 'she'll be right' philosophy, that nature heals itself, that mankind's actions are shrugged off. This is a dangerously mistaken doctrine, and it should never be spread about.[91]

Wildlife Australia has been a troubled publication for a long time. Former editor Anthony Brown, who headed up the magazine from 1989 to 1992, described his time there as 'always a struggle to keep going'. In a 1997 interview Brown, who arrived in Queensland from Ireland in 1984, said he had tried to give the magazine a national focus and wider appeal but money was always a problem. 'Colour photographs really stretched our little budget,' he said. 'We tried to do more serious work, too, to do the big stories, but people wanted the local angle and they wanted "furries" as we called them, stories about koalas and cute little animals, and we needed the support so that's what we gave them.'[92]

By 1999, *Wildlife Australia* was run by a committee and it showed. Each issue seems unplanned and prepared in haste—there are far too many typographical errors for a quality publication—and its philosophical underpinnings are unclear. It tries too hard to please its readers and there is often an air of confusion about where the magazine is heading. For example, in the Winter 1997 issue, a new column was announced to be called 'Animal Hospital'. When readers actually turned to the page they found 'Animal Clinic'—a small error but an annoying one. The piece was written by Dr Geoff Pye, from the Currumbin Sanctuary on the Gold Coast, and gave some sensible tips for dealing with injured wildlife. But the regular feature disappeared after one go. What happened? We never find out.[93]

AUSTRALIA has a rich variety of magazines with environmental themes—and something to appeal to every taste and mood—and they've survived against long odds. Putting out a magazine is expensive business, public taste is hard to measure and sometimes fickle; Australia's market is small, and advertisers are hard to woo to any publication that's not absolutely in the mainstream.

Many magazines fall by the wayside—new ones come out all the time but cannot make a go of it. Even a publication as apparently successful as *Simply Living* couldn't go on forever. But most of the magazines featured in these pages have been around for years. They have large, loyal subscriber bases, increasing sales and manage to turn a nice profit for their owners.

The lesson seems to be that Australians want to read about their environmental heritage; they want to learn about the wildlife, the history, the geography of their unusual land—and to learn to live lightly upon it. These magazines fulfil this desire—and they do it with elegance and style.

6

FILMS AND PLAYS
Scripting the landscape

MOST of our greatest writers for the screen and stage have addressed environmental issues. The significance of their contribution has been deftly defined by acclaimed film director and screenwriter Bill Bennett, who said movie-making would be 'flaccid and impotent if we lost our passion and our zeal to tell a story which we arrogantly believe the rest of the world would want to see'. Such zeal, according to Bennett, 'stems from an anger—a sense of defiance, a sense of outrage, a sense of injustice, a desire to right what we think is wrong'.[1]

Throughout the often illustrious history of the Australian cinema, our film-makers have felt outrage about the devastation of our ecology, and they have translated their feelings into every genre of cinema.

Prominent writer and reviewer Scott Murray has noticed a 'sensitivity' to environmental and other issues evolving in the Australian film industry over the past thirty years in contrast to the early local cinema, such as in the 1940s when the cutting down of trees was depicted as a 'heroic moment'. Even as late as the 1980s a film like *The Man from Snowy River* (1982) showed ecological disaster in a heroic and glorifying context.

Murray, a co-founder in 1973 and long-time editor of the influential *Cinema Papers* magazine, commented in 1999 that the environmental consciousness in full-length feature films was partly inspired by the many documentaries of the 1960s and 1970s which were a significant attempt to 'define the relationship between Australians and their landscape'.[2]

From the mid-1980s onwards, as urban blight became the focus of more and more films, issues such as development, corruption and community anger were increasingly examined, according to Murray. He cited the popular *Mad Max* series as ecological fantasies which showed the outer suburbs of Australia's cities—and beyond—as a horrid wasteland. '*Mad Max 2 (The Road Warrior)* succeeds almost flawlessly as a visceral, exciting and powerful fable,' Murray said. He described *Mad Max* director George Miller as 'one of the greatest directors alive, and one of the best storytellers alive', whose several films were all 'profoundly interesting'.[3]

Murray also pointed to the anti-Vietnam War protests of the late 1960s as a catalyst for the radicalisation of Australian movie-making, which increasingly tackled such themes as the British domination and then the 'McDonald's-isation' of Australia and what these were doing to our culture and quality of life. Murray described the *Mad Max* trilogy and *Crocodile*

Dundee as classics which infused an environmental message into their plots.[4]

At the same time the Australian cinema became increasingly concerned with realistic urban settings, according to the Head of the Department of Screen Studies at Flinders University, Michael Walsh, who commented in 1999 that concern for the environment had been adopted by several film genres over the past thirty years.[5]

Walsh, an expert in Australian cinema from its earliest silent days, described several categories like the 'New Talkie', in which category he placed films such as Susan Murphy Dermody's *Breathing Under Water*, made by an academic film theorist who broadens her cerebral vision to reach out to a wider audience and deal with myriad ecological problems; then there are 'Mockumentaries' like Mark Lewis' *Cane Toads*, which combines a mainstream documentary style with comedy and offbeat approaches, and in which the boundaries between fiction and non-fiction become fuzzy as serious environmental issues are canvassed; and then there are mainstream documentaries, a field in which Australian film-making has always been prominent, with some of our most prominent and internationally recognised directors like Bruce Beresford and Scott Hicks having been documentary-makers.

The role of theatre in forging people's consciousness on issues like the environment was well put by Louis Nowra in discussing his successful play *The Temple*, whose theme was the corporate boom-and-bust of the 1980s that ravaged and destroyed much that had been good about Australia. Nowra said in 1993 that theatre had the capacity to 'actually tell you how society is' and that it was 'the art form that is the most susceptible to fashion and what society is interested in'.[6]

Throughout his career, Nowra has expressed concern about where humanity is heading, and in 1982 he described global politics in the twentieth century as 'murderous', culminating in nuclear madness. '*Dr Strangelove* is no longer a funny film,' Nowra said. 'It is becoming real.'[7]

Stephen Sewell, like Nowra one of the most influential playwrights of recent decades, sees the theatre as an outlet where 'realities are generated and questioned' and as 'a nerve centre of controversy'. To Sewell, theatre is and must always be 'a significant contributor to the way we perceive the world and our own positions within it'.[8]

The environment has been one of several potent strands in Australian theatre since it was revitalised and energised by the new political consciousness which accompanied the election of the Whitlam Government in 1972. Leonard Radic, long-time reviewer for Melbourne's *Age* and one of the nation's leading theatre critics, said the best dramas of the last three decades of the millennium 'directly challenged Australian society'. Radic said in 1999 that the New Wave of Australian playwrights presented political dramas which reflected the 'important movements in society and the political changes'. Radic noted several plays in which the environment was used both in its own right or as a metaphor for other ills increasingly coming under scrutiny. He said theatre had to be seen as part of a larger movement among

Australians seeking change; Australian plays actively questioned and challenged institutions, organisations and government, and they were not afraid to tackle burning issues like uranium, logging, land development, mineral exploitation and conservation. Radic recalled how a 90,000-strong anti-Vietnam march led by Labor frontbencher Jim Cairns in the centre of Melbourne was part of 'an enormous body of protest' which would soon spill over into wider political discussions which would influence the theatre, which in turn would breed groundbreaking companies like the Australian Performing Group in Melbourne and the Nimrod in Sydney.[9]

In the process of completing a book discussing the major themes in Australian drama, Radic said that although environmental issues were not always a 'major preoccupation', landmark works of playwrights like Dorothy Hewett dealt with the issues of salination, bad farming practices and ruinous get-rich-quick schemes. Radic described Hewett's *The Fields of Heaven* as a 'classical, definitive environmental play' which examined the continuing fragility of an already battered land.

The New Wave in theatre began in the late 1960s as opposition to the Vietnam War radicalised Australian cultural life, according to Radic, who noted that the myths of the Outback were being gradually replaced in Australian drama by urban issues. Simultaneously, there was a growth in consciousness about Aboriginal issues and feminism that would significantly widen the areas of interest. 'Australian plays were beginning to show that there was a world outside Australia, and they began to reflect the migrant experience and there were plays about Asia,' Radic said.[10]

Leonard Radic, the *Age*'s principal drama critic for twenty years, and for twenty-eight years a regular leader-writer, said the Australian theatre had dealt with specific issues like uranium mining, as depicted in John Romeril's *The Radioactive Horror Show*. Radic cited the importance of Australian plays in questioning the Federal Government's uranium policy. 'These plays, a by-product of the Vietnam protest movement, challenged the prevailing old-fashioned views and helped to promote discussion,' Radic said.[11]

Another important strand in Australian theatre has been the idea of Europeans adapting to a new and alien landscape in a strange continent, according to Sydney drama expert John McCallum. 'Playwrights like Louis Nowra and Stephen Sewell have depicted the environment as part of the overall picture that makes up the society that they so carefully explore,' McCallum said. McCallum, the Sydney theatre reviewer for the *Australian* and the Head of the School of Theatre, Film and Dance at the University of New South Wales, said in 1999 that varying attitudes to environmental issues were reflected in the proliferation of new works by playwrights from different cultural backgrounds. 'Many of the characters in Australian plays are displaced—be they immigrants or Aborigines—and cultural, social and environmental challenges are examined,' he said.[12]

For thirty years, the severe ecological challenges being faced by Australians have been at the core of a robust film industry and an imaginative theatre.

AS EARLY as the first years of the 1970s, Australian films already were depicting a world where resource exploiters were prepared to destroy entire areas—such as, for example, the Great Barrier Reef—just to extract the mineral and oil riches there. One of the earliest films with a strong environmental theme from that era was the thriller *Avengers of the Reef* (1973), in which scriptwriter Anne Brooksbank deals with the ruthlessness of resource exploitation—and the ability of humans to destroy fellow humans and anything else around them for the sole purpose of making money. Among the issues is pollution on coral reefs attracting the crown of thorns starfish which is destroying vast offshore areas in North Queensland. Coral reefs in Fiji are also besieged.[13]

Born in Melbourne in 1943, Anne Brooksbank was among a group of influential writers who would seek to impart a different set of values to the generation of Australians who had elected a new reformist government in 1972. Brooksbank, who has written novels and scripts for well-known television programs, shares many of the environmental concerns of her husband Bob Ellis, a well-known and controversial writer.

Much of Brooksbank's writing is based on her belief that 'the forces of the marketplace are damaging to most people' and developers can ruin the lives and aspirations of ordinary citizens. She also rejects economic rationalism, which she believes puts the rights of powerful individuals ahead of communities whose hopes and dreams are squashed.[14]

Many more films would follow *Avengers of the Reef*, whose gripping plotline puts forth the proposition that paid killers stand ready to eliminate the defenders of the environment.

Such defenders of the environment are also under siege in another film of the mid-1970s, *Storm Boy* (1976), based on the children's novel by Colin Thiele (see Chapter Eight). Scriptwriter Sonia Borg has crafted a warm-hearted portrayal set in various pristine, remote locations of South Australia. The majesty of these natural surroundings is contrasted with the evil petty-mindlessness of their despoilers, typified in the barbarous act of shooting a pelican.

Ten-year-old Mike Kingley is the Storm Boy with a fascination for pelicans and other wild creatures. Mike lives a remote and lonely existence with his father Tom Kingley (Hideaway Tom) and his vision is broadened by exposure to the wise and wizened Aborigine Fingerbone Bill.

'You kill a pelican and the sky comes up with a storm,' Fingerbone warns the lad. This leads to the poignancy of Mike tending to a young pelican in his bathtub, despite Tom's insistence that the creature's chances of survival are poor.[15]

Before too long Mike is confronted by thoughtless hoons hell-bent on trampling on the environment and killing all bird-life within it. As Tom is forced to deal with this menace, his clash with the local ranger powerfully shows the difficulties the defenders of nature have in being heard:

Ranger: They say you fired a shot at them [the intruders].
Tom: What if I did?
Ranger: There's a law against guns in the Coorong.
Tom: And when's there going to be one against lunatics in buggies? After the Coorong's ruined? When they've churned up every blade of grass?
Ranger: You could've killed someone!
Tom: So could they![16]

Storm Boy deals with many potent issues such as the rights of Aborigines to inhabit their ancient lands, even though European authorities decree these to be wildlife reserves, and the right of Tom to bring up Mike in his own way.

Born in Vienna in 1931, Borg arrived with her family in Australia in 1961 after travelling the world. She has written about Australian themes and issues, usually through collaborations or by adapting the works of authors like Thiele and Frank Dalby Davison. In *Dusty* (1982), she adapted Davison's novel about the journey into adulthood of a dingo pup. Another highly acclaimed collaboration was with Aboriginal writer Hyllus Maris, with whom she wrote the SBS TV series *Women of the Sun* (1985).

In the adaptation of *Storm Boy*, Borg reflects author Thiele's insistence that the pelican, Mr Percival, die at the hands of unthinking humans. 'The pelican had to die—Thiele believed all of us, including children, have to be able to face death,' Borg recalled in 1999.[17]

Borg's involvement in projects with a strong environmental theme was evident as late as 1998 when she wrote the script for a made-for-TV animated feature, 'The Way of the Birds' (1996), by children's writer Meme McDonald. According to Borg, her script is 'strong on conservation' in McDonald's story about a girl who believes that she is a curlew and is involved in such adventures as changing a rubbish dump into a nesting place.

Since 1983, Borg has lived with her mother Ilse and brother Hans on a property in the Otway ranges about fifty kilometres from Colac. The family chose this area in their desire to get away from city life, and they agist dairy heifers as well as keep ten horses and many pets. Borg described the region as one of rolling hills and untouched forests, and told of the family being involved in efforts to preserve it from loggers and irresponsible tourist operators.[18]

Borg said one of her moments of greatest satisfaction was receiving a postcard from Greece informing her that the local pelican was called Mr Percival.

THE most popular and best-known Australian films, from as early as the 1970s, have canvassed ecological issues. The *Mad Max* series, for example, is a disturbing trilogy which examines an Australia of the future which is facing dislocation and collapse. The first *Mad Max*, scripted by James McCausland and its director George Miller, was released in 1979 and quickly gained cult status with its cinematically scintillating portrayal of a futuristic, wild and dangerous place where the deadly chaos of the roads mirrored the anarchy throughout society.[19]

In *Mad Max* (created by prominent production house Kennedy Miller Entertainment) society is in total collapse, everything has become a deadly game and all concepts of decency disappeared generations earlier. Life and everything that sustains it became cheap a long time ago. 'People don't believe in heroes any more,' says one of the denizens of the future world. In such a society it is no surprise that mad motorbike gangs have taken over the highways.

Ordinary citizens are at the mercy of these thugs, and mindless violence is everywhere. It's not merely the survival of the fittest; it's laissez-faire at its unfairest. 'Anything you say,' is the hapless response of a harassed citizen. 'Anything I say,' is the thug's callous response, 'what a wonderful philosophy'.[20]

This is a society where the authorities have the most sophisticated and high-tech communications and weapons, yet the police are powerless against bikie hordes. No edifice is more ramshackle than the Halls of Justice.

One of the policemen is Max, who despite some good human qualities is a product of his brutal times. Like his colleagues, he has become blasé to every horror. Max may be king on his bike—scanning a vast, flowing and spectacular landscape—but all he sees is trouble and ruin.

The sequel, *Mad Max 2 (The Road Warrior)*, released in 1981, has been widely recognised as a classic of its genre. It represents what reviewer Adrian Martin has described as 'one of the first "high concept" movies of the 1980s' which showed the frightening scenario of 'a savage, post-apocalyptic landscape'. Martin described *Mad Max 2* as 'one of the finest films ever made in Australia . . . a dizzy height of cinematic invention'.[21]

In *The Road Warrior*, officer Max finds himself amid what the script describes as 'flurries of dust and sand . . . through an eerie, barren land'.[22]

There is increasing rage in the midst of the battles on the highways. The environmental degradation is such that scorching oil engulfs the combatants. The audience becomes gradually aware that the barrenness is one of the results of a monumental energy crisis. The world of international commerce is crumbling and we are in a sort of hell on earth, as described by the film's narrator: 'For without fuel they were nothing. They had built a house of straw . . . And in this maelstrom of decay, ordinary men were battered and crushed . . . in this blighted place.'[23]

A typical scene describes 'smouldering wrecks and dead bodies [which] litter the ditch. The animals in the corral have been slaughtered'.[24]

When Max helps Nathan, bleeding and wounded, it is purely for ulterior motives:

> **Nathan:** Thank you. Thank you . . .
> **Max:** Let's get this straight. I'm doing it 'cos I need fuel.
> **Nathan:** As much as you want. Take me back . . . Don't let me die.[25]

Fuel is the principal—indeed, the only—currency of any real value in a wasteland where everyone is scraping merely to survive:

> **Humungus:** Give the gasoline and I'll spare your lives. Just walk away. I will give you safe passage in the wasteland . . . Walk away and there will be an end to the horror . . .[26]

Of *Mad Max 2*'s seminal influences, reviewer Adrian Martin wrote: 'The cultural legacy of *Mad Max 2*'s high concept was extraordinarily extensive and diverse. In cinema, it created a veritable 'post-apocalypse' genre and has been parodied, quoted and ripped-off in everything from teen movies to music videos.'[27]

The Road Warrior was co-written by Kennedy Miller stalwarts George Miller and Terry Hayes and by Brian Hannant, who infused the script with his many environmental concerns and beliefs. When a sequel to the original *Mad Max* was being conceived, Hannant saw a 'natural progression to the next step', an even more distant future when law and order had broken down even further in a post-apocalyptic world.

'In the original *Mad Max*, which was shot just outside Melbourne, law and order had already begun to break down in what was a road movie set in the Australian context,' Hannant recalled in a 1999 interview. 'As *Mad Max 2* begins, the audience becomes immediately aware that SOMETHING TERRIBLE has happened to the world of the not-too-distant future as shown in the first film.'[28]

Hannant told of how the popularity of *Mad Max* had created an 'enormous pressure' for a sequel, and Kennedy Miller principals Byron Kennedy and George Miller returned to Australia to do it. Hannant became involved in the fledgling project through Jenny Day (later his wife), who was in the hub of the Kennedy Miller activity as a production assistant, providing feedback on the shoot as well as post-production.

Mad Max 2 (The Road Warrior) began as a ten-page series of notes by the trio, who imagined a world whose population had scattered and polarised, where roaming bands of marauders held sway in a rapidly decaying and destroyed civilisation. During this early note-taking and brainstorming between Miller, Hayes and Hannant, the plot evolved of hard-pressed populations living near a pool of oil which was central to humanity's ability to survive. 'We tried to show that in the post-apocalyptic world humans were concerned only in running their vehicles and nothing else mattered, so fuel was the sole currency,' Hannant said. Short-term and selfish aims dominated this world in which nobody cared what they were doing. According to Hannant, the audience was made to feel that 'we shouldn't go that way'.

As the script evolved, the social problems engendered by such a world were portrayed in the film by the feral boy. 'The feral kid was written as a direct reminder to show the extent to which societies can descend,' Hannant said. 'Such feral kids are a reality in the Third World today—look at Rio, and look at the orphans in the Congo and in Kosovo right now. These kids are traumatised and are forced to live on the streets whose desperation makes for the sordid world shown in *Mad Max 2*, which I saw as a cautionary tale.'[29]

Hannant, a Brisbane-born Baby Boomer who in 1999 achieved more than thirty years as a writer/director in Australia's mercurial film industry, began his professional career at the old Commonwealth Film Unit (later Film Australia), where he worked for ten years. He made many documentaries and docu-dramas, one of which was *Three to Go*, in which he directed one of three short films about youth (another of the young directors on that project was Peter Weir). Hannant also made a tele-movie, *Flashpoint* (1973), in which his environmental concerns first surfaced in the story about mateship on the iron-ore fields of Western Australia, filmed on location at Mt Newman.

After the Film Unit, Hannant went freelance, producing documentaries and docu-dramas in South Australia, and then he went to Sydney and got involved with *Mad Max 2*. Five years after *Max*, Hannant made *Time Guardian* (1986), a sci-fi love story about a time traveller in the Outback. Since then he has been writing scripts and developing projects, as well as teaching screen writing at the Australian Film, Television and Radio School.

Hannant recalled how in the making of *Mad Max 2* and many of his other films he was concerned about ecological issues such as the world energy crisis. 'During the oil shocks of the early 1970s, I believed that the car I owned would be my last,' he said. 'And everybody had to start thinking about how an energy shortage would create havoc in their lives.'

Only a major shock or a jolt to complacency (such as he hoped *Mad Max 2* had achieved) would wake people up to the need to rethink environmental issues, according to Hannant. 'Arseholes run the world,' he said, 'and they will only change their ways if they are given a huge shock.' Hannant said Australian governments needed to provide rebates to home-owners who put solar panels on their roofs, and he branded the Federal Government's plans to give farmers a diesel fuel rebate as a lunacy which would 'put even more diesel fumes into the air' despite warnings from the Australian Medical Association of the ominous health risks. 'All the logic is against current official environmental policy, but the energy companies are too powerful and the governments too frightened,' he lamented.[30]

On a philosophical level, Hannant described himself as a disciple of the Gaia Hypothesis, which regards the earth as a living and sentient organism. People could ignore the demands of Gaia at their own peril: 'If we pump too much shit into the atmosphere, the earth will turn against us in its bid to survive the onslaught; the earth will squeeze us.' Hannant described the pivotal British TV series 'Edge of Darkness' (1986) as having a 'dramatic effect' on his later thinking.[31]

The co-author of *Mad Max 2* said he regularly contemplated the catastrophic consequences of such things as rampant development and the future of cities. He described the 'gridlock' in the major conurbations throughout the world and said this could not continue. 'People will be forced to gather in localised regions where they will need to become self-sufficient, including in energy,' he said. 'This is likely to happen in what are now deserts, which will be the only places left.'[32]

As a result of such global trends, Australia is heading in a 'terrible' direction, according to Hannant, who stressed that in *Mad Max 2* the main aim was 'to tell a good, exciting story' but that the 'context' took over. He believes that its portrayal of mayhem and the fifteen minutes of chase scenes have never been bettered in world cinema. Pointing to the 'Italian rip-off merchants' who had duplicated many of the scenes created by Miller, Hayes and himself, Hannant said world audiences were able to relate to bleak and forbidding visions.[33]

Well beyond the apocalypse of *The Road Warrior* came the windswept, bleak and barren landscape of the third of the *Mad Max* series, *Beyond Thunderdome* (1985), which appropriately began with a fire in the midst of a wasteland. Max at first looks like a Bedouin, trying to somehow survive on the parched earth, under a blistering sun.[34]

By now a man with no name or identity, Max enters Bartertown, a scene both futuristic and resembling a Middle Eastern bazaar. Long-haired and scrofulous, he is someone barely subsisting among armed authorities who resemble Roman centurions. Surrounding him are enslaved gangs, rigidly controlled by an utterly brutal society. On sale at the flea market are precious commodities like water.

This horrid aggregation of human forms is being sustained by the efforts of even more frightful under-classes, trapped in an Under World whose main activity is to breed the pigs which produce the methane gas on which the entire stinking mess survives. This horrid lumpenproletariat is under the thrall of Master Blaster, a hideous and vicious being whose power among the low-lifes can enable him to place embargoes on the pig shit without which the 'upper' stratum of society cannot survive. By such means of blackmail and total control over the depleted energy sources, Master Blaster can terrorise the known world.

As Max is flung into this horrific environment, he is pitted into gladiatorial battle at the heartless venue designed for it: Thunderdome. At Thunderdome the fights are literally and definitively to the death and, as the master of ceremonies so delicately tells the combatants: 'I know you won't break the rules—there aren't any.'

If all this is so utterly repugnant, imagine the gulag into which Max is flung for transgressing the non-rules of Thunderdome. In this feral hinterland, mankind's technological thrust into the future has blasted it back into pre-history.

Between the unyielding landscape and the brutal society beyond, there is no hope for the ferals. 'If the earth don't get you, Bartertown will,' is the awful message.

But, even amid the worst horrors, there is a grain of hope. As Max mobilises the ferals he gives just a hint that their shred of decency will prevail over the divided and increasingly brutalised monsters of Bartertown.

THE apocalyptic mood of the *Mad Max* series is also evident in *Ground Zero* (1987), a collaboration by scriptwriters Jan Sardi and Mac Gudgeon. The film was inspired by the McClelland Royal Commission into British nuclear testing in Australia during the 1950s. The earliest visions of *Ground Zero* are of a degraded physical environment (a sparse desert) and a degraded cultural environment (the world's largest hot-dog accompanied by pneumatic hot-dogettes). In this decaying panorama—and in almost every indoor shot from a home to a TV studio—the shadow of Maralinga stalks.

Ground Zero's hero is a cameraman, Harvey Denton, whose apartment has large spotted rats, because of which he arrives home with six large rat-traps as a routine addition to his groceries. Denton, although a man of some substance, is consumed by the chaos and clutter around him.

Amid a growing mood of cover-up and paranoia, Harvey finds himself at ASIO headquarters, where he learns that his father, who had filmed the A-tests, might not have drowned as had been believed all these years but had died from a bullet in the head.

Two jogging ASIO operatives confirm the British attitude to Australia during the Menzies era when the tests were approved and conducted—and it becomes clear that things had not changed all that much during the three decades that followed. Trebilcock, a senior ASIO agent with whom Harvey makes contact, tells how the British thought of Australians as merely 'colonial rabble', an attitude which guided all their actions:

> **Tepper** (*a second ASIO agent*): But they're a friendly power . . .
> **Trebilcock:** Friends! The same friends who honoured us by testing their bombs on us?![35]

Harvey also tracks down a surveyor, Prosper Gaffney, who dramatically describes the horrendous scene at Maralinga: 'we saw the first one blow. It was like the earth cracked open and we'd released the fires of Hell. Smashed the bush. A million years of nature's work gone in a moment. Not a living thing for miles.'[36]

Describing 'that great black cloud' which hung above the test site, Gaffney tells of how the scientists callously and imperiously 'cheered and danced a jig, like the Devil himself'. Gaffney then sadly recounts the horrible legacy: 'One joyous flash and ya spend the rest of your life payin' for it . . . both now and in the hereafter and we'll suffer . . . for eternity.'[37]

Screenwriters Jan Sardi and Mac Gudgeon suffuse their film with the radicalism attained in their young adult years during the tumultuous late 1960s. Born in Melbourne in 1953, Sardi grew up in inner-city suburbia where he played in a rock band, was interested in acting and was a movie buff. Gudgeon, born in the New South Wales steel city of Wollongong in 1949, was a draft resister during the late 1960s, working under an assumed name on the waterfront as a member of the Painters and Dockers Union, one of the few organisations to provide employment to those actively opposed to conscription for Vietnam.

In the late 1970s, Sardi taught drama and English at a Catholic college, a cross-cultural experience which inspired his film *Moving Out* (1982), which he described in 1999 as one of the first Australian films to 'look at the world with non-Anglo-Saxon eyes'.[38]

When the idea for *Ground Zero* began to germinate, Sardi was seeking to express his concerns about several contemporary issues like US bases on Australian soil and the whole problem of atomic weapons and uranium mining—all expressed within the context of a 'paranoia thriller', a genre popular at the time. 'After reading the newspapers and magazines about what was happening in the world, I wanted to write something of a political nature,' Sardi said in 1999. As he and Gudgeon began work on the script, the evidence from the McClelland Commission began to mount and they 'sought to collate all the stories about the outrageous happenings at Maralinga'.[39]

Sardi's curiosity was drawn to one of the planes involved in the weapons testing during the 1950s being buried at Maralinga because it could not be decontaminated. 'What a way to begin a movie, we thought,' Sardi recalled. 'A plane is dug up from under the earth after 30 years; this had possibilities for an entertaining, page-turning thriller. What if there's a body in the plane, and if there is a body whose body is it, and is there anybody alive today who's related to the dead pilot? Our imaginations were also running wild with news that was hot off the press. We were arriving at a script that we hoped would be a little subversive, and relevant to Australia at that time. We badly wanted to create something important and relevant—not just popcorn escapism. We felt it was very important to convey our strong feelings that Australians had been deceived in the 1950s about what was happening at Maralinga, just as they were being deceived in the 1980s about the role of American bases on our soil.'[40]

The ability to tap into feelings and issues to which Australians can relate and empathise was evident in *Ground Zero*, as it was in Sardi's script for the phenomenally successful film *Shine* (1996), which won several Academy Awards. Sardi said *Shine* became an international success because of its 'universality in examining family relationships and how people cope with life and adversity'. Such ingredients are also found in other Sardi film scripts like *Secrets* (1991), about a group of Australian teenagers in the 1960s off to see the Beatles during their tour Down Under, and his various television works such as the series 'Embassy' and 'Halifax, fp', in which he sought to explore relevant contemporary themes.

In 1999, living in suburban Melbourne and still flushed with his *Shine* triumph and working on ambitious new projects (including a script about Dan Eldon, an American photographer killed at age 22 while covering the troubles in famine-hit Somalia), Sardi told of being concerned by many environmental issues, such as uranium mining and how the demands of big business still override community priorities. He said he looked back upon *Ground Zero* as a good entertainment which had important messages to convey.[41]

Mac Gudgeon had been a videotape editor in the USA before returning to Australia upon hearing that there was a renaissance in the local film industry. His experiences on the docks and his concern for social justice led to writing the script about the 1928 wharfies' strike, a story which became the TV mini-series 'Waterfront'. Then, sharing Sardi's concerns about the nuclear issues and his love of the political thriller, their collaboration on *Ground Zero* proceeded organically.

In 1999, living in rural Victoria, Gudgeon told of being 'constantly concerned' about the degradation of the environment and he was 'horrified' by the Howard Government's treaty on greenhouse gas emissions. Gudgeon said he was equally alarmed by genetic engineering of crops, especially at how 'multinational companies are buying up and patenting seeds, and conducting experiments on procedures about which nobody yet knows the consequences'. Gudgeon said he was concerned at any trend which reduced the diversity of species. Asked if he was in any way optimistic that the current trends would in due course be overtaken by more enlightened attitudes, Gudgeon replied: 'I can't imagine the next generation being any thicker.'[42]

In *Ground Zero*, Jan Sardi and Mac Gudgeon show how violence and murder prevailed at Maralinga and how there remain violence and murderous consequences for anyone digging too deeply into what happened during Australia's years of infamy.

THE unspeakable and unfathomable horrors unleashed by the Vietnam War and the shocking aftermath of chemical poisoning are described in Bill Bennett's *A Street to Die* (1985). This film is an angry yet poignant look at one dying man's raised consciousness about the deadly injustices done to him by his own government.

A Street to Die begins as a rash on Vietnam veteran Col Turner's arm is getting worse—a legacy of exposure to chemicals in that war zone in the late 1960s. At the same time Col gradually discovers that the Australian authorities don't care about how its soldiers have been disfigured, dismembered or disoriented. The Australian Government refuses to concede that chemical spraying in Vietnam was the cause of Col's cancers, and automaton-like bureaucrats dump vets like Col into a neighbourhood of service homes on bleak outer-suburban streets to die. The themes are explored with fine-tuned precision by Bennett, whose earliest film-making was in documentaries.

For Col, the maladies from chemical exposure never quite go away, and the symptoms multiply—his rashes become infected, with increasing regularity he falls asleep on the job, and his mental stability becomes more and more questionable.

The children of Vietnam vets are also affected. The film shows this in all its enormity, as well as the cover-ups and denials by successive Australian governments. The governments would never concede the shameful reality which Col observes in one household after another in his blighted street:

> **Col:** Number one has miscarriages, number three bashes his wife, Craig next door in number five has skin rashes and his boy's got something wrong with his brain, Tom in number nine's a loonie . . .[43]

Col himself is about as severely affected as anybody in his street. A dispassionate medical carer explains the harsh realities to Col's wife:

> **Clinic physician:** Mrs Turner, your husband has either leukemia, lymphoma or aplastic anaemia—we don't know which yet. We're going to have to take some more tests.[44]

Col is becoming increasingly outraged about the effects of chemicals on society, and he is especially appalled when a gardener takes no heed when told of the dangers of spraying 2,4,5-T in school grounds. Partly as an act of denial in refusing to believe that his Vietnam exposures will eventually kill him and partly through an effort to learn as much as possible about the deadly defoliants, Col has become energised and radicalised into fighting against the madness.

The greater his concern grows, the greater seems the stupidity of society's attitude to chemicals. When he phones the Education Department to complain about the gardener's spraying, his frustration is complete:

> **Col:** The stuff is poison. It causes cancer. Kids are playing in it. Isn't that enough to make you worried?
> *(Col listens for a moment, then hangs up in disgust.)*
> **Col:** No, of course not.[45]

As the Turners realise the inevitability of Col's fate, they mobilise all their resources to remedy the injustices:

> **Col:** It stinks. Why isn't somebody doing something?
> **Lorraine:** Because they're scared. That's why a lot of people wouldn't fill out those forms—'cos they thought it might get back to their boss or something.
> **Col:** I reckon it stinks. Blokes go off to fight a war for their country then their country turns around and kicks 'em in the balls.
> **Lorraine:** Yeah, but you imagine if the Government owned up—they'd have to pay out billions. No, they're not going to do the right thing.[46]

Such attitudes make it painfully clear that any government which could so heartlessly ignore the welfare of its citizens would not be interested in environmental justice anywhere.

Meanwhile, Col's continuing agony and physical deterioration are heart-rending, and *A Street to Die* encapsulates all the tragedies of how exploitative governments can allow their citizens to die in degradation and neglect.

A Street to Die reflected a range of concerns for Bennett, who was brought up in Brisbane. He has directed in Hollywood but has been quoted as saying that he prefers making modest productions in Australia reflecting his concerns rather than American blockbusters. *A Street to Die* is the work of a scriptwriter whose attention to detail stems from years as a TV current-affairs journalist, who while reporting in the early 1970s on the Bjelke-Petersen Government's ban on street marches was himself arrested and hauled away in a paddy wagon.[47]

A feature writer for the *Sydney Morning Herald*, Sally Loane, in 1993 described Bennett as a film-maker who 'has refused to make "bimbo" films or compromise the integrity of his projects'. Loane continued: 'He rejects being dubbed a social reformer, but aims to leave his audiences both entertained and pondering the issues—including tolerance, justice, racism—which he's raised in his various films.'[48]

ORDINARY citizens fighting to preserve their patch are the heroes of several powerful films which chronicle how only total vigilance can prevent the destruction of a neighbourhood. Prominent among them was *The Killing of Angel Street* (1980), whose first frame shows slogan-shouting demonstrators confronting developers about to move in. Their banners read 'People Before Profits', 'Hands Off Our Homes', 'Don't Steal Our Street', 'The People Are Winning' and so on.[49]

The script immediately gets to the heart of the matter as the film's principal hero is introduced: 'Angel Street does not exist as Jessica Simmonds knew it. Only two houses remain standing, whilst a third, between the two, is half-demolished . . . all this near her old home.' The scene of a neighbourhood in the final throes of total demolition is quite frightening and menacing, as it always is when the agents of destruction are circling. The script describes 'sinister men, loitering for no apparent reason, the demolition machinery and the demolition crews, and the odd pattern of demolition in the street with most gone or going and Simmonds' house standing alone in its garden debris'.[50]

In several powerful scenes, those things which make a home are seen being carted away from Angel Street. The demolition crews and the security thugs are removing every vestige of homeliness and domesticity. As a truck carts away a load of lace ironwork and debris, Jessica's father, Professor B.C. Simmonds, a classical and crusty radical, observes: 'See? Another load of our architectural heritage off to become razor blades.' But Simmonds won't allow such devastation and dispossession to continue without a struggle, and he is oblivious to the obvious dangers. 'Nonsense,' he replies to Jessica, who increasingly fears for his safety. 'I'm only exercising my democratic rights. I will not allow my home to become a concrete ant-hill 22 storeys high, my street turned into a giant cash register, and this beautiful city of Sydney . . .'[51]

It becomes increasingly clear that only the likes of B.C. have the inner strength, resolve and wherewithal to fight such seemingly overwhelming odds:

Jessica: You know you're going to lose. Cities grow and change.
Simmonds: Men make cities.
Jessica: Angel Street isn't one of the wonders of the world.
Simmonds: We could argue about that too. Don't you think it's worth preserving?
Jessica *(after a pause)*: Yes.
Simmonds: Then I'm ashamed of you. First you want me to quit so I won't get hurt, and now because I'm going to lose. It's not the winning or the losing, girl, it's the struggle.[52]

During this discourse, and indeed throughout every frame, a feeling is conveyed of menace and a sense of foreboding. We are reminded of Juanita Nielsen, who was killed in similar circumstances, in the cause of preserving Victoria Street at Sydney's Kings Cross against the developers and power elites.

Four demolition workers at Angel Street exemplify the attitudes which allow such atrocities to happen: 'I'd knock down the Opera House for time and a half.'

The Killing of Angel Street deftly shows the difficulties of different classes of citizens uniting in a cause—the suspicions by many in Angel Street about the 'quality' Jessica being manipulated by a blue-collar brigade. The events also reveal the power of publicity in any campaign such as this.

The main scriptwriter for *Angel Street* was Michael Craig, who had a long career as an actor and a writer in Australia after having been a regular in Rank Organisation films in Britain throughout the 1950s. Craig, born in 1928, was also a writer in Britain and among his scripts was a film recognised as one of the finest of the 1960s, *The Angry Silence* (1960), for which he was nominated for an Academy Award. In the 1970s, he was involved in the scripts for such celebrated Australian films as *The Fourth Wish* (1976) and *The Irishman* (1978).

At about the same time as *Angel Street* was released, *Heatwave* (1981) appeared, dealing with terrace houses and their occupants about to meet their fate. *Heatwave*, directed by the internationally recognised Australian filmmaker Phillip Noyce, begins with doors being broken through with an axe amid the smashing of glass and general pandemonium. From the earliest frame the developers treat these seemingly powerless tenants as some form of deplorable pest. As in *Angel Street*, there is an ongoing, ceaseless struggle:

Mary: Well the issues [of long-established neighbourhoods giving way to development] are the same . . . the only thing that changes are the battle lines.[53]

On the opposing side is the off-hand, uncaring attitude of the typically rapacious developer as he describes an earlier project:

Houseman: First one I ever put up mate, and that's as ugly as sin, but it's put a roof over a lot of people's heads and it's made me heaps . . .[54]

Heatwave proceeds to examine the interesting situation of someone like Houseman suddenly besieged by mysterious forces more powerful than he. *Heatwave* suggests how total solidarity among action groups can thwart even the biggest developer, whose bottom-line is often quite precarious.

MANY unashamedly popular and light-hearted attractions also embrace an environmental awareness, foremost among them the blockbusting *Crocodile Dundee* (1986), whose good humour and overflowing charm won worldwide acclaim. Comedian Paul Hogan and veteran scriptwriter Ken Shadie created Mick Dundee, who roams the back of beyond yet reveals strong feelings about his natural environment. These emotions are somewhat coarse and not always properly articulated, but his hero status—within Australia and among an adoring international movie audience—gives a new dimension to the appreciation of the Australian ecology.

At first it seems Mick's concerns do not stretch beyond that day's activities:

> **Sue** *(a hard-nosed New York reporter, asking almost sneeringly)*: . . . Where would someone like you stand on the nuclear debate? What are your thoughts about the arms race?
> **Mick:** *(shrugs)*: None of my business.
> **Sue:** None of your business. It's everybody's business. You should have an opinion. You should have a voice.
> **Mick:** Who's gonna hear me out here?[55]

With the charm that won over a cynical international audience, Mick then shows a sensitivity towards his harsh and remote environment not normally associated with the rugged individualists of the bush. He eloquently describes the Aboriginal experience. 'They don't own the land,' he tells Sue. 'They belong to it. It's like their mother.' In one of the more poignant moments in Australian cinema, Mick warms to his theme: 'Like all God's creatures [the Aborigines] just want the right to walk across the earth in peace . . . That's something you city folk don't understand.'[56] Mick, an earthy soul, points to a distant escarpment and describes it in his irreverent manner: 'These rocks have been there 600 million years . . . still be there when you and I are gone. Arguing over who owns them is like fleas fighting over who owns the dog they live on.'[57]

Paul Hogan achieved cult status for his Outback hero, who passionately questions the sanity of visiting urban hoons shooting kangaroos just for fun.

Hogan and Shadie created the character Mick Dundee and developed the film's plotline over a six-month brainstorming period, when they travelled to the Northern Territory and other locations. When it was time to put pen to paper, Hogan and Shadie had created, as Shadie put it in 1999, 'a character who was in tune with the Outback, someone who had great affinity with the Aboriginal people'. Shadie said Mick Dundee was portrayed as an innocent abroad, one who was a natural in his own environment but who was out of his depth anywhere else. As the plot was developed, Shadie said he and Hogan sought to grapple with Mick Dundee's conundrum as he faced new challenges in different surroundings and strange cultures: 'We were interested in "What would he do?", "What would he think?" and "What are his values?" '[58]

Shadie, born in the Sydney beachside suburb of Bondi in 1935, brought to *Crocodile Dundee* the experience of a writer who in 1957 joined the infant television industry. He had spent many years in radio as audio assistant to legendary personalities Bob Dyer and Jack Davey, whom he followed into the new medium. He subsequently wrote the scripts for some of Australia's best-loved TV shows such as 'Review '61 and '62' and the hugely popular 'The Mavis Bramston Show', both of which featured biting satire and topicality. Shadie began his association with Paul Hogan when he wrote for his early TV specials.

In 1999, while Hogan and *Dundee* producer John Cornell divided their time between the USA and the resort of Byron Bay in northern New South Wales, Shadie and his wife lived in Brooklyn on the Hawkesbury River on the New South Wales Central Coast, where he was involved in protests against planned developments at Kangaroo Point. Shadie said he was appalled at plans to construct 184 apartments, built on pylons, along the tranquil waterfront at Brooklyn.[59]

Despite the passion that often permeates *Crocodile Dundee*, it nevertheless is above all meant to be jolly escapist entertainment, poking fun at many of Australia's sacred cows and the preciousness which surrounds them—such as all the mystique and reverence surrounding the Dreamtime and other aspects of Aboriginal culture. This was the case in one of the film's funniest moments:

> **Nev** *(an urban Aborigine whose father is a tribal elder)*: Oh no, you can't take my photograph.
> **Sue:** Sorry. You believe it will take away your spirit.
> **Nev:** No, you've got the lens cap on.[60]

The charm of Paul Hogan/Mick Dundee created a realisation that time-honoured stereotypes could no longer apply in a world changing beyond recognition, where even the Outback could no longer remain isolated from the outside world.

ARMAGEDDON as comedy is the essence of *One Night Stand* (1984), in which John Duigan takes whimsical and gentle digs at the foibles of young people and their preparedness—or lack of it—to confront imminent extinction. The end of the world, as described in sound bites on TV screens, is a peripheral matter to Brendan and Tony, whose more pressing aim is to try to shag Sharon and Eva.

The global crisis and a decaying planet are of no concern to two lads more keen to embellish to each other their supposed sexual performances, and the

Cold War—yes, at that time there *was* a Cold War—had no place in their narrow consciousness. They are intent on tomfoolery while their nation is plunged into peril as Australia's US bases are hit by nuclear devices and all citizens are ordered to remain indoors.

From the earliest credits, the nuclear threat is made clear as a US warship sails in to the usual anti-nuke horde of demonstrators. Carnal lusts preoccupy the demonstrators even though the main topic of concern is presumably global peace or war. We learn, very gradually, that global politics is facing an international crisis from which Australia cannot remain aloof.

Duigan—best known as the director of popular and much-loved whimsical films like *The Year My Voice Broke* (1987), *Flirting* (1990) and *Sirens* (1994)—asks sardonically in *One Night Stand*: who needs nuclear destruction when people already live surrounded by the debris of half-empty wine flagons and leftovers of Kentucky Fried Chicken, with rock music constantly blaring white noise?

The philosophy which underpins *One Night Stand* had its genesis during Duigan's formative years which included studying history and a Master of Ethics degree at the University of Melbourne. He was a student radical and member of the peace movement (taking part in Vietnam Moratorium marches), as well as head of the university's student theatre. He performed anti-war street theatre, taking the message to shopping centres, beaches and country towns.

Early in his career, Duigan—born in Hampshire, England, in 1949 to a British mother and Australian father and who arrived in Australia at age 11—also made films of gritty realism such as *The Winter of our Dreams* (1981) and by the late 1990s had directed major international releases such as *The Leading Man* (1996) and *Lawn Dogs* (1997).

Duigan began his career as a director with seven small films on a range of issues and revealed his social concerns in the feature *Mouth to Mouth* (1977), about unemployed teenagers. In the mid-1980s, he joined other prominent personalities like rock star Peter Garrett (see Chapter Seven) and ABC TV's well-known Sydney newsreader Richard Morecroft in seeking to spread the anti-nuclear message, and at the same time he wrote press releases for Garrett's unsuccessful campaign for the Senate.

As Armageddon approaches amid the chaotic and disordered domestic scenes in *One Night Stand*, the full extent of what is happening sinks in very gradually:

> **Eva:** Millions of people could be dead.
> **Brendan:** We don't know that—could be all hype. They exaggerate everything—maybe there's going to be an early election.
> **Sharon:** They wouldn't go that far.
> **Brendan:** Well they wouldn't tell us what was going on, would they?[61]

The state of humanity that would react so scattily to any sort of environmental disaster or political crisis is lambasted by Sam, a deserting pacifist US sailor being sheltered by the Aussie quartet: 'If there was a kid from outer space doing a school project on human beings he'd have to say we were all crazy.'[62]

Here is a gentle yet incisive film which humorously questions the morality of today's youth and how they would react to a nuclear threat to themselves or to a holocaust happening somewhere out of sight and even better still out of mind.

Looking back on *One Night Stand* in 1999, Duigan said from his London home that he wanted to take a satirical look at the average young person's lack of awareness of what a nuclear war would mean. 'I tried to suggest that Australia wouldn't be removed from a global nuclear struggle because of the frequency with which US ships, usually nuclear-powered, made routine visits to Australian ports,' Duigan said. 'I was not trying to make light of the immense issues associated with nuclear war; I felt it important to satirise the complacency, lack of awareness and blinkered attitudes of young Australians.' Duigan said the film's unconventional approach went against the stark realism prevalent at the time in the Australian cinema.[63]

In London, Duigan remains concerned with many global environmental issues, and he is a financial contributor to Friends of the Earth. In 1999, he cited the state of the world's oceans as his main concern and he was saddened by 'the number of species of marine life which have been decimated by pollution and global warming'. Duigan's other main concern was genetically modified crops; he said their rapid introduction seemed out of control and nobody was aware of the consequences of tampering with nature.[64]

IN THE offbeat and unconventional *Breathing Under Water* (1991), scriptwriter Susan Murphy—writing under the name Susan Murphy Dermody—injects a dose of horror into her voyages of fantasy and magic. Murphy takes us underneath Sydney, which she describes as 'below the maze, below ground zero'. The senselessness of much of human endeavour is both hinted at and made quite explicit; we are shown, for example, archival footage of University of Sydney scientists removing the shell of a snail with tweezers.[65]

There is only a thin line between offhand and seemingly trivial little cruelties and the most massive and widespread of holocausts. The inevitable retribution for all destructive enterprise is described by a Narrator overwhelmed by civilisation's direction: 'The fathers of the bomb have become its trembling sons.'[66]

Breathing Under Water describes how many modern technologies are at best frivolous and at worst the harbingers of disaster: 'Each new device turns out to be a demanding riddle that doesn't shrink but expands the field of the unknown, the field of catastrophe.'[67]

Many of the scenes are mad and disjointed, and confusedly flick between colour and black-and-white. There is superimposition upon superimposition—each more bizarre than the one just before. In *Breathing Under Water*, there is not much difference between the real and the fantastic.

Susan Murphy's work is an amalgam of mystery, imagination and dream sequences. It warns of the unspeakable savagery and nihilism of humankind's dark side. Yet there is an acknowledgement—and a glowing hope—that a civilised society can yet flourish if people can somehow be focused into turning away from mean-spiritedness.

In *Breathing Under Water*, a mother and daughter have explored, then thought and reflected, then undertaken new explorations to discover other ways to live which are more palatable and sustaining. The next generation is given a chance to 'understand where heaven has to be'.

Also adopting this semi-documentary tradition, interspersed with dramatisation and fictional characters, is Martha Ansara's *The Pursuit of Happiness* (1987), set amid the backdrop of all the America's Cup '87 hoopla at Fremantle and the continuing prickly issue of visiting US warships.

As a docu-drama, *The Pursuit of Happiness* traces a typical couple—Anna and her husband John—as they are affected by the American presence. Anna has doubts about how the locals benefit from it while John's mind is in a whirl about the money to be made.

Anna, a rookie newspaper reporter, has the impossible assignment from her editor of conveying a serious nuclear message without getting overly emotional. Her editor—and the citizens of Fremantle—are deaf to the things they don't want to hear. They recognise the potential harm of the American presence but ignore it in favour of their pocketbooks.

As Anna and John wrangle, a TV newsreader is talking about the Cup, but his words could also apply to wider global concerns: 'The pace is fast and furious and only the superfit will make it. In the end, it's not how you play the game, but whether or not you win.'[68]

Surrounded by the greed and boosterism of the locals, Anna is pessimistic about whether people's values can be changed:

> **Anna:** . . . it makes you wonder whether it's worth it.
> **Dorothy:** Well, everything's been an uphill battle, hasn't it? You know, an uphill battle to get the vote for women. An uphill battle fighting child labour. An uphill battle fighting against slavery and all the people who started those battles off were all regarded as loonies. None of my friends call me Dorothy, they call me Dottie![69]

Young people also reflect this pessimism, and where there isn't nihilism there is a depressing and overwhelming cynicism and selfishness, as discovered time after time when Anna hits the streets to conduct interviews for her project:

> **Anna:** . . . I'm getting young people's ideas on how they see the world in the future. Can you tell me what you think the future of the world is?
> **Girl:** We're going to die. I don't know—umm—I think Russia's going to blow the bomb soon . . . I think there's going to be a nuclear war soon because they keep making weapons and more weapons. I reckon they should stop

> making all those weapons and just start giving the money to people in Africa because they're dying.
> **Anna:** How do you see the world in another 30 years' time?
> **Boy:** I don't worry about the world, I worry about myself. I don't care what happens to the world. I worry about myself.[70]

In a creative mix of dramatic and documentary techniques, Ansara presents real people and actors to show the diverse range of people who guided historical developments at the time—everyone from the WA Inc business tycoons in close cahoots with the Americans, to those working tirelessly to change the system.

SOME Australian documentaries have managed to attain a cult status—and thereby spread important ecological warnings to a wide audience. Foremost among them is *Cane Toads* (1987), whose quirky and irreverent approach contains a serious message about the spread of introduced species. Alternating cuteness and scariness, scriptwriter/director Mark Lewis shows how *bufo marinus*, a toad imported from Hawaii, is running as wild as anything found in the pages of Eric Rolls.

A sparkling, fast-moving and often funny film, *Cane Toads* is a major wake-up call to those who believe our environment can sustain and recover from almost anything.[71]

Mark Lewis, born in 1952 and raised in the country, graduated as a Bachelor of Economics from the University of Sydney. He studied film direction at the Australian Film School and became at first a sound mixer, then established his reputation as a director with *Cane Toads* (on the strength of this film and another about dogs Lewis would go to Hollywood). In 1999, his fascination with nature, the environment and living things continued, and he was filming *The Natural History of the Chicken* at various locations throughout the USA. He had just completed *RAT*, about the spread of rodents in New York, and *Animalicious*, about unusual animal incidents.

In *Cane Toads*, he tells the story of how the importation of the species into Australia resulted from a world conference on sugar technologies held in Puerto Rico in 1930. The film reveals how cane toads eat everything—except the cane grub and beetle they were brought in to eliminate. At the time of filming, cane toads had spread to 40 per cent of Queensland and were making incursions into other States and the Northern Territory.

Mark Lewis pointedly remarks on how all living things (with the exception of cane beetles) are potential prey as the toads are ever ready to devour native animals, including marsupial mice, and will eat anything which fits into their mouth. At the other end of the cycle, snakes and ibises are killed by toads which they try to swallow and instead are knocked over by something not unlike a cyanide pill. They can even be harmful to humans, as a spray from their poison glands can cause pain, paralysis and even death.

In the film, a cross-section of highly articulate experts and observers are interviewed, for example writer and cartoonist Patrick Cook who debunks the 'perverted reverence' with which many Australians treat this marauding pest. The toads may indeed be cute, but their ecological destructiveness must not be ignored.

The idea for *Cane Toads* came during the late 1970s when Lewis read a small item in a newspaper about the spread of the pests. 'I thought it was a good story,' Lewis said in 1999. 'I did some research and found many interesting stories about how the toad was ingrained in the culture and finally in 1987 received the money to make the film. Although it had a very strong environmental theme I was also attracted to the characters and the great stories that I found. Unfortunately, the very bleak environmental future and potential the film depicted of the cane toad and its spread is being realised.'[72]

THE spectrum of environmental issues is canvassed in documentary films and one of the most evocative is *Wild* (1991), a cinematic essay based on Eric Rolls' *A Million Wild Acres* (see Chapter Three). With the Rolls classic as a catalyst, scriptwriters Ross Gibson, John Cruthers and James Manche approach their documentary with a lofty sacredness, 'to provoke, cajole, entertain and lead its audience into an adventurous encounter with a set of urgent questions about our responsibilities in the land, questions which require brave and inventive new responses from today's population if we are to live long and well here'.[73]

Filmed with fabulous flair in the Pilliga Forest of north-western New South Wales, from where Rolls had written his masterpiece, *Wild* explores the multitudinous influences on a wilderness ecology. All the guilty parties responsible for the destruction of a rare ecology are named and brought to task: timber-cutters, cattlemen, charcoal-burners, bee-keepers, wild horses and pigs, cats and rabbits and foxes, rubber weed, thistle and lantana. Just like Rolls in his book, the film-makers see a larger purpose for their narrative, it being 'the story of a forest, but it could also be the history of a nation'.[74]

Wild discusses, as did *A Million Wild Acres*, the horrendous damage of the Acclimatisation Societies which introduced inappropriate species across the continent. The script is scathing about the stupidity of the European settlers of every generation: 'The Society was a club of English newcomers who saw Australia as a playground for the plants and animals of the world.'[75]

For example, there are stark descriptions of the introduction of a Jamaican rodent called the agouti:

> . . . we've had no first-hand experience of the rodent's special habits: it is nocturnal, prolific and voracious; it eats almost any vegetable matter but it also thrives on the flesh of unwary animals. It will bite through a wooden fence and burrow under a metal one. In the bushland of Australia it would have been privileged, mobile and unthreatened by predators. It would have had the run of the land. But the land would have reacted.[76]

As did Rolls, the film also focuses on the rabbit. The commentary is quite succinct: 'They ran wild. A legend of breeding. They burrowed into the land, churned it and moved it, turned it inside out.'[77]

Wild contains dramatic archival footage which shows the manic and frenzied destruction of eucalypt forests. The vivid black-and-white film footage and stills show the axemen as mythological, brave heroes whose violence and lust for fire were regarded as admirable.

Down through the millennia Aborigines had managed their environment with great adeptness, and the film compares the treatment of the land by the European arrivals and by the original inhabitants. The commentary is at once poetic and soaring:

> For thousands of years the Kamilaroi kept their country glowing with careful fires—some to germinate special plants, some to rejuvenate the soil, others to shepherd animals with hunting grounds. It was a type of farming synchronised to the rhythms of the country; it was a science unrecognised by white settlers who disparaged the Aborigines for never learning to hoe the soil or work in rectangular fields.[78]

The film, focusing on the wide-ranging concerns of Eric Rolls, is a collaboration between three intrepid film-makers and scriptwriters—Ross Gibson, John Cruthers and James Manche. Gibson, born in Brisbane in 1956, began his practical involvement in film-making in the early 1980s as a founding member of the Sydney Super-8 Film Group, and as a co-organiser of the Sydney Super-8 Film Festival. In 1985, he and Cruthers made *Camera Natura*, described as a 'history of white Australian landscape'. Gibson gained much of his expertise and insights after studying at the University of Queensland and the University of London, followed by courses in film and general theory at the London Polytechnic and the Sorbonne. As well as film-making, Gibson has also been a writer, curator, publisher and teacher and one of his aims has been to speculate on possible new models of knowledge and how to depict environmental issues. Cruthers, born in Perth in 1953, since the early 1980s has worked as a producer, production manager and sound recordist on a range of independent films. An honours graduate in Australian literature at the University of Western Australia, Cruthers has also worked as a senior project officer, consultant and assessor for the Australian Film Commission. Manche, the associate producer and editor of *Wild*, is an honours graduate in political science from the University of Melbourne and has a Bachelor in Education from Victoria College.

In *Wild*, Gibson and his fellow writers warm to the theme that white Australians must learn a new respect, a sacredness, for the land which nurtures them and which is a part of their very being. The reward for 'listening' to the forest is 'a catalogue of splendours', which in turn is part of a 'grandness', a 'mythic' experience.[79]

Wild is a voyage of discovery through history, a look at the present and a speculation about what lies ahead.

AN EQUALLY powerful documentary is *Give Trees a Chance* (1980), in which Jeni Kendell examines the anti-logging protests at Terania Creek a year earlier. Setting the scene is narrator Jack Thompson, one of Australia's best-loved actors and personalities, who warns the audience that the story of Terania Creek in northern New South Wales is the 'story of rainforests everywhere'. Describing rainforests as the 'cradle of evolution', Thompson solemnly and sadly tells of how only forty per cent of the world's rainforests remain and 30,000 hectares a day is disappearing. Furthermore, by the year 2001, about a half of existing species are likely to be extinct.[80]

Give Trees a Chance, amid the stirring songs of protest including a clever reworking of *Give Peace a Chance*, has dramatic images of violence within the forest as chainsaws and heavy machinery move in. Glorious footage shows Terania's creatures and vegetation rudely invaded. Police shift blockaders so the bulldozers can move in over several days in August 1979.

Kendell's cameras are rolling as local landowner Hugh Nicholson, an anti-logging activist, confronts some of the senior police officers, questioning the right of the constables to be on his property. Other demonstrators stand in front of bulldozers, chanting 'We're not as gutless as sawmillers think'. Meanwhile, the police ignore an ABC reporter's question as to why they're not wearing identification numbers.

One protester yells to a policeman who hit him in the testicles: 'What sort of an animal are you?'

Then come the images of destruction as tree stumps and the debris of the logging operation affront the senses and send shockwaves of further anger among the protesters.

Give Trees a Chance gives many examples of how a well-orchestrated and committed campaign can be effective beyond the resources of those participating in it. For example, it is pointed out how on every day of the blockade a hundred police (most of them from the Lismore area) were needed, with up to thirty reinforcements sent up from Sydney. The Terania protesters devised daily tactics to slow down bulldozers, causing sufficiently long delays so that the public could 'think about the issue'.

Jeni Kendell shows how such concerted action turned Terania into a national issue, which in turn forced New South Wales Premier Neville Wran to confront the powerful pro-logging elements within his government and to proudly and publicly proclaim his green credentials. *Give Trees a Chance* documents Wran's sense of powerful responsibility to protect environments for future generations.

As the battle for Terania Creek continued under a media spotlight, a State Government committee went to investigate the matter in September 1979, and after a second delegation went to the area the demonstrators were successful in their key demand—an environmental impact statement on how logging would affect the rainforest.

The film is one of the finest and most inspiring song-rich and colour-filled documents of locals fighting for their environment and way of life, as is a subsequent work by Kendell, *Earth First* (1987), which also formed the basis for an illustrated book released by ABC Enterprises (see Chapter Four). *Earth First* was co-written by ecologist John Seed and Jack Thompson, who as in *Give Trees a Chance* is the familiar face sharing Australians' grave concerns for the future.

Music and songs also form the backbone of *Earth First*, which begins with the powerful and pulsating rhythms of didgeridoos and the sounds of nature. A spectacular rainforest emerges from the mist—and here is a complex natural system of interlayed vegetation. Waterfalls, streams and rivulets begin a journey into the rhythms of life.

We learn that these are the wet tropics of North Queensland, where remain the last vestiges of such wilderness, which is (for the moment) just as it was millions of years ago. As Thompson wanders through this glory, he holds a seed which he says was part of the first flowering plant when the region was still part of Gondwana.[81]

Thompson describes the lush vegetation in which he treads as a 'living museum'—and he points to a cassowary which he says walked here at a time when all of Australia was covered by rainforest. With a passionate and intense delivery, Thompson tells of how such a rainforest has a 'stabilising effect on climate and soils' and how rainforests represent the 'most diverse and complex ecosystem on earth'.[82]

Just as the sights and sounds of all the creatures in such a glorious setting enrapture the audience, the peace is cruelly disturbed by heavy and destructive machinery. Loud and hideous bulldozers toss trees aside as Thompson gives the ominous statistics: destruction is happening at the rate of 22 hectares a minute, a football field a second. Thompson tells of the 'catastrophic effect' of destruction which, if it continued, would see the end of all rainforests in the space of one human lifetime.

Among those interviewed for the film is US ecologist Paul Ehrlich, who warns that such destruction is 'cutting off our own life' and that it is tearing away humankind's 'support system'. The destruction is further likened to a nuclear holocaust.

Earth First also features footage, as in *Give Trees a Chance*, of the fight for Terania Creek, including Wran telling of how this struggle convinced him of the need to protect forests and trees. Thompson then describes how the 'Spirit of Terania' was then transported to south-west Tasmania and the fight to save the Franklin River.

In what is almost the defining moment of the film, Ehrlich warns that if the disappearance of rainforests continues at the present rate, within 100 years mankind will not resemble what we are now.

DEALING with many of the same issues—but in Sydney rather than the bush—is Pat Fiske's *Rocking the Foundations* (1985), which deals with the New South Wales Builders Labourers Federation (NSW BLF) and its Green Bans to stop the widespread developments threatening to change forever the city's landscape. In a vivid and exciting documentary (in which file footage mingles dramatically with Fiske's interviews of the principal BLF players in the Green Bans struggle) the incredible story unfolds of how this particular union was a most unexpected source for social reforms, as well as environmental activism.

Rocking the Foundations, which took six years to make, tells how under the leadership of the troika of Jack Mundey, Joe Owens and Bob Pringle, the BLF did the previously unthinkable and put the environment and the needs and aspirations of citizens ahead of the job prospects of its own members.

Fiske, once a builders' labourer herself, shows the incredible transformation of a union which had not been 'active or viable' even in dealing with the needs of its membership. She shows how until Mundey, Owens and Pringle came along (Mundey was secretary of the NSW BLF between 1968 and 1973, and Pringle had been prominent in the campaign to save Kellys Bush) this had been a corrupt union which didn't care about the many abuses and the appalling conditions on building sites.[83]

With evocative black-and-white footage, Fiske shows how massive building development started during the affluent and free-wheeling 1960s, with office blocks getting higher and higher in an increasingly dense Sydney skyline. The Sydney CBD became a series of wind tunnels owing to the irregular and unfettered designs of the structures amassing. Meanwhile, inspired by the anti-Vietnam War movement of the time, Mundey and his comrades became dedicated to the idea that unions had a right and a duty to be involved in social, political—and environmental—matters.

The next phase, as exquisitely dissected by Fiske, involved the incredible coalition of wealthy and privileged people with a Communist-led union to fight for the preservation of things that were precious to them. Events came to a head at Kellys Bush, a public parkland in the well-to-do harbourside suburbs of Hunters Hill and Woolwich. The affluent residents of that area were opposed to a planned housing project, and so evolved the rare spectacle of opulent matrons joining BLF activists in fighting the State Government and developers.

Interviews with Mundey, Pringle and Owens reveal how BLF bans on Kellys Bush stopped the bulldozers which would have ripped it asunder forever. Simultaneously, the BLF activities made people realise that other parkland throughout Sydney was similarly at risk.

Then came the momentous struggle at The Rocks in Sydney Cove, birthplace of the nation, where five to six generations of families were threatened by the rampant high-rise development in the district. The residents were seemingly powerless and faced insurmountable odds as they confronted the might of the State Government led by Sir Robert Askin, who gave carte blanche to powerful development groups. The commitments of the BLF were the only hope for the residents. This was a time when scab workers were prepared, without warning to those who lived there, to demolish houses which the union had decreed should—for the moment at least—be saved. The demolition crews moved in as police looked on. At the very time that these demolitions were occurring, other communities became inspired by the whole Green Bans concept in what Fiske describes as an era of 'real change'.[84]

The Green Bans movement spread to all vulnerable areas and sites across Sydney in what Fiske, born in the USA in 1946, sees as the first successful land-rights struggle in Australia. It all started in the inner-Sydney working-class suburb of Redfern, and among the campaigns was a fight against plans for a giant sporting complex at Centennial Park which would involve one of the local residents there, Patrick White, who in turn would be inspired to write his play *Big Toys* (see below).

This struggle was won as plans were quickly shelved to build the complex in time for a mooted Olympic Games in Sydney in the 1988 Bicentennial year.

Then came one of the supreme struggles—that for Victoria Street at Kings Cross where Green Bans were set up to protect local residents from having their lives changed forever by a sea of high-rise horizons surrounded by freeways.

One company building, a major office block in Sydney, was placed under Green Bans because it also had links to a planned woodchip plant in Queensland. Thus, rainforest protection was defined as just as important as preserving urban heritage.

Mundey the unionist and Communist explains to Fiske that organised labour's struggles for such things as the 35-hour week are meaningless if workers have to live in polluted cities and in dwellings where they can no longer afford the rent.

The greatest achievement of *Rocking the Foundations* was summed up by Fiske in a 1985 newspaper interview. She reported being gladdened when audience members told her: 'Thank you—I didn't know.' As Fiske explained: 'They had read about it [the BLF's campaigns], and thought they understood the period of the Green Bans, but they didn't comprehend it as a whole. They were genuinely surprised at how advanced the NSW branch of the BLF had been in other issues such as employing women on sites, gaining limited tenure of office for union executives, and in their effective strike tactics.'[85]

In the film, Fiske records the dramatic events leading to the deregistration of the NSW BLF in June 1974. But she also reveals, and exults in, the fact that, like the Lake Pedder issue in Tasmania, this was a defeat which became an inspiration for future fighters for the environment.

THE cream of Australia's playwrights, like the screenwriters discussed above, have become increasingly alarmed at the state of the environment and have explored their concerns in their works. It is significant that when the highly respected *Theatre Australia* magazine in 1981 listed its top 27 Australian playwrights at the time, the top five—Dorothy Hewett, Alex Buzo, Jack Hibberd, John Romeril and David Williamson—had all dealt with environmental issues, and would all continue to do so for the next thirty years.[86]

One of the earliest modern playwrights to directly address environmental issues was John Romeril, who has spoken dramatically of the 'imminence of the deluge, the collapse of the planet, chaos, the end of the road'. Born in 1945, Romeril began writing at the University of Melbourne and then was part of the radical ferment at the Pram Factory and the Australian Performing Group. It is his view that plays need to reflect that 'the continued existence of life on the planet is at stake'. His green consciousness was evident from the early 1970s, and he has proudly claimed to be the author of probably the first plays produced in Australia on a solar-powered computer.

Romeril exemplifies a group of New Wave radical playwrights who converted their anti-Vietnam War anger of the 1960s and early 1970s into rage against what was happening to the environment. Indeed, Romeril many times states this explicitly in *The Radioactive Horror Show* (1977): 'Yesterday Vietnam. Today uranium mining. The Australian civil war is hotting up.'[87]

The Radioactive Horror Show began as a play-by-committee, in which all participants in the original Pram Factory production had an input, with the actors improvising scene and image ideas, and sharing in the research. Then, after incorporating information from newspapers, pamphlets and the distillation of viewpoints, Romeril 'wrote' the play.

'In practice, many people contributed directly to the project,' wrote Romeril for the poster released for the Pram Factory's production. 'Given the subject matter theft was deemed an appropriate procedure. Unbeknown to themselves, luminaries like Einstein, Teller, Gore Vidal, George Bernard Shaw, Asimov and Doug Anthony have had a hand in shaping the text.'[88]

The Radioactive Horror Show, which followed Romeril's highly successful and path-finding *The Floating World* (1974), comprises 16 scenes with such provocative subtitles as 'The Hiroshima Panels' ('Any debate that fails to acknowledge both the terrible consequences—and imminence—of nuclear warfare doesn't begin to be relevant'); 'Ted and the Energy Mafioso' (in which we meet The Fixer, 'a member of Rockefeller's SS'); 'The Nuclear Days of Our Nuclear Lives' (in which Sarah, 'housewife and Optimist, sees a light at the end of every tunnel, even if it's the headlamp of the train rushing towards her'); and 'The Activist Pays Us a Visit'.

The last mentioned tackles many of Romeril's concerns, such as the rampant and uncontrolled growth of uranium mining. The play's characters express environmental concerns which in later years would become part of a growing national consciousness:

> **Roger:** I see a time coming when because of thermal pollution—because we've altered its climate—broken down the world's defences—raised its temperature—the polar ice caps'll melt—sea level will rise and giant tidal waves'll leave a city like this a hundred feet under water.[89]

This is a very early, albeit exaggerated and quite over-the-top, prediction of global warming, one of a range of concerns evident from the Romeril Papers, held at the Australian Defence Force Academy Library in Canberra. The papers contain Romeril's original source materials on which this play and many of his other works were based. They reveal Romeril's mammoth reading and research into all aspects of nuclear and related scientific research.

The Romeril Papers include the full gamut of influences on *The Radioactive Horror Show*: publications by the Australian Conservation Foundation and Friends of the Earth, popular magazines like *Newsweek* and *Rolling Stone*, plus documents revealing the damage to reactors around the world, especially in the USA. In his personal papers, Romeril reveals other influences on his play, including Doris Lessing's cataclysmic novel, *The Memoirs of a Survivor* (1974).

Horror Show deals with all kinds of ecological threats, not merely radioactive ones. The play also presents visions of waste being channelled into Lake Eyre, which becomes empty and degraded by salination. The destruction of forests is also mentioned with reference to the One Tree Hill tale, which will also feature in other prominent plays.

The increasing destruction of the environment is getting madder and madder, according to Romeril, and it is, both literally and metaphorically, a cancer on an already depleted planet:

> **Roger:** The ozone layer—a thin shield miles above the earth's crust . . . it keeps lots of ultra-violet rays at bay . . . you get too many ultra-violet rays you get cancer . . . They use [freon, a propellant] in spray cans—every time you kill a fly, up it goes.
> **Sarah:** The fly?
> **Roger:** The freon. What does it do? Battle with the ozone layer, that's what it does . . . it drifts off . . . dissipates . . .

The crumbling morality which has inflicted so much suffering on a fragile ecology is also explored in another Romeril play, *Bring Down the House* (1992). It describes how the co-existence between all living things, such a prerequisite for survival, has been shattered. One of the main characters, Will, muses how 'morality, in troubled times, is a luxury' and the others speak of 'the world . . . what's left of it'. Their descriptions are stark:

> **Morg:** We had riches.
> **Rita:** We had trees.
> **Morg:** We turned the riches into paper.
> **Rita:** The paper wasn't worth the trees it was written on.

The conclusion is best expressed by Morg: 'Nature fought us, and lost.'[90]

Similarly, quality-of-life issues are the concern of *Lost Weekend* (1988), whose broad sweep includes the pollution of aircraft noise at an idyllic rural retreat. This play is described by Romeril in his papers as 'something I wanted to write about: the town and the country; the experience of the town coming to the country, the country colliding with the town'.

Lost Weekend is as much a celebration as an expression of concern about bucolic values, as Romeril reveals in a discussion of the play. 'I dig, I plant,

I breathe unpolluted air, enjoy fairly simple pleasures, watch birds, listen to silence, hear my own heart,' he says.[91]

A plethora of concerns, based on Romeril's widely expanding fields of interest, imbued his later works, such as *Red Sun—Red Earth* (1996) and *XPO—The Human Factor* (1998). Romeril has been described by leading critic Katharine Brisbane as 'the most political writer of all the playwrights'. Brisbane has also acknowledged Romeril's use of humour and burlesque to convey extremely serious concerns, especially 'the position of the small man, the individual in a world which is dominated by multinational powers which take very little account of the individual.[92]

Katharine Brisbane has acknowledged the efforts of New Wave theatre to tackle environmental problems and to counter the prevailing conventional wisdom which insists that the Australian landscape is 'too big to fight' and 'alien'.[93]

John Romeril has angrily expounded on the complacency and cravenness which has allowed the depredations, pointing to how, despite Australia's dizzy wealth of resources, 'there'll be nothing to fight over, nothing to own, nothing to appropriate'.[94]

ALSO tackling the nuclear theme was Alison Lyssa's *The Boiling Frog* (1984), inspired in the late 1970s when Lyssa was in San Francisco and learned at first hand about the overwhelming community anger surrounding the death of US campaigner Karen Silkwood, murdered while uncovering and publicising wrongdoing and cover-ups at the nuclear reactor where she had worked.

The play spans death and destruction over three centuries, beginning with the Great Plague in London in 1665, then fast-forwarding to more modern cataclysms like nuclear holocausts and reactor meltdowns.

The Boiling Frog was first performed at Sydney's Nimrod Theatre on 11 July 1984, under the direction of John Bell. In a multi-faceted work, Lyssa examines the one-man-against-many theme and shows how in the nuclear Cold War age there is not much future for someone like Paul, a gifted scientist with a conscience outraged by the cover-ups and deceptions, especially the public being told that plutonium was strictly for industrial and not military purposes. His more pragmatic (and famous) colleague Magnus tells Paul that it is futile to challenge the irresistible forces of a military-industrial complex; that conformity is what is expected from the best and the brightest, whose star can very easily and effectively be snuffed out:

> **Magnus:** . . . if you want to be the Left's unemployed hero, suit yourself. Go to the press if you want to. Go on your peace march this weekend. Hold hands with the old ladies with their 'Ban the Bomb' badges pinned to their knitting. But don't come back to us. You won't get research funds. You won't get a posting, you won't get students, you won't get published. In five years you're a has-been living in a caravan.[95]

Such warnings and examples of suppression have been noted in Chapter

Three, especially in how the chemical industry has been able to crush the aspirations of any scientists who dared to challenge it. Lyssa reveals how powerful and sinister forces have always been on hand to crush the voices of conscience. In Lyssa's play, not much changes over the centuries in mankind's ineffectual responses to environmental destruction.

One of Lyssa's time-travellers across the centuries is Joan, who has witnessed one horror after another and provides a poignant assessment of vicious thefts and lost belongings:

> **Joan:** . . . the biggest hurting [from radiation exposure] will be here, in my heart, because it wasn't necessary. And if you bury me in the earth, or scatter my ashes, that bit of plutonium will not have gone away. It will blow about in the dust in a hundred years or a thousand years, and some other creature will breathe it, and die hurting, and not understand why . . . I feel cheated, and I hate you for it.[96]

The fate of Karen Silkwood was that of countless quixotic warriors down through the ages but, although crushed, they heighten people's consciousness which, in turn, will inspire bravery to fight the outrages afflicting our lives.

The Boiling Frog was two years in the making, during which time it underwent a thorough workshopping process. When *The Boiling Frog* was first conceived, Alison Lyssa, was a promising playwright in her mid-30s. She had already won recognition with *Pinball* (1981), but the aftermath of *The Boiling Frog* would test her innermost strength. Whereas *Pinball* had dealt with feminist issues, such as lesbian mothers being allowed custody of their children, *The Boiling Frog* confronted wider social and political concerns and was thereby deemed too presumptuous by many.

A prominent Australian political thinker, Hester Eisenstein, put it succinctly to Lyssa: 'As long as you're writing about what they see as women's issues, [critics and others] can be patronising and pat you on the head and say, "That's a nice curiosity." But you dare step into the political arena and write about the whole world! You dare challenge the entire political structure! No wonder they got angry with you.' Such a piercing appraisal of the theatre establishment by the internationally respected New York-based Eisenstein, author of *Contemporary Feminist Thought* (1983), hit Lyssa hard.[97]

The unkindest cut, though, was the blowtorching of *The Boiling Frog* by H.G. Kippax, at the time the most influential theatre reviewer in Sydney. He summed up the event as a 'dismal evening' and the play as 'a yawn'. Writing in the *Sydney Morning Herald*, Kippax said *The Boiling Frog* was 'unrelieved tedium' that insulted the intelligence of anybody watching the play, which was an 'abuse of science and technology'. Kippax accused Lyssa of 'silly anti-intellectual dogmatism' which would alienate any right-thinking person against the anti-nuclear cause. 'At the end we are urged to revolt and, if I follow the playwright, to climb back into the trees,' Kippax wrote.[98]

A more sympathetic review by John Moses in the *Australian* described the play as 'lucidly written' and 'a very thoughtful piece, based on a neat enough theatrical device, and with an important message about survival in the nuclear age'.[99]

AN ALLEGORICAL approach to nuclear issues such as Maralinga and the wider concerns of Australians' relationship with their land and culture is taken by Louis Nowra in *Sunrise* (1983), a highly creative and dramatic work. *Sunrise* begins simply by showing a dysfunctional family coping with growing pressures. Each family member is seen as having a very ominous inner demon to contend with, or is grappling with the contradictions of youthful dreams—with nuclear threats hovering in the background.

The patriarch of this motley collection is Clarrie Shelton, master of a property called Honey Ant Dreaming, whose varied life is gradually revealed as having been complex, inexplicable and sometimes quite bizarre. At first we know Clarrie as a fairly nondescript and inoffensive radio astronomer and only later is it revealed he had links with the atomic tests at Maralinga.

These flawed individuals as portrayed by Nowra are a microcosm of Australia's cultural dislocations that result in callous treatment of the land. The theme is developed by Clarrie's brother-in-law David who, in a supremely ironical metaphor, is a cancer specialist whose own health and lifestyle have self-destructive qualities about them. As David describes it, 'Australians never take root . . . we never have and never will', and his wider analysis is scathing and bitter:

> **David:** We follow European and American culture as if we were lip-readers trying to make out what's being said. Other countries have their roots in something, that's why they have ideals and ideas, they fight for them . . . Our torture is to have no vision worth dying for. We're a silly hybrid, like the head of a monkey sewn onto a dog—and have no soul.[100]

Nowra describes the need for Australians to come to terms with the devils inside them, to control their hatreds and resentments, and to cast aside simplistic, naive notions.

All the characters are united by a disturbing element, and their conversations highlight Australia's vassal-status on almost all matters, including the degrading land around them. Nowra portrays Australians as complicit or apathetic as the Menzies Government allowed the British to irradiate Maralinga. The exchanges between Richard, a lefty State MP in his mid-30s, and Clarrie, his father-in-law, reveal the British secrecy and Australian officialdom's subordinate role:

> **Richard:** I didn't know you were part of the Australian atomic tests, Clarrie . . . It's not something you've been voluble about.
> **Clarrie:** I observed one test. I was an Australian scientist, it was an English bomb. Christ, we were bloody lucky if they allowed us to polish it.[101]

Nowra suggests that Australians then and later could not come to terms with the enormity of what was being done, and they adopted a simplistic and child-like response to cope:

Clarrie: . . . You have no idea how magnificent it was.
Richard: Like cracker night?
Clarrie: A white light so intense you felt like a blind man first seeing the sun. Or the dawn of creation. Majestic, god-like, terrifying . . .[102]

But, as part of the painful self-evaluations by all members of the unhappy clan, Clarrie is no longer blind to the horrors, including the effects on the Aboriginal population, many of whom lost their sight or suffered horrendous life-long disabilities. 'Men are barbarians with the creative gifts of gods,' sighs an increasingly depressed Clarrie.[103]

Sunrise builds to a shocking and surprise ending, with the spectre of bush-fires a constant, searing image of disaster, of which a nuclear blast is the supreme example. Born in Melbourne's struggling housing-commission belt in 1950, Louis Nowra has written plays, novels and films with themes as varied and controversial as his own life. He was born a Doyle but, in a bid to rid himself of all traces of bog Irishness, in the early 1970s changed his name, as did Tennessee Williams, to a geographical location, choosing one of the major centres on the New South Wales South Coast. As he humorously explained many years later, this choice could easily have been one of the State's rivers but he figured that Louis Murrumbidgee somehow didn't sound quite plausible.

THE uranium debate and the activities of foreign-owned miners are among the concerns of *Big Toys* (1977), in which Nobel Prize-winning novelist Patrick White (see Chapter Two), who had originally intended to be primarily a playwright, explores what Katharine Brisbane describes as the 'moral and spiritual bankruptcy' of attitudes which allow environmental despoliation.[104]

Also described as a black comedy which angrily reflected White's growing concerns about nuclear power during the last decade of his life, *Big Toys* features a series of vignettes lampooning the trendy middle classes who might be in a position to do something about environmental vandalism but either cannot or will not do so. The leading characters are more concerned with mind-games and trivialities—and cynical posturing and hidden agendas—than challenging powerful forces.

There is an air of resignation in the portrayal of Sir Douglas Stannard, the new chairman of Intermond Mining (with its almost unlimited capacity to 'play' with uranium) who has in his possession 'the biggest, gaudiest toy that ever escaped from a child's hand'.[105]

The play's forcefulness partly stemmed from White's long-held disgust at the manner in which complacent Australians tolerated such outrages as the British atomic tests at Maralinga. 'Puppets of the English earlier in history, we found ourselves cast in the same role for the Maralinga experiment thanks to [Prime Minister Menzies'] infatuation for his monarch,' White told a symposium on nuclear issues, 'Imagining the Real'. This symposium was held in Canberra on 11 November 1986, when scores of writers gathered on

Armistice Day to swap concerns. 'In the scramble after mineral wealth in the nuclear age, our country's beauty will be scarred, her spirit estranged by developers and their touts, the political non-planners.'[106]

A major strength of *Big Toys* is that, just like within society there are mixed and contradictory responses to serious issues, each character has some good and some bad—especially union leader Terry Legge who, although he allows himself to be manipulated by the corporate interests he is opposing, will not budge or betray the core issues in which he passionately believes.

The character of Terry Legge is loosely based on Jack Mundey (described above), and through the medium of *Big Toys* White hails the BLF fighter for his vision and incorruptibility. Among his campaigns, such as fighting to save Sydney's Centennial Park, White would be part of the Green Bans process led by Mundey.

Big Toys is consistent with White's feeling, as expressed during the Bicentennial in 1988, that 'the earth is angry . . . so, I am angry'. In an article for the *Sydney Morning Herald*, White wrote: 'If the earth is angry, the human beings who inhabit it have cause to be angry too. Unless a superpower or hidden talent of a lesser nation succeeds in destroying us, we must all, in the years to come, work towards a civilisation based on humanity.'[107]

The wide-ranging article was a seething attack against the ills of Australian society, reflecting the attitudes which inspired *Big Toys*, with White expressing the hope that 'we may feed, not only the starving nations, but the starving in our midst'—both literally and spiritually.

The corporate types caricatured in *Big Toys* were partly drawn from real life, stemming from White's friendship with Klari Daniel, whose anecdotes about her nephew Peter Abeles reinforced White's notions about how corporate development had an insatiable edge. White's biographer David Marr relates how Klari Daniel, many of whose family fled Nazi persecution, 'deplored the vast accumulation of wealth by Jews'. When she asked her nephew, one of Australia's foremost captains of industry and significant political power players, 'Why do you go on? . . . You have more than you need,' Marr chronicles Abeles's reply: 'Tante Klari, you don't understand. You have to go on. You can't stop.' Such talk radicalised White significantly and his inner burning rage is evident in plays like *Big Toys* and his novels of that era.[108]

Another major inspiration for *Big Toys* was what Marr described as 'The Blow' which would fall on White's neighbourhood of Centennial Park—plans for a massive State sports centre. Under this plan, announced on 16 March 1972, parts of Centennial Park and nearby Moore Park would be covered in concrete if Sydney won the right to hold the Olympic Games during Australia's Bicentenary. Had the plan proceeded, according to Marr, White's 'own house and 35 others along the spine between the two parks would be bulldozed to make way for the Olympic stadium'.[109]

As White's anger grew about the looming desecration he was equally concerned at the hapless and inept manner in which the government and corporate plans were being challenged, and he incorporated his disdain into the words and actions of many of the characters in *Big Toys*.

White developed an almost religious and messianic fervour towards saving precious things, and he told a rally at Centennial Park on 18 June 1972 that plans to build an Olympic sports complex there would spell 'the ruin of a park which compares with the most beautiful in the world'. He compared Centennial Park to Hyde Park in London and the Pincio in Rome. White said of those famous parks: 'That they have endured for centuries is because generations have respected them as a delight to the eye and for many other homely, more down-to-earth reasons. As I understand it, a park is certainly not a Luna Park, with a music bowl, public address system, cycle tracks, traffic whizzing along raised roadways, and a great concrete amphitheatre looming over it, with a crowd roaring monotonously inside.'[110]

White's disdain and utter contempt for multinational corporations was so all-embracing that this attitude permeated every facet of his private life, even affecting long-time friendships. According to Marr, White was 'distressed' when Geoffrey Dutton agreed to write the text for a book commissioned by the Mobil Oil company. White was so affronted that 'he complained even to Dutton's children that their father had sold out to the multinationals' by becoming part of a book of photographs of Australia. White believed Mobil was among those destroying parts of the continent, and he could not fathom how any self-respecting man of letters could accept work from such a source.[111]

All environmental issues were of equal concern to White, who spoke and wrote just as fervently about the Sydney monorail as he did about nuclear war. He believed all environmental atrocities, local or global, usually stemmed from the greed and unjustifiable power and influence of spellbindingly powerful corporations.

THE controversial Dorothy Hewett has written plays with ecological themes: such as her celebrated *The Man from Mukinupin* (1979) and *The Fields of Heaven* (1982), as well as the lesser known *Golden Valley* (1981), and *Songs of the Seals* (1983).

The Man from Mukinupin, set between 1912 and 1920 in a fictitious 'typical West Australian wheatbelt town east of the rabbit-proof fence', was commissioned for WA's sesquicentennial celebrations and Hewett used the play to express her growing outrage at environmental despoliation. In the play, she recalled many of her personal and family experiences in the early years of her life spent on an isolated wheat farm near the township of Wickepin.

Born in Perth in 1923, Hewett has lived on both sides of the continent and personally learned about every environment—from the remote areas of Western Australia to the bustling inner-Sydney suburb of Darlinghurst to the tranquillity of Faulconbridge in the New South Wales Blue Mountains. Her surroundings are the material of her prolific output of plays, poetry, novels, essays and an autobiography.

The Man from Mukinupin, amid its vaudevillian atmosphere and clever dual role playing, is the play through which Hewett achieved her most consistent national success. In *Mukinupin* emerge visions of a merciless sun beating down on a degraded countryside from a cruelly clear sky. 'The alarm clocks are ringing . . . across the salt lakes' is the ominous warning early in the play.[112]

Beyond the Hewett trademarks of jovial repartee and good-naturedness—buttressed by classical allusions ranging from Shakespeare to Tennyson to Henry Lawson's 'The Roaring Days'—a realisation grows of crops failing as a summer's intense heat shows no respite. There is young love, gaiety and joyousness, but the townsfolk must also deal with ominous world developments and preparations for war. As their land becomes increasingly degraded, other problems arise.

> New Holland is a barren place,
> in it there grows no grain,
> nor any habitation
> wherein for to remain
>
> she drowns me in the winter
> and she bakes me in the sun.
>
> I'll plant her and I'll rape her . . .
> Cut her plains with sheep pads,
> mine the black beach sand . . .[113]

As the 'New Holland Song' (inspired by Randolph Stow's *Tourmaline*) indicates, this is desert country, beset by salination, a place where regularly it is 114°F in the shade and all around are droughts and rabbits. Incapable of dealing with the range of crises, the locals are strangled by their increasingly irrelevant conventional wisdom, and the viciousness of their racism is matched only by the breathtaking hypocrisy which fuels it.

The land is degraded. The people are degraded. Amid boisterous theatricality, Hewett presents sometimes poignant and sometimes bitter moments showing how the townsfolk's callousness towards each other cannot reap any fertile fruits.

But the play's denouement is not entirely without hope. As unlikely marriage alliances occur, there is a chance that 'now . . . we could perhaps rearrange ourselves'.[114]

Golden Valley, based on an unpublished novel for children, features such characters as Jane, a giant high-stepping crane; Em, a huge wombat; Nee, a great possum; and Di, an enormous mopoke.

The inability of humans to properly tend their environment is quickly put by Hewett to her young audience as Nee tells of how 'humans are strange creatures . . . they can never be like us'.[115] Through Marigold, a red-haired orphan girl adopted by the denizens of Golden Valley, a deep love is evoked, a humanity and a sense of caring, as first seen by her concern for a feral cat suffering in the jaws of a trap. There follow humorous touches such as Nee the possum, though proclaiming that the dreaded wild cats have no place in the bush, helping tend the feral feline's injuries. Hewett, with a justified reputation for irreverent humour, never allows the serious scenario to become too weighty for the young audience:

Nee: Cut it out, Di. Don't be such a galah.
Di *(indignantly)*: I'm a mopoke . . . mopoke . . . mopoke.

Hewett depicts the play's most unsympathetic (human) character, Jack Swannell, whose villainy surfaces in his every action, doing such positively inhumane deeds as skinning rabbits alive. He proclaims his odiousness in verse:

I'm Swannell, Jack Swannell, I'll skin 'em alive,
Sittin' 'ere pluckin' the wings orf the flies.
Sometimes it strikes me, nobody likes me;
It hits me and shits me that nobody cares.

But when I'm downhearted I like givin' pain.
To a wombat, a possum, a mopoke, a crane.

Swannell is the dreaded overlord of Golden Valley and is constantly threatening to evict its hapless inhabitants. When Marigold pleads for his mercy, he can only sneer: 'Why should I care what happens to a crane, a wombat, a possum and a mopoke?'[116]

Delightfully, Swannell will get his comeuppance, and Hewett even presents the exquisite notion of putting those who harm the environment on trial, to be judged by the creatures who have been harmed.

Hewett's other play for children, *Song of the Seals*, features Marlin Prawn, a local can-do developer and captain of industry, who sees limitless potential for opening everything up, with the devil taking the hindmost. 'What's in a fish, what's in a tree?' Prawn callously asks. 'Lots of lovely dough-ray-me,' he replies to his own question, unconcerned about the consequences of his deeds.

To Marlin Prawn, all the glories of nature are reducible, and expendable to the cause of One Big Deal. But, for the children in her audience to retain faith in the integrity of their futures, Mr Prawn also will be victim to a most just revenge.

In *The Fields of Heaven* (1982), Hewett most vividly encapsulates environmental concerns built up throughout her life, especially her youth in Western Australia's rural areas. Spanning 1929 to 1951, the play reveals the tragedies of broken dreams and the futility of enlightened attitudes being able to prevail against stolid conservatism and self-interest.

In the typical Western Australian rural surroundings of Jarrahbin, Hewett introduces a bleak landscape which the well-to-do land-owning gentry describe merely as 'desert'. But in the beginning there are still hints of optimism, as by the visionary landowner Tom Barrow, although they are resisted on every front:

> **Tom:** . . . the deserts shall bloom again.
> **Emily** *(his wife, in bitter reply)*: Stinkwort and poison bush.
> **Tom:** Everlastings and purple clematis, bottle brush in the creek bed.
> **Emily:** The dry creek bed.[117]

Tom, a character based on Hewett's father who also fought a relentless and quixotic fight on environmental issues, accepts the hardships inherent in improving his precious acres, and he has an unflagging, unchanging and unchangeable commitment to make something of his 'perverse' land. Hewett has commented that her father was 'the first conservationist I ever knew', and in a 1984 interview told of how 'he had that dream about trees and had the same reaction from the Western Australian Agricultural Department as Tom Barrow has in the play'.[118]

The Fields of Heaven emphasises the futility of Tom's intentions, Hewett also rails against the inflexibility of public administration on rural matters, showing how the rigidity of rural folk ensures toleration of the worst land practices which inevitably result in soil degradation:

> **Tom:** I've got such dreams [for his property, Marvell Locke]. But they won't listen, those fools in the Agriculture Department.
> **Louie** *(his feisty daughter)*: I know.
> **Tom:** They think I'm impractical, a dreamer. I told them, 'Give the farmers the trees to plant and the salt problem will disappear. If you don't, in 20 years the land'll be useless.' I told them that, Louie . . . And do you know what they said to me . . . 'Go home, in 20 years we won't even be here.' Can you imagine it? 'In 20 years we won't even be here.'[119]

The apathy which Tom Barrow so abhors leads to tragedy, as Hewett stresses with bitter forcefulness. Hewett contrasts the humane and progressive father and daughter with the (often unseen) unmovable power-brokers whose decisions are criminal, short-sighted and utterly callous. Louie, one of the most sympathetic and impressive characters in the Australian theatre, is forceful and inspiring:

> **Louie:** . . . Marvell Locke is beautiful with trees.
> **Tom:** But it isn't enough, Lou, can't you see? If we all don't plant the salt'll run down into the creeks and we'll all be destroyed.[120]

Hewett shows how there is the thinnest of lines between human creativity and exploitation, and between sordidness and pristine romance. There are many heart-tugging moments as dreams and ideals are swept away by practical imperatives, especially for the star-crossed lovers Rome Bodera and Louie. Over the next few years Louie will find wisdom and depth, while Rome will come to epitomise the destructive and uncaring exploiter.

Prosperity is the result of Bodera's endeavours, and the manic path to wealth will inevitably devour him. In his total zeal to be a man of substance instead of a 'dirty dago', Bodera crushes anyone and anything that gets in his way. If putting another sheep run on Marvell Locke at the expense of the ancient trees means more prosperity, Bodera will crash through regardless of anyone else's feelings or ambitions.[121]

The Fields of Heaven was described by Sydney critic Elizabeth Webby as one of Hewett's 'most realistic and relentless' plays with its emphasis on 'a land destroyed through selfish materialism'. Webby also elaborated on Hewett's belief that the play 'proved too confronting for Australian audiences'.[122]

The deteriorating urban environment is Hewett's theme in her musical *Pandora's Cross* (1978). This tuneful entertainment begins with the deafening sound of demolishers at Kings Cross, the inner Sydney hub which has been targetted by developers increasingly running rampant. In the seamy, varied bohemian life of human vices and foibles it is instantly clear that 'somehow things are not the same'. As the characters tell of their dreams and aspirations, the sound of the wreckers gets louder and louder.[123]

An empathetic and warm-hearted Hewett tells how many of the characters may be deemed by 'polite' society to be the dregs of humanity, but at least they *are* humanity. Even fictitious creations like the notorious poet Ern Malley have more soul than those directing the wrecking crews.

The bohemian community is helpless against the full force of corporate power, and this is sadly recognised by Primavera, an ex-stripper and now club proprietor who realises the overpowering potency of the corrupt links between the State Government, the 'bloody developers', the 'crims', the 'cops' and the 'hoons'. 'It's awful to watch people movin' out. Have you ever noticed? It's like a war-torn city. They come outa the woodwork, dazed, helpless, like refugees, with fifty bucks in their hands to store their lives away in a taxi truck.'[124]

The entire community then joins in a funereal chant to pay their last pathetic homage to the world they had known: 'Under concrete and glass Sydney's disappearin' fast,/It's all gone for profit and for plunder . . .'[125]

This community—like all communities powerless against rampant urban development—will never be the same again. Both the people and their environments will change irrevocably in an atmosphere of futility and degradation.

Dorothy Hewett's environmental concerns have imbued her vast body of work and she has delivered a forthrightness which often has proved too confronting for smug and comfortable Australian audiences. Elizabeth Webby has lamented how 'it is regrettable that her sort of wide-ranging "total theatre"—one which appeals to the senses and the intellect, one which aims to enlighten and provoke as well as entertain—has so rarely found directors, theatres, productions or audiences equal to its challenges'.[126]

SIMILAR sentiments are expressed by Marxist playwright Stephen Sewell, who in *The Blind Giant is Dancing* (1983; revised 1995) shows how Australian politics, society, culture and ethos have all combined to destroy the land as well as all noble human aspirations. Australians are portrayed by Sewell as having no notion of independence, meekly acceding to horrors ranging from the nuclear threat to the obliteration of individual neighbourhoods.

Several of the characters in *The Blind Giant is Dancing* discuss how Australians are aimlessly and thoughtlessly destroying their natural heritage for short-term gain, turning once fertile land into desert:

> . . . the desert coming in over the farmland . . . that intensive farming and overstocking. They're destroying the soil . . . There's no cover on the soil now. All it needs is a big wind and it'll blow it all away, and that'll be the end.[127]

The causes of such degradation are macro-economic and can be solved only by a powerful political will and a total overhaul of Australians' sets of values. But the Australian political structure—across all parties, across all philosophies—is incorrigibly corrupt and beyond salvation, according to Sewell. Conversations throughout the play reveal an arid and hopeless place:

> **Bruce:** But—how can you blame the drought on the farmers?
> **Louise:** I'm not blaming the drought on the farmers. All I'm saying is that the methods they use to maximise their profits are destroying the land's ability to withstand drought.[128]

The necessity of chasing worthy ends via a corrupt political process is recognised by everyone in the play—from the leftist revolutionary to the wealthiest of conservative elements. A conversation between Bob Lang, a Tory economist and banker, and his wife Janice, shows how Australians have an innate capacity to turn a blind eye to whatever is happening around them:

> **Janice:** What's it got to do with us?
> **Bob:** It depends on what kind of country you want to live in, doesn't it?
> **Janice:** I don't care what happens to the country! I care about what happens in my own house![129]

A society that destroys also has an innate capacity to achieve wondrous things, and this contradiction gnaws at many of the characters:

> **Allen:** . . . this society is the product of human decisions.
> **Rose:** What kind of humanity is it that spewed out this monstrosity?
> **Allen:** The same humanity that produced the Sistine Chapel and the Theory of Relativity.
> **Rose:** Atom bombs and ashtrays sold by the Vatican.[130]

The Blind Giant is Dancing reflects Sewell's stated belief that Australians allow any and every desecration to occur because of their 'gutless refusal to confront'. Sewell says Australians need to 'confront in a serious way the terror and majesty of life and to weep and laugh and scream and exult in it'. As have Patrick White and most of the other playwrights who explore environmental issues, Sewell calls on Australians to assert themselves and not be dictated to by overseas powers: 'It is perhaps time to think that it is common sense to fight your own enemies and not the ones chosen for you by an international bully preparing for nuclear war.'[131]

When *The Blind Giant is Dancing* was revived twelve years after it first appeared, the *Bulletin*'s theatre critic Brian Hoad acknowledged its importance as an environmental testament by commenting that it had become 'a dance of death that seems more terrible now than it did 12 years ago . . . but all a playwright can do is warn'.[132]

Many of these potent themes are discussed even more powerfully in *Dreams in an Empty City* (1986), which depicts the greedy 1980s at its most blatant, the madness of this frenzied decade at its most insane, and its violence and rapaciousness at their most deadly. Sewell, born in Sydney in 1953, saw the '80s from the prime vantage point of his late 20s and early 30s. He began writing and producing plays immediately after graduating in Science from the University of Sydney in 1975, also a watershed year in Australian history with the dismissal of the Whitlam Government.

Dreams in an Empty City is a savage and mocking critique of the pretensions of foreign interests, especially American, moving into Australia and taking over the local economy—and the local ecology. These foreign interests can, and do, destroy anything in their path to pay for paper deals back home. Sewell describes such cavalier treatment of Australia as a route to ruin:

> **George** *(an American banker)*: The new strategy is to diversify, go for the developed countries.
> **Harry** *(Kaufmann, the floor manager of Interbank, a money-market trader)*: That why you're here?
> **George:** Yeah. I'm just sniffing around, checking the place out. There's not much of it, is there? Two big cities and a few ports. Someone told me the other day the capital used to be a sheep station . . . But you wait. They've got a national debt of 46 billion now. What'd they spend it on? The same thing jackasses all over the world spend it on: real estate speculation, company takeovers, having a flutter on the sharemarket.[133]

Such disdain for local things can be achieved only with the collaboration of local accomplices, and typical of these is financier Simon Wilson, who is prepared to destroy anything worthy for the perceived big picture. Wilson can cynically justify his indefensible actions with a simultaneous revulsion for how things in Australia have turned out. 'If there was any morality, there is no more,' he says in self-justification. 'Modern war has destroyed what once seemed immortal; that nothing is sacred, nothing unthinkable.' Talking late at night to a call-girl in her apartment he tells of how 'ordinary men and women like you and me are at this moment calmly prepared to unleash the holocaust as soon as they receive the signal.'[134]

As the play reaches a denouement, a soliloquy by Wilson crystallises Sewell's observations about how precious belongings can so quickly and irretrievably be lost:

> Let me show you the power of evil! The city . . . one of the most beautiful cities in the world. A city sprung from greed and ambition, built from human flesh, every building the grave of a labourer whose life was less important than the profit its construction meant. The city . . . thrown up by finance and speculation, conceived in bribery, corruption and murder. A sparkling, empty edifice of dreams and nightmares. Listen to it . . . listen to it sigh; listen to its misery, its glory, its hunger. Feel the hum as power flows through it, organising it, animating it, connecting every part of it from the magnate in his penthouse to the single mother in her room . . . The city of death, glistening in the light of its own conceit.[135]

Dreams in an Empty City failed at the box office and had very mixed reviews when it was staged at London's Hammersmith Lyric theatre. Sheridan Morley, writing in *Punch*, was impressed by *Dreams* as an 'urban nightmare of the Sydney underground'.[136] But it was damned in *The Listener*, where Jim Hiley described the play as 'the most lamentable export ever from the antipodes . . . I can only imagine this venture is an act of post-imperialist revenge, randomly targetted on West London'.[137]

Another *Bulletin* theatre reviewer, Jeremy Eccles, described Sewell's 1988 play *Hate* as 'undoubtedly the most important . . . of this Bicentennial year [with the theme that] Australia is a country that's never believed in anything, has never achieved anything, has just staggered on in the hands of self-seeking power-mongers'. Eccles further notes how Sewell's ecological concerns are at the core of much of *Hate*, in which the western districts of Victoria are variously described as 'loveless' and of an 'arid emptiness'.[138]

ONE of the earliest plays to directly confront environmental destruction was Jack Hibberd's *A Stretch of the Imagination* (1973), which captures an atmosphere of human decay and desperation. The play, first put on at Melbourne's Pram Factory by the Australian Performing Group in 1972, is a monologue by a man at the end of his tether and at the end of the road.

This wreck of humanity, Monk O'Neill, appears as an unkempt and disjointed misfit unable to cope with the crumbling world around him. 'Nature shrieks,' he laments as he greets the audience. As he looks about him, O'Neill sees nothing but 'a panorama of dust, ant-hills, and dead grass. Nought else'.[139]

Here is Monk O'Neill, at the start of the 1970s, decrepit and irrelevant as he falls victim to the surging world around him. How can the earth thrive when such as he walk about, however haltingly, creating mischief?

The world has passed him by, and both O'Neill and this world are equally delinquent in attempts to make a better fist of things. About all O'Neill is capable of is whingeing. 'A dry hot day,' he moans, on and on and on:

> Still. Not a breath of wind. Heat. Hot. The sun. An emphasis of red sand. No moisture in the sky. Nor moss beneath the feet. Grey grass. Bleached-out weeds. Granules. And sticks. Fire. Fire. *(Pause)* Subsist. The rest is silence.[140]

To be present at O'Neill's demise is to face a most unpromising future, with mankind's stupidity at the heart of the malaise. With the callous and uncaring manner he displays even in these dog days of his life, it is no wonder the world is stuffed. O'Neill realises very accurately the monstrousness of such attitudes, but he selfishly drives all doubt or conscience from his mind. Total destruction is such a simple, mechanical and reflex thing:

> When I first came to One Tree Hill there was this one tall tree. Nothing else. A hairy tree. Conspicuous, almost outlandish. Not wishing to advertise my presence, I took an axe and chopped it down. Thud.[141]

Even from within his haze of indifference and thoughtlessness, O'Neill realises the enormity and short-sightedness of such deeds, but his petty and mean-spirited mind still manages to hold sway:

> I looked at the tree . . . and the uproar was immense. Every cockatoo, crow, emu and rosella in the kingdom had taken to the heavens filling them with spleen and indignation. Lesser men would have regarded this as a harbinger and knelt down in supplication. Not me. I seized the old shotgun and fired salvo after salvo at the demented pricks.[142]

O'Neill's obliteration of one tree is consistent with what Hibberd has described as Australians' attitudes towards vegetation and the land. 'Australia's native vegetation seemed entirely foreign to arriving colonists,' Hibberd wrote in a 1996 essay for the Melbourne *Age*. Hibberd described how 'contempt' was fostered towards trees which were regarded by Australians as merely fodder for 'clubs, fires, houses, bows and boats'. He further described an 'Australian arboriphobia' whereby 'electricity authorities, who employ gums as poles and hack trees for wire access, are the exemplars. [The Keating Government's] failure to halt all wood-chipping and force all television cables underground showed that it had been seeped in the poltroonish culture of arboriphobia. The cults of the stump-jump plough and chainsaw bite deep'.[143]

Hibberd's strong environmental concerns were recognised by theatre critic H.G. Kippax when reviewing a revival of the play in 1977. Harry Kippax noted in the *Sydney Morning Herald* how in *A Stretch of the Imagination* the land was portrayed as 'a victim of aliens and predators like Dr Hibberd's ancient anti-hero, Monk O'Neill, the hermit of One Tree Hill whose first action on taking up residence was to cut down the tree'. According to Kippax, O'Neill is crafted in such a way by Hibberd that the character 'emerges . . . as a compulsively destructive element, in human relations as in nature'.[144]

In *A Stretch of the Imagination*, an angry and compelling play, Hibberd insists on how we all must at all times husband and nourish our little acre, and any belief which suggests that any other way is possible is truly a stretch of the imagination.

Hibberd's forthright theatrical statements stem from his being an integral part of the burgeoning New Wave of the late 1960s and early '70s whose Melbourne contingent revolved around the Australian Performing Group at the Pram Factory (with John Romeril and David Williamson among the other notable alumni). Hibberd, who became a medical practitioner after graduating from the University of Melbourne, was brought up in similarly humble surroundings as the other playwrights who found a new voice via the Australian Performing Group. The son of the local plumber, young Jack was brought up in rural Bendigo and educated at the Marist Brothers College.

A Stretch of the Imagination, which Hibberd wrote at age 27, was a significant offering during the radical theatre of an era in what Dorothy Hewett has cleverly described as 'a form of folk theatre without being folksy . . . a theatre of gum without being gumnut'. The play was, according to leading critic Margaret Williams, 'our first unmistakable theatre classic'.[145]

A Stretch of the Imagination enjoyed many revivals throughout Australia, and its environmental concerns became widely recognised. Of the 1991 revival at The Playhouse in Canberra, *Canberra Times*' reviewer Ann Nugent described Monk O'Neill (as portrayed by Max Gillies) as 'a symbol of Australian failure, the often unsuccessful struggle by the [European] "pink" tribe to establish any meaningful culture in the vast continent'. According to Nugent, Monk is depicted as someone who 'lives out his dying days uncomprehendingly and still hating the things around him'.[146]

Seventeen years earlier, in reviewing Harry Schmidt's interpretation of Monk at the Australian Theatre Workshop at Canberra's Childers Street Hall, the *Canberra Times*' Paul Barron described this play as 'one of the major works of contemporary Australian drama' in which the playwright examined 'a bleak, barren landscape and a civilisation based on hypocrisy and a people who have lost contact with the "real" world'.[147]

In 1987, *A Stretch of the Imagination* became the first Australian play to be professionally staged in China with a successful season in Shanghai. It became recognised internationally as a potent and provocative work that dissected the Australian ethos. 'Monk O'Neill' became part of the language for what he represented. According to the *Australian Financial Review*'s theatre critic Paul McGillick, 'Monk, in his values and relationship to the land, is all Australians. The greatness of the play, though, lies in its universality—a fact proven by the many performances around the world in different languages'.[148]

ENVIRONMENTAL themes are central to many of the plays by Alma De Groen, who won widespread recognition and critical acclaim for works which dealt with other important issues like feminism. In 1983, a decade after *A Stretch of the Imagination* was staged by Brisbane's La Boite Theatre, it presented De Groen's *Vocations* (1982), first performed in the previous year and revised for the Brisbane production. Hibberd was the dramaturg on *Vocations* at the Playwrights' Conference.

This play represented an early indication of De Groen's growing concern about environmental issues, which would blossom with a series of powerful plays in the late 1990s. In *Vocations*, De Groen is scathing in her depiction of how someone committed to fighting and caring for the ecology is a hideously flawed and ineffective human. This character is Ross Hepple, a zoologist in his late 20s, who is shown by De Groen as someone concerned about the fate of his planet in a purely scientific, abstract fashion—someone without the merest glint of a soul who treats fellow humans, especially his companion Vicki Bloom, an actress, with total shabbiness and unconcern. Ross believes he stands for the loftiest ideals but little realises how these are diluted by his simplicity, immaturity and fanaticism.

De Groen is savage against the pretensions, feyness, gullibility, self-delusion and ridiculousness of the likes of Ross, even though he refers to the Gaia Hypothesis, a concept close to De Groen's heart:

> **Ross:** That's hopeful to me—the fact that even though we've produced the neutron bomb we still know what Plato said.
> **Vicki:** What did Plato say?
> **Ross** *(cornering her)*: He said the world is an animal. A living breathing soul . . . An organism, capable of healing itself! What it means, Vicki, is that the earth we're inhabiting is alive! . . .[149]

Ross displays an innocent and naive optimism which is sadly out of place in this drama of sordidness, manipulation and arrogant power moves dictated by a careless and apathetic society. In this study of the priorities of life and what defines a successful and satisfying existence, Ross is found sadly wanting. Even when it comes to balancing relationship and career, Ross is out of kilter. The environment awaits its fate without him. This is despite his breathtaking presumption of what his own powers are and what they should be:

> **Ross:** . . . what I want . . . a stupendous discovery that says Yes, optimism, not pessimism, is the driving force of the universe. But you have to look, you have to find that order in what you do, and ignore people being negative about what you're trying to achieve.[150]

Some of De Groen's environmental consciousness stems from her upbringing in New Zealand's North Island, where she was born in 1941. Alma Mathers spent her girlhood in what has been described as 'the largest pine forestry plantation in the world' and she learned early in life the consequences of living in a 'small shrinking township, founded to serve a hydro-electric power station . . . destined to be whittled away as the surrounding area was inundated'.[151]

Vocations is an often witty warts-and-all play featuring uncompromising dialogue by a playwright who holds none of her rage in check. It is believed to include the first mention in Australian drama of the Gaia Hypothesis, a fact of significance to De Groen. 'I'm sure it preceded "Edge of Darkness",' De Groen said in 1999. 'I was writing *Vocations* in the late '70s before the Gaia Hypothesis became common currency.'[152]

De Groen, who for two decades has lived in places as varied as bustling East Sydney and the comparative tranquillity of Leura in the New South Wales Blue Mountains, has an all-encompassing concern about Australia's environmental future, and she believes no human action can be separated or quarantined from wider consequences. If nature is degraded, De Groen has stated, then so will everything else be degraded.

Discussing her new play *The Woman in the Window* (1998), De Groen told Peter Robinson in an interview for the *Canberra Times* that it was based on the premise that 'if nature disappears then art as we know it will disappear'. She told Robinson of her extensive fears: 'My play portrays a future when the night sky is destroyed by an array of space stations.' In that night sky, the size of the moon has shrunk, and the unclear causes for this are ominous.[153]

The Woman in the Window has several important concurrent themes, flitting between the plights of celebrated Russian poet Anna Akhmatova at age 62 still living in the dying days of Stalin's regime, and Rachel Sekerov in her early 20s in the Australia of the future. Akhmatova, having survived early Terrors and developed a fatalistic attitude towards survival, is strong and street-smart; whereas Rachel is confused and subjugated and quite at the mercy of a high-tech Stalinist State of a futuristic Australia.

The nature and beauty that Anna knew and experienced are but vague concepts for Rachel. What was once glorious reality is now cynical artificiality. Sandor Voss—Rachel's lover who is resident of one of the millions of space stations that travel around the earth—makes the occasional pathetic attempt to rekindle a once exuberant, bursting and life-filled planet:

> **Sandor:** There should be birdsong . . . something relaxing. I'll turn on the birds.
> *(He touches his tracer: the harsh and deafening sound of kookaburras and cockatoos. Rachel cringes, covering her ears.)*
> **Sandor:** Sulphur-crested cockatoos, and kookaburras! Don't know which is which!
> **Rachel:** Stop it! Please![154]

Meanwhile, in Stalin's Soviet Union, the environment was utterly degraded, as was the environment of Australia generations before Rachel's era. Anna's friend Tusya, although only in her early 20s and the same age as

Rachel, is well aware of the devastation—and the consequences of it—happening around her:

> **Tusya:** People are getting sick at Chelyabinsk so they've stopped dumping nuclear waste in the river, and now they're diverting it into a lake. The lake is self-contained. Yuri asked what would happen if there's a hot summer and the lake dries up. There'd be radioactive dust blowing for hundreds of miles![155]

A constantly deteriorating environment is happening simultaneously as the last stages of individuality and uniqueness are being snuffed out in Russia; they have been non-existent for generations in the Australia of the future.

In the two societies depicted by De Groen goodness and decency have been obliterated, and the most sordid kinds of behaviour prevail. Akhmatova, unlike the utterly subjugated and conformist citizens of Future Oz, still believes despite all the persecution that she is partly to blame for what is happening:

> **Akhmatova:** We danced and made love on the edge of an inferno. We tumbled into Terror in our Carnival masks and our satin shoes. There were people starving, who had no shelter, who couldn't read, and we didn't care.
> Are we part of humanity—or just some awful lesson?
> No one knows the era they're living in until it's too late.[156]

The future of Australia as envisaged by De Groen is enough to make one weep, or laugh—it's so horrendous and outrageous. Both tragedy and comedy are the stuff of a radio play, *Stories in the Dark* (1996, first broadcast 1995), co-written with Ian Mackenzie.

Stories in the Dark, Australia's entry into the 1996 Prix Italia, Drama Section, is also set in the next millennium. An outfit called Emu Tours has an office in Beijing, where there is a pollution alert with 'particulates exceeding 1,500 times the safety limit'. With such conditions, interest in Fun Park Australia is very intense, and the Great South Land has become known as 'the Earthly Paradise'.

Despite its serious intentions and the bleakness of the future it presents, the play turns into a delightful spoof. Society has degenerated to the extent that cyberspace rules OK and human input and influence are a distant memory. As Chloe, a classic child of this brave new cyber age, tells her mother Basia: 'Face it, ma, you've been out-evolved by technology.'[157]

As in *The Woman in the Window*, the death of culture is the death of everything. The lowest common denominator prevails, and a popular artist like Ken Done is remembered as 'one of our great Masters from the 20th century'.[158]

This utopian Australia has arisen on a planet being destroyed by global warming and depletion everywhere. We are told that the dome of the Sydney Opera House was built at huge expense when the ice caps starting melting . . . The artificial island and the elevator shaft were built up in prefabricated sections, to keep pace with the rising ocean'.[159]

The play shows a degraded and poisoned world in chaos and confusion, to the extent that Lapstone Wharf had become 'one of the few places in the world where the fish are safe to eat'.[160]

De Groen and Mackenzie have also collaborated on the film *Deadly Access*, which was in pre-production in 1999. At the same time Mackenzie was working on a treatment for an environmental film, *Green Zero*, and he was commissioned to write another for ABC Radio called 'The Taxi Files'.

Mackenzie, born in 1966 at Dunedin, New Zealand, described his main environmental concerns as 'short-sighted economic rationalism' which ran counter to long-term survival, as well as 'appalling religious beliefs inflicting suffering and damage on not only this generation and its environment but as many future generations as they can'.[161]

With plays like *The Woman in the Window* and *Stories in the Dark*, De Groen has confirmed her reputation as being ahead of her time with a range of social and environmental concerns.

A MAN standing up against the mob to fight the degradation of his environment is a central theme of Alexander Buzo's *Batman's Beach-head* (1973), a powerful variation on the theme of an inexorable slide into ecological destruction. Based on Henrik Ibsen's *An Enemy of the People* (1882), Buzo's adaptation is set in a fictitious—but alarmingly typical—country town on the Victorian coast, whose local burghers are prepared to tolerate—and cover up—every environmental outrage for the sake of short-term profits and their own positions within the community.

The play accompanied the twentieth anniversary of the Melbourne Theatre Company, at a time when Buzo was becoming a recognised and popular playwright who had tackled a variety of themes beginning with *Norm and Ahmed* (1968), his searing study of racism.

Although based on Ibsen, *Batman's Beach-head* is in many ways a play ahead of its time, discussing a range of environmental issues years before they seeped into the broader popular consciousness. Critic Peter Fitzpatrick noted that there had been fears that the Melbourne Theatre Company's 'cautious' subscribers might not be impressed by the play's themes, but these proved unfounded as 'conservation was by 1972 not just modish but thoroughly respectable'.[162] Just as Ibsen had challenged Scandinavians almost a century earlier, Buzo urged an Australian audience to cast aside complacency about the increasing pollution around them.

As we meet the leading citizens of Batman's Beach, they tell of an idyllic spot—which is even better with riff-raff stringently discouraged—a tourist hub far from the madding metropolitan throngs of noise, carbon monoxide, advertising over-kill and an exploding population. These local well-connected boosters do not take kindly when Tim Stockman—a local GP, conservationist and brother of the mayor—warns that, rather than being a pristine wonderland, Batman's Beach is facing an environmental crisis. 'Our whole community's polluted, not just the water,' Stockman warns.[163]

The locals are forced to defend their various vested interests against such threats of exposure, and a web of corruption and gross mismanagement emerges. The only 'logical' response by the city fathers is that pollution is happening everywhere, and that Batman's Beach is not nearly as sullied as are many other places. The attitude becomes clear at the beginning of the play, as Bruce Halstead, gun regional TV journalist, tells the influential locals: 'Batman's Beach is the place to be. Clean, beautiful, everything on tap. The air's fresh, the water's pure, the beach is immaculate. This isn't a seedy resort town like a few others I could name.'[164]

As the play progresses, Halstead's observations prove to be pure wishful thinking. Dr Stockman's research has uncovered that muck of every variety is harming the area, the fish are full of mercury and, because of the outpourings of the nearby oil refinery, water purity is at risk.

But such truths are inconsistent with the township's immediate economic health, and mayor Peter Stockman leads an orchestrated and intense campaign to discredit the scientific findings. Buzo cleverly describes the eternal conflict between people with different priorities and agendas in an environmental crisis:

> **Tim Stockman:** I'm the one who has the good of the town at heart.
> **Peter Stockman:** You want to wreck us, destroy our chief source of income.
> **Tim:** But the source is poisoned! Our town is based on a lie!
> **Peter:** Anyone who wants to destroy his own town must be a real ratbag, as far as I can see.[165]

As the criminal and grindingly efficient machinery of cover-up gets into full swing, Tim Stockman in utter desperation calls a meeting at the local greyhound track where he intends to inform the local residents about the perils they are facing. This event sinks under his own overwhelming intellectual pretensions and because of his inability to prevent his brother the mayor and his stooges from manipulating and derailing the increasingly farcical proceedings. As happens at all levels of government, vital information is stifled under points of order, and pollution continues to ravage Batman's Beach as debating niceties prevail over effective action. Dr Tim gives an eloquent speech about the planet being in mortal peril, but the townsfolk ignore him.

Throughout *Batman's Beach-head* are instances of one brave and committed person standing up against the many, but meanwhile the sludge makes a greater sweep across the waterways and more fish are being poisoned.

Another Buzo play performed at about the same time, *Tom* (1972), is set in Sydney 'now or soon' and deals with the corporate ethics—or lack of them—which allow environmental outrages to happen. *Tom* shows a corporate world embarking on a path of destruction; there is no place for rational debate and all human effort must be suitably adapted to meet the needs of development. The ethos, as heartily endorsed by the citizens of Batman's Beach, is succinctly explained by oil executive Stephen:

> Controversy and argument are totally outmoded and incredibly uncool. Opinion and polemics are of no consequence! People must be taught to understand, even to investigate for themselves—but never to question. It's out of date in the conglomerate of the universal metropolis.[166]

This is pure hyperbole, but the attitude it reveals is extremely widespread, as Buzo warns. All the characters in this play, especially Stephen, have a manic zeal to subvert everything and subordinate it to the corporate cause. In one of the more outrageous lines of the Australian stage, Stephen states what he believes to be self-evident: 'When you see an oil slick on water, it's the water that's the intruder. Oil is the ultimate and should be spread.'[167]

(If this seems somewhat farfetched, it was matched by reality. 'We care about water pollution, but it isn't as important as a $100 million industry . . . Pollution of minds is a more important problem than air or water pollution,' said Victorian Premier Henry Bolte in 1970.)[168]

Lurking behind Buzo's entertaining parody of human foibles and pretensions is a serious exposé, also years ahead of its time, of citizens coping with drastic technological changes beyond their control. 'To survive we need to change mankind completely,' Stephen says, stressing the need for a 'new behaviour'.[169]

Tom has been described by critic John McCallum as a picture of 'the nadir of Australian materialist, hedonist society' which provides 'a valuable antidote to the hype which surrounded the natural resources boom of the '70s'.[170]

McCallum praises Buzo for being able to savage the pretensions and self-delusions of people who see themselves as movers and shakers—and, in the process, destroyers—but who, in the end, will inevitably be spat out by people even more vicious and ruthless than they are.

Much of Buzo's concern about the environment—and of the failure by city planners and civic leaders to properly address issues—sprang from his schooldays at Armidale in the New England region of northern New South Wales. As Buzo recalled these times, he told of how town planning in Armidale not only displayed all the tenets of philistinism but also was utterly impractical. 'Locals cannot recall a day that has passed without two cars running into each other at one of Armidale's hundred blind intersections,' Buzo wrote in his contribution to a collection of memoirs, *Country Childhoods* (1992). 'The city is laid out on two remorseless grids running east–west and north–south, with the creek providing the only geometric relief . . . the highway runs right through the town.'[171]

In his non-fiction writings, Buzo shatters the illusion of Armidale being a tranquil and contented place. He is even vitriolic about the town's setting around Dumaresq Creek, which he describes as a 'winding, spiteful little tract of what purports to be water [and] has long been a town joke . . . for years it has been a grudging conveyor of oil and effluent'. The scene is rounded off quite starkly: 'In floodtime [the creek] swells across roads and parks and forms a wide brown band through the valley, stopping just short of Beardy Street, the main drag and centre of Armidale's café society.'[172]

Buzo also spent much of his early life in Sydney's Lower North Shore, and in later years he became concerned at how once tranquil and vegetation-lush suburbs like Cremorne and Neutral Bay—with gracious old houses and a profusion of bird-life along the tree-lined streets—were being changed forever by developers ripping down what was there and replacing it all with home units and town houses. Philistine middle-class attitudes and their disdain for precious belongings which allows old glories to vanish are analysed in two Buzo plays of the late 1980s, *Stingray* (1987) and *Shellcove Road* (1989).

Both plays deplore the greedy incursions of developers, but Buzo notes that these would not have been possible without 'the code of silence which exists among the middle class'. He said in a 1989 interview: 'What I regret is that the things I want to turn into subjects for art—our suburbs—are disappearing before my very eyes. They're gone before they've been immortalised.'[173]

FIGHTERS for the environment are held in similar disdain in *Money and Friends* (1992), by one of Australia's most successful playwrights, David Williamson. *Money and Friends* deals with a range of issues including how well-to-do and ambitious people with selfish agendas deal with materialism and corporate interests. It is set in the township of Crystal Inlet, a beachside settlement of weekenders where we meet four couples getting away from it all.

Early in this play we meet trendies full of moral outrage who do a lot of talking about the environment but who seem unwilling to actually do anything about it. For example:

> **Conrad Wesson** *(a TV personality specialising in the environment)*: What the developers don't seem to realise is that if they bulldoze the mangroves to put up their hotels, the fish the tourists come to catch won't be there in 10 years' time. Greed, ignorance, apathy—the deadly troika.
> **Peter** *(a professor of pure mathematics)*: It's depressing . . .[174]

The effectiveness and the very integrity of Conrad are further questioned by Margaret, a university history lecturer, who in a soliloquy to the audience questions people's motives:

> **Margaret:** Turn on your television and there's Conrad taking us off to yet another Environmental Outrage. I'm not saying Conrad's not sincere. He is, but unlike many of us with similar feelings, he had managed to turn Environmental Concern into a nice little earner . . .[175]

Conrad may have a high opinion of himself, but his environmental message is being submerged by his arrogance and shortcomings, dismissed as 'that big-mouth media greenie'. Margaret, described as in a 'sudden full-on rage', defines the hopelessness of it all, with environmental concerns extremely low down on the priorities of the trendy set:

Margaret: When I come down here the only motives on display are making money, getting one's photo in the social pages . . . I'm sorry, but it is utterly, utterly depressing![176]

David Williamson, born in Melbourne in 1942, during the early 1960s became the first graduate in mechanical engineering at the then new Monash University and later taught thermo-dynamics at the Swinburne Institute of Technology until he began his glittering career as Australia's most popular playwright.

Much of Williamson's characterisations are based on his own observations and experiences from his time at Monash and Swinburne. During this time, according to his literary biographer Brian Kiernan, Williamson had already concluded that 'human beings are essentially selfish creatures who seek to further their own ends and win approval by pretending to altruism and compassion, but who are more motivated by irrational sexual, territorial and aggressive instincts than they care to admit, even to themselves'.[177]

Williamson regarded the theatre and his own plays like *Money and Friends* as direct efforts to influence people's thinking and to be a vehicle for change. Throughout his career, Williamson has regularly lent his support to environmental causes and has shown an abiding concern to prevent wanton destruction of precious things. Various environmental issues were of equal importance to him, and he was 'fanatically' opposed to the pollution and unhealthiness of tobacco smoking.[178]

In June 1998, he joined a cavalcade of showbiz personalities in signing an open letter in support of the Wilderness Society which at the time had come under attack from several fronts, including the Queensland Premier Rob Borbidge. The letter criticised the Queensland Government for its environmental record, especially the planned Port Hinchinbrook resort and residential development.

Williamson was a long-time peace and anti-nuclear campaigner, beginning with opposition to the Vietnam War. He and Patrick White were among marchers on Palm Sunday in 1983 in support of nuclear disarmament and he also participated in peace rallies over the years.

During the '90s Williamson lived firstly at Pearl Beach on the New South Wales Central Coast, then at Noosa on Queensland's Sunshine Coast—both places where he could observe the threat of rampant development which would engulf and overturn communities that had been there for generations.

THE influence of plays on the green consciousness is incalculable and they transcend what appears on the Australian stage. As Leonard Radic has noted, Australian plays are 'everywhere—in print, on videotape and film, in school libraries and on university courses, and in the repertoire of every subsidised company in the country'. Discussing the extent to which Australian plays have attained their own unique identity, Radic, writing in 1991, goes on to observe: 'It is a very rare week in Sydney, Melbourne and the other

State capitals where there is not at least one new Australian play in production'.[179]

With some hyperbole, Radic has hailed the ground-breaking and nation-building qualities of Australian drama and likened some of the works of White, Williamson and Romeril as on a par with the classics of Shakespeare, Ibsen, Chekhov and Brecht. But few would dispute Radic's touchstone claim that since the early 1970s Australian drama has 'come excitingly into its own'.

Many of the plays have deep within them a hope which can conquer the bleakness and destruction they portray. In a London review of Stephen Sewell's *Dreams in an Empty City*, *The Guardian* critic comments on how 'unchecked greed and ambition will in the end destroy themselves and . . . human goodness can never be extinguished'.[180] This assessment, coming in the year of Australia's Bicentenary, expands an optimism that the ecology will somehow prevail against the most horrendous of threats.

But any thoughts of a bright future ahead must be significantly tempered. In another review of *Dreams in an Empty City*, the London *Observer* noted how this play had revealed how language—and concepts—can become a 'devalued currency' whereby 'words such as "principle", "civilisation" and "humanity" are held up to the light and shown to have no value'.[181] And Louis Nowra, the prolific critic of theatre and culture, has described Australia as a place without moral or ethical questioning. 'Compared with Americans, who came like us from a frontier heritage and who have a continual spirit of self-questioning, we seem to be confronted by some spiritual anaemia,' Nowra said in 1983 in a discussion of his *Sunrise*.[182]

The films and plays discussed in this chapter have extolled the importance of giving value and credibility to all the vocabulary of fulfilment and survival, and of inspiring their audiences to overturn the madness of the real world outside the theatre and cinema.

In Australian films and plays over the past thirty years there are messages of hope in a variety of contexts. These messages are often complex and must surmount much of the horror and doom that also confront us. But for the sake of our lives, and the lives of the generations to come, we must heed the ominous warnings and take heart from the glorious optimism.

To paraphrase Louis Nowra, plays aren't simply produced and films simply made, they *demand* to be produced and made by civilisations concerned about their fate.

David Williamson, in a personal letter to author Peter Carey written in 1993, mused about his growing love for Australia. He wrote, 'As I get older, the love of this country, like your own, starts to focus more and more on the landscapes.' He spoke of his gratitude that life had given him the chance 'albeit briefly, to marvel at the interlocking weave of life forms around me'. And he spoke longingly of his dream of getting away from

Sydney and moving to the tropical regions of Queensland, which he and his wife duly did. He writes: '[T]his is a wonderful country . . . I sometimes feel as if a benevolent but mischievous god gave our forefathers the use of a land they tricked up as hellish, then slowly revealed to us, their descendants, that in fact we had been given paradise.'[183]

7

SONGS AND POEMS
Troubadours and troubled times

THE earliest settlers of Australia still had their hearts and minds in England, and they saw the local landscape as hostile territory. Most of the voyagers of the First Fleet and the first generation of European settlers saw the antipodean land as nothing but desert or scrub. Unlike the beauty of England, there was nothing here worth preserving. Such attitudes would change only when songs and poems inculcated a new set of values.

Environmental poet Mark O'Connor probably put it best at the start of the 1990s: 'Judith Wright's great environmental poems written 30 or 40 years ago are today stopping more bulldozers than any single group of activists.'[1]

In another interview in 1990, O'Connor also recognised the importance of literature in instilling respect and love for a nation's environment—and in this regard saw Australia as singularly lacking. 'You can't bulldoze the British Lake District, for instance, because the great English writers have written about it,' O'Connor said. 'But what Australia got lumped with was an English-speaking culture that didn't see any part of the continent as sacred—in the way the Aborigines see it. Maybe that's why it's not seen as a crime to chop down a rainforest—and people confess if they lusted after their neighbour's wife but not if they "sinned" against the environment.'[2]

O'Connor expanded on this theme in 1997, stating that if citizens sought to fight destructive developers and miners in treasured places it was necessary to 'give them an opera, a novel, a poem or a symphony' to instill a conservationist ethic. O'Connor has championed the cause of poetry as creating a 'national awareness' of what was happening to the continent.[3]

As for songs, music has been part of every revolution, both the successful and the unsuccessful. It has often been quipped that although Franco's Fascists won the day in Spain, the Internationals had the best songs. Both sides of the American Civil War produced songs to inspire the troops and keep the home fires burning. And the spirituals of enslaved African–Americans inspired a people to hope when all hope must surely have seemed lost. People need music; people fighting against long odds need songs to give them courage when they're feeling fainthearted; to crystallise their beliefs and to remind them what they're fighting for.

Songs have been sung at all of Australia's environmental battles—at Terania Creek, the Franklin and at Jabiluka. The songs have not made it into the mainstream and most of them are known only to the protesters who sang them. These days they are being passed on at workshops and festivals in the hope that they will inspire others at the other battles which are sure to come.

But in the wider picture, the songs a society sings are a reflection of what that society cares about. It gives great hope for our nation that so much of our mainstream music (the music played on the radio and sung in concert halls) is thoughtful, provocative, stirring—and that our Australian environment has become a worthy subject for song.

JAZZ singer and trumpet-player Vince Jones has long been a voice of conscience in Australia. He has performed at fundraisers for numerous good causes over the years—for the Wilderness Society, for Amnesty International—and has taken stands on issues as diverse as freedom in South Africa, saving the ABC, the shenanigans of Richard Nixon and the troubles of youth. One of his greatest passions is for the environment.

Jones—who sometimes baulks at being solely identified with jazz, preferring the term 'world music'—has said that Australia, with its wilderness and wide-open spaces, has always presented for him a sharp contrast to Europe. He said in 1996 that his boyhood in Glasgow had given him an appreciation of nature because he had seen so little of it. Describing Glasgow as 'so industrialised', Jones commented: 'There were factories everywhere and the rivers were so polluted. We lived in 10-storey tenements. There was nowhere wild; everywhere had been walked on. You come to Australia and it's wild.'[4]

The experiences while touring in Europe in the mid-1980s brought this home to Jones again—and strengthened his commitment to the bush. Around that time, green themes began to appear in his music.

Jazz is not usually regarded as a comfortable place for social activists. The image of jazz is that it is an exclusive medium, infinitely subtle, and which speaks powerfully to its initiates but is often closed to outsiders, offering them just tantalising glimpses of a private country. But, unlike most jazz performers, Vince Jones is engaged with society at large.

In *On the Brink of It* (1985), the title track adds images of world destruction to the standard 'life is short' of musical seductions:

> Don't talk about moral convictions
> Cold wars or atomic predictions . . .
>
> We're on the brink of it.[5]

In the track 'Boys on the Corner', Jones laments the fate of young people in a ruthless, mechanised society. Jones unleashes a swirl of vivid images of urban decay.

On the album *Tell Me a Secret* (1986), the hauntingly lovely 'Sensual Item' laments a particular lost love—a woman changed by time and circumstances. But the narration expands that sense of loss to suggest that the singer feels lost in the world, that he has lost his personal power over his own fate:

> No one ever asks my point of view
> Still you make the decisions that you feel

Convincing others we need the new
Murals painted everywhere in sight.[6]

In the song 'Too Much Too Soon', Jones lambastes the mindless conformity of yuppiedom, where everyone thinks the same way and where the goal is to accumulate as much stuff as possible:

Words that you just cannot use
Fruit that nobody can eat
Places we don't go to
Doors that you can't go through
What have they done to this place?[7]

Jones fans have never been in doubt about where he stands on issues. When performing in clubs, he intersperses his musical numbers with what has been described as a stream-of-consciousness patter, underlining the subtle message of the music with bolder interpretations. Although some reviewers have criticised this practice—saying the audience fidgets and that the comments are naïve, intense or cryptic—when the authors saw Jones at Sydney's The Basement on 19 January 1999, the audience was very much on his wavelength. He kicked off the concert by saluting the fact that it was smoke-free (at his request), which got a round of applause and calls of 'Thank you' from the crowd. The audience, on that night, was an interesting mix—from hard-core jazz enthusiasts (who responded with murmurs of appreciation to any difficult or impressive riff) to middle-aged suburbanites and great numbers of young people, including one group who looked so young that they had to sweet-talk their way past the doorman.

Between songs he told little stories about his boyhood, how he felt then about this 'beautiful blue planet of ours' soaring through space and lamented the fact that greedy people—who would have known better when they were little boys—were spoiling everything. Late in the concert, which consisted mostly of very jazz-oriented love songs (Jones said the theme of the concert that night was Love), he dedicated a piece to the '600 or so gentle souls up at Kakadu who are protesting against the expansion of uranium mining'. He hailed the protesters as 'not doing it for the money, you know, they're doing it for you and me—and for your children'. Jones expanded on this theme for a few minutes, joking that maybe some uranium should be dumped in the backyard of Senator Robert Hill, the Federal Environment Minister, because he thinks it's so safe.

The remarks were applauded vigorously, and the song—a very jazzed-up version of 'If I Had a Hammer'—brought the house down. Jones dedicated the last song of the evening to three Aboriginal women whom he'd met at Jabiluka. They were three of the traditional owners whose lives and works he found so inspiring. The song, 'If One Woman Awakens', was a lovely ballad about personal empowerment and how we can all influence each other and the course of events—for good or for ill.

Perhaps Jones had honed his patter a bit—or he had a particularly sympathetic audience on the night the authors heard him. In a 1994 interview, Jones said that sometimes people didn't like it when he talked between numbers and that perhaps it didn't pay to be too 'hard-edged or aggressive'. He said that sometimes it could be risky, talking to people like that, 'but I can sneak it in . . . the public are pretty cool about it'.[8]

Jones knows something about risk. In 1994, the engine of his private plane was apparently tampered with, in what he believes was either a warning or an attempt on his life. A year later, he said the incident had happened because he was speaking out against logging. 'Someone tried to kill me,' Jones said. 'And that was because I stood up and said what I thought, supposedly in a country of free speech.' The day before the incident, someone had written 'Greens cost jobs' on the rear window of his car.[9]

The life of Vince Jones has mirrored his environmental concerns. A vegetarian, a non-drinker since 1991 and a passionate opponent of tobacco, for about five years he left the city to live in the forests of East Gippsland in the Australian Alps. From early 1993 to 1998 he lived 800 metres above sea level in a cottage overlooking the Murrindal Valley, near the village of Buchan, commuting to gigs in Melbourne and Sydney, usually flying himself there in his Cessna 180. He said at the time that playing in smoke-filled clubs had begun to affect his health. In fact, at one Sydney gig in 1990 he did the then-unheard of thing in demanding that audience members put out their cigarettes before he would continue singing. Since then, he has been doing an increasing number of smoke-free gigs and has been in the vanguard of artists calling for tobacco-free workplaces.

At the time of his move, Jones frequently spoke out against the artificial and unhealthy life of the cities. He told the *Age* in 1994 that he'd liked it in Melbourne, where he lived in the bayside suburb of St Kilda, but had realised it was a very unnatural existence. 'I have a lot more control over my life than would people in the city,' he said 'They're on the track some economist has directed them.'[10] And he told the *Sydney Morning Herald* that he was appalled by the pollution of the cities. 'We are the first generation of petrochemical kids, and it's going to rear its head somewhere,' he said. 'You can already see it with all these old people and kids falling over with asthma, building up toxins.'[11]

Jones rejoiced in the unspoiled grandeur of the wilderness and in the freedom he found there, but as he began to spend more and more time in the country he found a new awareness about the speed at which wilderness areas were being gobbled up by 'progress'. 'The whole ridge across from my farm has been logged totally,' he said in 1993. 'Now they're bulldozing a road into a brown mass and in five years' time they'll subdivide it and sell it as farms.'[12]

In early 1998, Jones moved away from his tranquil eyrie for the New

South Wales South Coast so that he could be within striking distance of Sydney. He told jazz writer John Shand that although the sojourn in the wilderness had been good for his lungs, he had missed being amongst the musical scene. 'The only thing you can really do in the country is go to local talent nights,' he said. 'There's not much music around . . . In the city I can at least go to poetry nights and I can hear the reflections of young people. I get more interested in reading about the social condition when I'm living in the city. I like getting into protest, too. There's none of that in the country.'[13]

Jones' most recent protests have been against the Jabiluka mine in Kakadu National Park. He has joined the protesters on the barricades, singing his songs and lending his famous voice to swell the rising tide of anger at the developments there.

SONGWRITING veteran John Schumann hit the headlines in the aftermath of the 1998 federal election when he, as the Australian Democrats' candidate for the seat of Mayo in South Australia, gave Foreign Minister Alexander Downer a scare. The singer-songwriter and former frontman of the band Redgum (from 1976 to 1986) came close to ousting the prominent Liberal from his blue-ribbon seat and becoming the first Democrat to sit in the House of Representatives.

After his failed political bid, Schumann, long an active Democrat, rejoined the staff of party leader Meg Lees. As her media adviser, Schumann became embroiled in controversies far removed from the years when he was lead singer for Redgum, which is still regarded as a major cultural force of the period.

Essayist Humphrey McQueen has praised Redgum for their contribution to Australian life and letters. McQueen has written of their use of the Australian idiom, of the quality of their verse and of their musicianship. In an introduction to a collection of their works, he wrote: 'You have to listen more than once [to their songs]; concentrate; think through the logic of their stories. It is in making these demands that Redgum approaches art.'[14]

John Schumann claims Redgum was the first band in Australia to sing environmentally aware songs. 'We were the first,' he said in 1999. 'We were singing about the environment when the Oils were still surfing at Narrabeen.'[15]

Be that as it may, Redgum's work seems all but forgotten. In 1999, discographic information about Redgum's songs was sketchy and unreliable. Most Redgum albums were unavailable by the end of the 1990s—even in libraries, where they sometimes appeared in the catalogue but not on the shelf. And when people think of the vanguard of environmentally aware popular music, it is Midnight Oil that comes first to mind. Speaking of this, Schumann said: 'The music industry has no sense of history. If they haven't sold any of something this week they just say "Let's get rid of it—erase it, throw it away".' But Schumann himself has a vast collection of Redgum-iana, 'buried under the last 10 years of my life'. He said: 'I've got it all—records, reviews, notebooks. I've been meaning to do something with it, but I haven't looked at it in years.'[16]

Redgum was formed on the campus of Flinders University in Adelaide in 1975—almost by accident. Schumann, Verity Truman and Michael Atkinson, all students in a politics and art course, volunteered to work together on a music project. They wrote a few songs, sang them for the class, and were so encouraged by the reception of their work that they wrote a few more songs and started singing them at pub nights. For the next few years, they sang on campuses, for community radio and eventually for the ABC. Along the way they picked up violinist Chris Timm and later a bass player and a drummer when a rhythm section seemed like a good idea. They toured the clubs of Sydney and Newcastle and made an album, *If You Don't Fight You Lose* (1979), with the help of friends who worked at the ABC. The album sold very well, surpassing the expectations of everyone connected with it, and on the strength of those sales and some good critical response, the band members found the courage to quit their day jobs.

Redgum were always committed to doing socially relevant music. Their early songs deal for the most part with the usual concerns of the Left—unemployment, the exploitation of workers, the subjugation of women and, above all, US imperialism and the loss of Australian cultural identity. Environmental themes are there but often seem almost incidental to the main point of the songs. For example, 'So Goodbye', by Schumann, tells of 'chimneys spewing thick black smoke' but is mostly about a poor young bloke ending a love affair with an upper-class young lady.

In their early days, Redgum was very much an Adelaide band and they sang about the part of Australia they all knew best. Songs such as 'Peter the Cabby' and 'Killing Floor' show us the lives of the working poor, immigrants mostly, in that city. In these songs, you get a real sense of what Adelaide was all about—its sights and sounds and the people who lived there. McQueen also praises them for that regionalism, saying that in such regional and particular voices a culture finds its strength.

Hugh McDonald joining the band broadened Redgum's horizons and their concerns grew to encompass national and rural issues. (Redgum was always fairly fluid with regards to membership—every photo of the band seems to show a different group of faces.) Two of McDonald's songs, 'Spirit of the Land' and 'The Diamantina Drover', have become classics which celebrate the courage of rural people and the issues with which they have to contend. 'Spirit of the Land' is evocative:

> The rivers are dry across the land
> And the farmers' fields have turned to sand

> The topsoil's gone with the hot north wind
> The crops won't grow and the rust's set in . . .[17]

Schumann has said that the role of music is to make a positive contribution to the intellectual, cultural and spiritual life of society. Speaking to a gathering of songwriters and musicians in 1990, Schumann said there was a 'huge obligation on the part of a singer who makes a living from his/her songs' and he sought to do this. 'That is why I write the songs I do,' he said. 'People like me are supported by society . . . Therefore there is an obligation on us to contribute seriously and intelligently to improve the conditions under which we live.'[18]

He went on to say: 'Even the standard love song is politically active because it takes our minds off injuries, inequities, oppression and war in our society and in the larger world'. In his talk, Schumann lamented the fact that nearly all songwriters were content with empty love songs, and he raised a smile from the audience by telling the story of being recognised by an air hostess once as the 'guy who wrote all that stuff about war and death and cancer'.[19]

The classic Redgum song is 'I Was Only 19', written by Schumann as a tribute to two friends who went to Vietnam 'and came back Agent Orange victims'. First released as a single in 1983, 'I Was Only 19' touched a chord with the Australian people. The song shot to No. 1 and stayed near the top of the charts for months.[20]

In his 'The Last Frontier', Schumann again draws from his personal experience—a trip to Alice Springs. His narrator, expecting to discover the very heart of Australia, finds that it has been sold off to a foreign power. He sees the uniforms of the US military, the barbed wire fences around Pine Gap and wonders why he has been shut out of his own country. This is a song about degradation, about the selling out of Australia, about militarism and all its evil works and about the fragile environment of Australia's heartland (the country shimmers through the windscreen).[21]

'Nuclear Cop' combines serious intent with a touch of levity, having a rompin', stompin' bluegrass sound and audacious rhymes 'Landscape looks like Salvador Dali/We all spend our summers on Bali'. In it, Redgum presents the case for uranium mining as if it were a viable proposition:

> Load it on a jumbo, send it overseas
> We got lots of money now, who needs trees?[22]

In what is perhaps the most moving song from the Redgum period, 'Where Ya Gonna Run To', all of the band's themes come together, with the environment the main focus of their concerns. This song lays out Schumann's rage about what has been done to this land in the name of commerce and militarism:

Southwest Tassie's got a dozer track
We'll never ever get Lake Pedder back . . .[23]

He cries out against the savagery and the cold, brutal detachment of modern warfare—and the strident urgency of the music here underscores the power of the words:

Shake me, wake me, tell me it's a dream
I got a B52 on my TV screen
There's a man in a tie
Pointing to the sky
Where ya gonna run to?

Post-Redgum, Schumann had an independent career in music, producing albums for Eric Bogle and recording some solo albums. *Etched in Blue, True Believers* and an album for children have disappeared from view despite brisk sales at the time and a warm critical reception.

In 1999, Schumann had opted for the slow and steady progress of professional politics and appeared to be out of the music business forever. Schumann says that he has no regrets about his time as a folksinger or about getting out when he did. He says he feels that he still has the best part of it—the music. Lucky for us, we have the music too. The album *Caught in the Act*, the only Redgum effort still readily available commercially and probably the only one ever to make it to CD, is a recording of a live performance in a small Sydney pub—the Rose, Shamrock and Thistle in Rozelle—in 1983. It is still a great album to listen to. It has a freshness and a vibrancy even though some of the political references almost need to be footnoted, and it captures the spirit of what a Redgum performance must have been like.

PERHAPS no other entertainer has identified himself and his music so strongly with the battle for the environment than has Peter Garrett of Midnight Oil. In 1999, Garrett was the president of the Australian Conservation Foundation for the second time and was very much involved in all its works, particularly the fight against uranium mining at Jabiluka.

Garrett has assisted the Cape York coalition of Aboriginal land councils and environmental groups seeking to preserve this wilderness area. He has written and narrated the film *Shoalwater Bay Up for Grabs*, and co-written the ABC radio drama 'Outlanders'. As president of the ACF, he has contributed articles to newspapers, and when there is a media event for an environmental cause, Garrett is frequently on the podium. He has fought to keep a munitions depot out of Jervis Bay on the New South Wales South Coast and to have that area declared a national park. He has been on the board of directors of Greenpeace International, and in 1984 ran for—and very nearly won—a seat in the Senate as the candidate for the Nuclear Disarmament Party.

While many performers write songs about their convictions of one sort or another, creating lyrics in support of environmental causes, or feminist causes or a particular political point of view, with Midnight Oil the message is the raison d'être of the band. Their songs—all their songs—are political in nature and one cannot imagine Midnight Oil singing any other sort of music. Garrett himself concedes this, saying: 'We're the moles of Australian rock—the political conscience.'[24]

An expert in contemporary Australian music, Michael Hannan, said in 1999 that Midnight Oil have been an important voice on a whole range of issues—and their success has been a direct result of their political messages. Hannan, an associate professor at Southern Cross University in Lismore, New South Wales, said of the Oils: 'They've inspired a global audience. They've drawn a huge contingent of fans—people who are concerned about their issues—and they've kept them involved over a long period. I think this continuity is hugely important.'[25]

Midnight Oil are so identified with their songs that they rarely—if ever—sing anyone else's. Neither can one readily imagine anyone else singing their songs. Garrett has on occasion mused about his role in the band, saying that he is not so much a singer as a performer. 'I see myself as someone who is able to work with songs that have come through the Midnight Oil thing,' he said, 'and it is out of those songs that I am able to make some kind of sense and make them work as vocal performances.'[26]

The band has come in and out of fashion over the years, with some albums selling superlatively well (*Diesel and Dust* in 1987 sold more than five million copies, most of them overseas) and others not making much of a mark. They've packed out the big venues and toured with the big bands, but have happily come back and played in small RSL clubs in out-of-the-way towns. This fluctuating fortune truly does not seem to bother the Oils. They've experimented with different sounds and musical ideas over the years—*Earth and Sun and Moon* (1993) and *Breathe* (1996), for instance, have a softer sound that probably baffled many of their hard-core, hard rock fans. And when they did have hits, they've refused to do publicity tours or the standard barrage of interviews.

Garrett, when asked by a reporter while they were walking along a beach, how he felt about sales dropping, pointed out a bird in flight, skimming lightly over the water. The reporter seemed to think that Garrett was wool-gathering or attempting to change the subject; but it seems obvious that Garrett's answer was meant to be something like 'I'm as worried as that bird is out there'.[27]

The Oils' longtime manager, Gary Morris, said in 1996 that the band had 'started out as a by-product of our lives'. He laughed when asked by a reporter if he was surprised by their huge success, saying: 'I thought we had hit the big time when we started packing out the Narrabeen Antler (a pub in the heart of Sydney's northside beach culture).'[28]

Midnight Oil—then called The Farm—started playing the pubs and clubs of Sydney's northern beaches in the mid-1970s. They soon became one of the many bands that gravitated downtown to French's Tavern, a dingy basement venue on Oxford Street that used to pack in young rock-and-roll fans. Even then, Midnight Oil weren't the usual new band full of wannabes—the members were accomplished musicians, for one thing, and had just graduated from various universities with degrees in arts and sciences—and, in Garrett's case, law. And they were, to a man, well-educated and politically and socially aware. At French's they became something of an underground hit, with the passionate Garrett belting out lyrics until he nearly dropped with exhaustion. 'They were the most full-on thing I'd ever seen,' said a friend, Peter Farnan, guitarist with the pop group Boom Crash Opera.[29]

After Garrett shaved off his blond hair and perfected his menacing lurch across the stage, the look was made. (Garrett said in 1999 that his trademark bald look resulted from having his head shaved because his hair got in the way when he was surfing, and the band insisted it stay off.)

The band recorded its first album, *Midnight Oil*, in 1978. Produced on the cheap, it was so successful, selling more than 70,000 copies, when they were thinking it might sell about 1,000, that all thoughts of careers in business or the professions were abandoned. The next two albums—*Head Injuries* (1979) and *Place Without a Postcard* (1981)—were also quietly successful and contained well-written songs that touched on social and environmental issues.

But the next album, *10,9,8,7,6,5,4,3,2,1* (1982), took the band to amazing new heights. The album marked an end to the band's 'pub rock' sound and introduced complex soundscapes with synthesizers, drum machines and white noise overlaying the acoustic instruments. And it contained the most overtly political material to date with a strong anti-nuclear message. Drummer Rob Hirst described the album as 'the right combination of attack and anger and melody and technology and spirit'.[30] It was recorded in London and produced by Englishman Nick Launay, who went on to work with INXS, the Church, the Models and later Kate Ceberano. Buoyed by *10-to-1*'s phenomenal success, the band became more and more political—in their music and in public statements.

Like most Midnight Oil albums, *10-to-1* has a wondrous unity of theme—both musically and textually. Most modern albums seek to do this—to create a single artistic statement rather than be just a collection of songs that are ready about the same time—but Midnight Oil is superlatively successful at this when many other bands fall short. Musically, *10-to-1* is full of jagged edges, sudden disconcerting key changes and chord progressions, and weird electronic sounds; it is powerful music, disturbing music that very much highlights the album's disturbing message. Each song is a statement on its own and several of the band's all-time hits are on this album—for example, 'US Forces' and 'The Power and the Passion', the latter described by *Rolling Stone* magazine as 'one of the few summer anthems that'll have you thinking and dancing at the same time'.[31] But as an album, each song builds on the last to set up a powerful series of images of madness, rage and despair. The whole album is a cry from the heart against the insanity of the arms race, but the Oils' argument is never simplistic 'ban the bomb' stuff. They create a picture, for example, of the alienation of the individual from the seat of power:

> The bosses they can sense your mood
> All in place to a hand that rules
> They all want to deal you out[32]
>
> When I'm locked in my room
> I just want to scream
> Now I know what they mean
> One more day of eating and sleeping[33]
>
> The bosses want decisions
> The workers need ambitions
> There won't be no collisions when they move so slow[34]

The theme of power and what it does to people is taken up over and over again. 'Short Memory', for instance, is a litany of historical abuses of power, from colonialism to economic imperialism. 'Read About It' is about the suppression of truth by the forces that control the world's media. 'US Forces', of course, is about military might, and 'Maralinga' is a reminder of what domination by a foreign power can lead to. The album is angry, certainly, and it is a plea for individuals to take back their power, to resist intimidation and to fight off apathy and cynicism.

While *10-to-1* raged against the dark and established the band as angry social reformers, Midnight Oil cannot be lightly dismissed as populist rabble-rousers. Their musicianship is undisputed and their lyrics are first-rate poetry, with a range of moods and a complexity of imagery that bears careful study. And they are capable of changing tacks and introducing new ideas into the arena of debate.

The album *Earth and Sun and Moon* sold just over a million copies in 1993, a total which by that time was considered disappointing. Much was made at the time of it being a departure for Midnight Oil—of its lush sound, more personal songs, softened rhetoric. Garrett at the time explained this change of emphasis, saying: 'We didn't sit around at a meeting and say "approach must be broadened". I think we just wanted the songs to be direct and they just happened to be personal songs. We're a little warmer, a little wiser, a little more open about things.'[35]

Earth and Sun and Moon is lyrical and lovely, paying homage as it does to the great bands of the past. Drummer Rob Hirst acknowledged this debt, saying: 'The bands you hear are the ones we grew up with. I think the influences we grew up with—The Who, The Doors, The Beatles—were fantastic influences.'[36]

The overall feeling of the album is one of hope, compassion and lightness. The truly lovely title track, by guitarist Jim Moginie, a musician of rare talent whose contribution to the band over the years has been phenomenal, gives us an astronaut's eye view of the planet. The narrator is standing in space looking down on the world with love, watching the weather, the mountains and the seas. He sees what is wrong—oil slicks, fires in the Amazon—but he calls on us to see the great beauty, the majesty of our world:

> I wish you could see this great mystery
> Earth and Sun and Moon, human tribe, thin blue line
> Earth and Sun and Moon will survive
>
> An axe upon the broken ground, the sigh of the trees
> It's floating in the ether, it brings me to my knees[37]

This vision of compassionate oversight—that the world is small, unified, that there's a bigger frame of reference—is picked up on many of the other tracks:

> Cool wind, clear my head
> Bright sun, grow my food
> Strong rain, clear my heart
> Fill my land
> It's a new day[38]
>
> The backbone of this country's broken
> The land is cracked and the land is sore
> Farmers are hanging on by their fingertips
> We cursed and stumbled across that shore[39]

Despite its 'big picture' perspective, the album is very Australian. There is a great sense of place, of love for the Australian landscape, of compassion for its people and their struggles, and of trying to come to grips with its sometimes shameful history.

Earth and Sun and Moon was not entirely a new departure for Midnight Oil. The earlier album *Blue Sky Mining* (1989) is also full of hopefulness and reverence and awe at the beauty of the world. When this album was still in its planning stages and as yet untitled, manager Gary Morris referred to it as Midnight Oil's 'green record', to which Hirst added: 'We've always been green. It's just a slightly darker shade this time.' And Garrett said: 'It has a definite green element to it, but then it's almost impossible not to have considering the world's environmental status.'[40]

Blue Sky Mining was recorded after a long and gruelling tour promoting *Diesel and Dust*, with the band spending nearly all of 1988 on the road in Europe and America. Hirst said many of the songs are 'a bit strange' largely because of all the travelling. Of the US, he said: 'You don't see anything except concrete and bitumen, and you see these enormous reflective surfaces like the melting car parks in Phoenix, Arizona, reflecting heat up into the atmosphere. Part of the travelling business was to see what havoc countries have wreaked on their environment.'[41]

The title track begins as a straightforward piece about exploitation. It's about mining, of course, and the damage that it does—and about the folly of the pursuit of wealth at the expense of the earth and of human beings—'there's nothing so precious as a hole in the ground'. It's also about the struggles of the common man, who works so hard to make a few men rich. The song was inspired by Ben Hills' book *Blue Murder* (1989) about the blue asbestos mine in Wittenoom, Western Australia. This mine was started by magnate Lang Hancock in 1943—when the dangers of asbestos mining were beginning to be known—and it was later taken over by CSR, the sugar refining company referred to in the song. The people who worked this mine have suffered greatly over the years, their lungs 'turning to cement' in a slow painful death. But, of course, the song is about all such mines and the people harmed by them. The title works on another level, Hirst explained, saying that a blue sky mining company was one that 'shoots for the stars without assets and then crumbles, like a few have done lately'. Hirst described such companies as being run by 'people who give nothing back to the community and have tax schemes in the Dutch Antilles'.[42]

The song is much more than a rallying cry against the evils of asbestos mining. The last lines talk about the rain coming, falling softly on this Blue Sky town. Taken literally, this image would not be one of hopefulness or joy—rain falling on a place like that just pollutes the waterways and creates even more problems. But in the context of the song, rain falls like a benediction:

> We've got nothing to fear
> In the end the rain comes down
> Washes clean the streets of a blue sky town[43]

There is a sense that even in such a place, where there is so much evil and ugliness, the earth is blessed, that in spite of our sin and folly, everything will one day be clean again.

In the next track, 'The Stars of Warburton', it is the stars that rain down blessing. The narrator describes being 'shaken down in a toy town'—a town full of air-conditioned malls, surrounded by degraded farmland and with its culture ruled from America—'Don't want to talk about Elvis Presley/Don't want to see his white shoes walking around and around here'. Yet when he gets away from the clutches of this town, he sees the stars:

> I couldn't believe the stars of Warburton
> Were waiting for me
> I couldn't believe the stars of Warburton
> Were living in me, raining down on me
> I couldn't believe the stars of Warburton
> Were washing down on me[44]

This is a joyous song, and it again calls our attention to the bigger picture. It describes evil as man-made, local, small; whereas the universe is vast, with a goodness and a majesty that will surely prevail.

The well-known song 'King of the Mountain' is also on this album. This is one of Midnight Oil's great sing-along choruses, the rock anthem at its best. But it too is about much more than first meets the eye and ear. The words here are powerful, the images complex. There are strong echoes of the work of T.S. Eliot, with references to wasteland, rustling grass, red rocks and cactus, dryness and rain, shadows and light. In the poem, Midnight Oil have brought this imagery home, as it were, because despite its literary feel it is also a pretty straightforward description of the Australian landscape. But in this song, the wasteland isn't a dead place, representing the darkest impulses and deepest fears of mankind, but rather a place teeming with life and beauty that brings joy to one who encounters it in the right spirit. The title 'King of the Mountain' in Australia is given to the winner of the annual Bathurst 500 car race. As such, it's a macho image of power, dominance over nature, sound and fury. In this wonderful song, Midnight Oil have turned this image of crude power into something far more subtle and beautiful.

By the late 1990s, Garrett was scathing about the state of music in Australia when Midnight Oil first hit the scene. 'When we started to make music and talk about stuff that we thought was really important, Little River Band was singing songs about America, the Bee Gees were dominating with disco, 'Countdown' (the ABC's televised pop showcase) was a continuous succession of one-hit wonders, boy-girl groups,' Garrett said. 'And nobody had a sense of rage. Nobody had a sense of yearning. There was no grief. There was no celebration. There was nothing. It was dead. And it was real sugar-coating stuff. So, we put on our Doc Martens and away we went.'[45]

Novelist Tim Winton (see Chapter Two) concurs, and for the cover notes of the Oils' CD *20,000 Watt RSL* (1998) he wrote about what the band meant to him, growing up in Western Australia. 'The music contained an unmistakable atmosphere of the suburban Australian life I was part of,' Winton said. 'Underneath the bland, safe surface, a jerky agitation, an itch I recognised. This new music connected to all that restless energy and hope, the dismay, the paranoia. Australia seemed about to stop thinking and just go shopping and here was a band anxious about our communal future. Someone had finally got beyond the easy nihilism of the time. The music sure as hell made you want to dance, but it carried ideas, raised pressing questions. At last, there was an Australian band with something on its mind. Well, with a mind at all, for that matter.'[46]

David Leser, writing in *HQ* magazine, said: 'For the vast majority of young Australians, who live with their backs to the Dead Heart of the continent, Midnight Oil has been the symbol of an awakening white conscience. A group of former private school boys, they have raised the banner of Aboriginal rights more effectively than their backgrounds could ever have hinted at. At the same time, they have managed to invoke a reverence for the land—and a despair over the way it is being ransacked—that few have achieved.'[47]

Playwright Stephen Sewell (see Chapter Six), quoted in the *HQ* piece, said the band had 'plugged into the passionate commitment to human values that is at the heart of left politics. That they are there and have an audience is a significant breakthrough for the left. They have achieved something for social change.'[48]

The election of John Howard's Coalition Government in 1996 and everything that flowed from this brought the rage back to Midnight Oil. *Redneck Wonderland* (1998), the angriest album the band had produced in years, marked a return to the hard-edged rock for which they are famous. Drummer Rob Hirst said: 'It's a very apt and angry record and I prefer Midnight Oil when it's operating like that.'[49] Hirst said that while they were making the record, 'the morning papers would come in and we'd see the political developments in Australia unfold, we got angrier and angrier and the album just toughened up.'[50]

Garrett, too, was openly contemptuous of the Howard Government and the direction in which it was taking Australia: 'The current government are not equipped to lead the country into the 21st century and this reflects itself on their approach to the environment as much as anything else,' he said in late 1996.[51]

Two years later, Garrett was equally aghast at the way things were heading. 'As long as politicians buy into the big lie of perpetual and inevitable growth at all costs then we can see things going pretty much the same way, continuing to run down the environment while speeding up the conveyor belt that we're all on,' he said.[52]

The album's title came from a piece of graffiti the band saw on a wall in Melbourne which denigrated John Howard and rainforest destruction. There was a sketch with a map of Australia and the words 'Redneck Wonderland' scrawled across it, which seemed apt for the way they were feeling about the state of the nation.

The apocalyptic vision of this album transcends any sort of petty political squabble or a particular narrow ideological debate. It speaks rather of the despair one feels watching a nation lose its way, becoming less than it was, and of the sense of loss and outrage which that begets. Musically, *Redneck Wonderland* is dramatic and confronting, with the acoustic, orchestral and electronic instruments working together to produce a powerful and thrilling mix of sounds. The music squalls and howls yet it is also strongly melodic and magnificent in its noisy splendour. The words of the songs, also, are confronting and upsetting and filled with images of decay and despair:

> In the city the heart still whispers
> Flaking metal a silent witness
> There's chemical fields and cathode clouds
> The Milky Way is emasculated as exhaust fans and smart cards shrink-wrap the coloured air and send it coughing into eternity[53]

In *Redneck Wonderland* we are 'through the looking glass' into a world of chaos and disintegration, where nature is a threat to be kept out, where other people are to be feared, where the air is poisoned, where peace and security are under constant threat.

Rob Hirst has spoken out about the importance of music having a political conscience. He said events such as Live Aid had made a huge difference to people, describing it as 'perhaps the most significant altruistic adventure of our age'. Hirst also cited 'the huge Mandela marches around the world, U2 and Sting's efforts with the rainforests and Amnesty International, and the nuclear marches and concerts which people like the Oils were involved with'. Stressing the value and importance of their stance, he said: 'There are those who think change can only ever be achieved through the strictly political protest. (Then-Prime Minister Bob) Hawke told us once "You can write as many songs as you like, you can scrawl your slogans on the wall, but it won't change one iota the decisions of my government." This was at the time of Pete's run for the Senate—and he only said it because he was terrified.'[54]

Midnight Oil are among the longest-running acts in Australia, and they have stayed together, with only one change in personnel, longer than just about any rock band. The band members have managed to maintain stable families and sensible lives in a business where that is unusual. And the glue that holds the band together is that rock-solid commitment to the issues that are dear to them, to the music and to their audience.

Peter Garrett, certainly, is an avowed Christian, a Bible-reader, and a believer in service to one's fellows. It is perhaps this commitment more than any other that keeps him and the band focussed on what they do and keeps them from the burnout that besets so many environmental activists.

Garrett has said many times that he doesn't get discouraged, that he sees signs of hope that things are getting better, that progress is being made on many fronts. At the Sydney launch of *An Environmental History of Australia* (1999) by his friends Drew Hutton (leader of the Queensland Greens) and Libby Connors, Garrett gave a brief address in which he said something else that helps to explain why he is able to soldier on so apparently tirelessly. 'The environmental movement isn't a mass movement, and it shouldn't be understood as one or expected to be one,' Garrett said. 'The huge campaigns of the early '80s when thousands of people were mobilised to march on the beaches—that was a spike, not the norm. No, the environmental movement has always been small, a few people working quietly to change things gradually.'[55]

THE grand old man of Australian folk music, Eric Bogle, is a transplanted Scot who has made his name singing about Australia. In fact, in 1987, he was voted 'Most Australian Australian of the Year' by the Newcastle and Hunter Valley Australia Day Society, an award of which he is very proud. Bogle is a much decorated performer—he has been given the Order of Australia Medal (1987), and in 1986 he was awarded the Peace Medal by the Federal Government in recognition of his efforts through music to promote peace and racial harmony.

Despite his long residence in Australia—and his declaration that he now feels a stranger in his home country—Bogle still sings his Dreamtime songs in the thickest of Scottish accents.

Bogle arrived in Australia in 1969 from Peebles, Scotland, where he was born in 1944. Determined to 'do well' in his adopted country, he got straight to work in a real job 'in those idyllic, gone-forever days of full employment', eventually becoming an accountant.[56] He took on all the trappings of a serious life—and in Canberra, where he says he settled because he didn't know any better. (He often jokes, too, that he acquired a pair of spectacles and a wife—and that though he soon ditched the spectacles he has kept the wife.) But Bogle had always been involved in music, and once in Australia he soon found his way to the folk clubs. He was amazed by the depth of talent and commitment he found there—'all notions of being the premier traditional singer in Australia went right out the window' he told music writer Robert Childs many years later.[57] He was surprised, though, that most of the music seemed to be Irish or Scottish, with little original Australian material.

Bogle started singing at the clubs—Scottish songs first, then gradually introducing his own material. 'I never fancied myself as a singer,' he joked once to music writer Stuart Coupe, 'but when I started writing my own songs I was the only one who was game to sing them.'[58] He continued working as an accountant, singing a little and writing songs to 'keep his sanity' until 1980, when he had finally had enough of his respectable job and chucked it in.

Bogle has a great and abiding love for this country. His wonderful descriptions of its baked earth, its golden sunshine and his frequent evocations of the Southern Cross reveal a man in love with the landscape. This love for the land and of its indigenous peoples shines through in much of Bogle's work and forms a good part of his ethos of conservation.

He sings about what is important to him—about humans and their foibles and their pain and their dreams. He is well known for his sensitive and emotion-filled songs about his parents—'Scraps of Paper' and 'The Day that Nancy Died'—about handicapped children ('Rosie'), about the casualties of divorce ('Harry's Wife'). In his concerts and on his albums, there is always a leavening mix of funny songs ('Nobody's Moggie Now', about a deceased cat; 'Little Gomez', a chilling tale about an oversexed chihuahua; and the very pithy 'I Hate Wogs').

Many of Bogle's songs deal with the displacement of the Aborigines. The theme often recurs that the land is sick because its proper custodians have been shut out of it by the brutal, ignorant conquerors. In 'Hard, Hard Times', he sings:

> Poor Jackie's a stranger in his own land
> Living in a world he can't understand
> Wrecked and drowning in a sea of grog
> His pride and self-respect's been robbed
> Worse off than a white man's dog.
>
> Spirits have fled from the rocks and trees
> The land is sick from white man's disease
> Their lust and greed does them betray

They raped the land to make it pay
The black man just got in their way.[59]

Bogle has spoken and sung often of his great admiration for the Aboriginal people. In 1999, he said: 'I believe that we—white society—haven't done the right things by this land. The Aborigines lived here for thousands of years—60,000 years is the accepted figure by most people—and they shaped the land. They created so much of what the first white settlers saw, with their fire-stick farming. That landcare method was so beautifully perfected over the centuries—it worked so well. But white settlers have been degrading this country from the beginning. We're still doing it too, with our intensive farming and our huge population sprawling all over. The land can't sustain us like this—not long term. The Aborigines knew they depended on the land—that's the difference. We don't seem to know that anymore.'[60]

Bogle's songs generally show great compassion and understanding for country people. It is clear that people who live and work on the land, who struggle through drought and cope with isolation and fear are not the enemy. In the song 'Bushfire', he tells of what happens when a family's property is engulfed. To the insistent droning of didgeridoos, he paints the truly frightening picture of the fire's approach—the distant smoke plumes, then its raging on far-off mountaintops—a fire that comes closer and closer to the farm while the people wait helpless before it. The song is really about nature's revenge, how we can abuse the earth up to a point, but nature will have the last word:

Arrogant man
He squats upon the land
He buys it and he sells it and he zones it.
He buys his seeds and cuts down his weeds
And imagines that he owns it.

Despite these words about 'arrogance', the song shows great compassion for the family's fear, for their ruin, for their sheep which are incinerated in the angry blaze. It continues:

For 20 years our land we cleared
We ploughed and we sowed and we tamed it
But where the bushfire has passed
There's only black ash
And Nature has reclaimed it.[61]

In 'When the Wind Blows', he again sings of nature's revenge, invoking the imagery of the Last Judgment to express his rage at the works of men. He sings of an earth swept away, of a million souls swept into a hell of their own making:

This good Earth we did inherit
We shall leave a smoking desert
A headstone for the heedless human race
To mark our final throes
When the wind blows.[62]

Bogle said this song was inspired by a book of cartoons by Raymond Briggs. 'He's a great social commentator; I think he's a genius,' Bogle said. 'He did a wonderful series of drawings about an old couple—just ordinary folks—who were going through a nuclear winter. It's quite funny—though horrifying too, of course, as that's still a very big threat to our world—it could still happen. I wrote to his publisher and told him I was going to pinch his title. I never heard back so I guess it was OK.'[63]

In 1990, Bogle and Kev Carmody did a three-month tour, which included more than 50 performances, on behalf of the Australian Wilderness Society. A part of the proceeds of the tour went to support the society's work, little Wilderness Shop booths were set up in lobbies of the various venues and a goodly number of 'wilderness' songs were sung on that tour. (Although, Bogle assured audiences at the time that it wouldn't be 'three solid hours about trees'.)[64]

Around that time, Bogle gave hundreds of interviews on the subject of his environmental concerns. He was accused of jumping on a trendy bandwagon, a charge he hotly denied. He spoke to the *Sun-Herald* newspaper's Susan Jarvis, saying: 'If anything it's the Stings and the Kylies who are jumping on the bandwagon. I—and indeed the whole folk movement—have been talking about environmental problems for two decades. I wrote my first greenie song 19 years ago, before the term had even been invented. I got sick of all those people suddenly developing an interest, so I decided to reassert the original environmental voice.'[65]

When he was reminded in 1999 of his earlier comments, Bogle said: 'Oh, I didn't mean to stick it up Sting. I think he's done a lot of good. But what I meant is that folk music, by its very nature, is concerned about how we treat the earth. That's what it's all about—that's what it's always been all about. Folk songs, even the bush ballads from the last century, were asking the questions we're talking about now. And the sort of people who are attracted to folk music have cared about the world around them—how we treat each other, how we behave. Basically, we were all green before it was fashionable. We'll be green still when if it ever stops being fashionable.'[66]

Bogle laughed as he remembered the Wilderness tour. 'Yes, it was good,' he said. 'I was glad to do it. I'd been a member of the Wilderness Society for years—I still am. But I don't think any of them actually came to the concerts. They may have—I don't know. But they're terribly driven people, conservationists. They don't always have time to listen to music.'

Out of the concert tour came the album *Voices in the Wilderness* (1990). On the title track, Bogle sings about what the idea of wilderness means to him—not just saving the world's resources, but of saving that place of wildness and freedom within ourselves:

> Deep inside each one of us
> There is a wilderness
> A wild and secret place
> Where our souls find release

Where we are children once again
Living in an unspoiled land
Of earth and trees and sky and peace

Save the wilderness
If it fades and dies then so do we.[67]

In 1999, Bogle said that he still felt this way—that what's most important to him is the 'green and quiet place within' which we all need to find. 'That's the real wilderness,' he said, 'and we need those green places in the world to be able to refresh the inner wilderness.'

He got to the heart of the matter: 'For me, I found that place up in Kuranda, on the Atherton Tablelands in Queensland a few years ago. I was at a festival there and I happened to be out in the bush by myself. I sat in a little glade in that forest and just listened to it, just experienced the peace. Now, whenever I want to, I can shut my eyes and I'm back there. I can feel that peace again.'[68]

ABORIGINAL activist and folksinger Kev Carmody writes songs of such anger and power that they are sometimes painful to listen to. *Rolling Stone* magazine refers to the 'black light, black heat' of his solo performances, but the dark vision of his lyrics frequently touches raw nerves in white society and he has always had difficulty with airplay.[69]

Words such as 'Just enough money to go get pissed/Thieve off the rich here just to exist'[70] and his spoken poem celebrating the communist tendencies of 'Comrade' Jesus Christ[71] have been considered too explosive for the radio. Carmody himself says that his songs are 'not entertainment' and he is not surprised that the powers that be in society don't care to hear his criticism. But, he says, 'you can screw the lid on truth for as long as you like—eventually it will come through'.[72]

Carmody's songs confront the hypocrisy, greed and vandalism of white society and decry what has been done—and what continues to be done—to the original inhabitants of this land. And they deal with issues of justice. 'If you look at it in the broad sense, I'm mainly concerned with a person's humanity to another person; and, second, our respective responsibility towards the Earth,' Carmody said in 1994. 'They're pretty basic bloody things. They're in all indigenous cultures. It's only within the industrialised societies that we've lost those dimensions that are part of our human nature.'[73]

Carmody is one of the most outspoken advocates of the poor and the marginalised in our society. He sings about urban violence and the hopelessness of our youth—and takes his message into the nation's jails. He has worked with street kids in Logan City, Queensland, one of the poorest, most dangerous and desperate urban areas of this country, and he is one of the foremost proponents of the issue of land rights for the Aboriginal people.

His own feeling for the land, as the source of his spirituality, informs his attitudes towards the destruction and desecration of the environment by white industrial society. On the issue of land rights, he explained in 1989: 'You know this is not just about giving some bloody blackfellas a block of dirt. The whole culture is based upon the spirituality of the land. It's almost as if the whole land was a cathedral. When I was a young fella in the early '50s, I saw old people actually weep when they first started bringing in bulldozers with big anchor chains between them. Have you ever seen them pull scrub? They pull down 300 acres a day. It's barbaric. If they do that to the land, we can understand why they do it to the people.'[74]

Carmody was born into a droving family in western Queensland, the son of an Irish father—whom he has on several occasions described as 'mad Irish, fighting Irish'—and an Aboriginal mother. He says he spent most of his first ten years on horseback before he was, as he puts it, taken by white society and forced to go to a Christian school in Toowoomba on south-east Queensland's Darling Downs. When he left school, which was as soon as he could, he shunned the whole academic experience, working as a shearer, a labourer, a welder, as need required, and playing the guitar a little on the side. In his early 20s he was beginning to establish a name for himself as a Rugby Union player, but his football career came to an abrupt halt when, as a member of a Queensland country side, he refused to play against the touring South African Springboks.

When he was well into his 30s, he became a student again at the Darling Downs Institute of Advanced Education, and because there was no guitar teacher, the college suggested he concentrate on Australian history instead. As an adult he took readily to education, eventually completing a PhD from the University of Queensland.

Carmody used to bring his guitar to tutorials—he says to cover up the gaps in his written English—and would sing about history rather than writing it.

This way of thinking about the music continues on into his professional career. 'This is all part of the Aboriginal way,' he said in 1990. 'My people were drovers—so-called illiterates. So their way of recording oral history was to make up a poem or sing a song. I've just continued on in that tradition. I've been writing songs for as long as I can remember.'[75]

In 'Thou Shalt Not Steal', Carmody takes on the hypocrisy of white society, which teaches Aboriginal children its 'commandments' and yet does not live by them itself. With the back-up choir—the Teddas—the song has a Gospel feel to it. The song is about land rights, but there is a green message here as well:

> With your science and technology
> You can make a nuclear bomb
>
> You should have found out long ago
> You better keep it in the ground.

And in the next verse:

> Well you talk of conservation
> Keep the forests pristine green
> But in 200 years your materialism
> Has stripped the forests clean.[76]

Carmody makes his point that the white settlers are not the rightful custodians of the land and nothing good can come of an original act of barbarism and theft.

Carmody's music has often been compared to Bob Dylan's, both in its message and its musical style, complete with banjo and harmonica playing. Carmody doesn't deny this influence, although he says he looks further back to the influences on Dylan—to Woody Guthrie and the Carter Family. And, he says, the main difference between himself and Bob Dylan was that Dylan 'was good at telling you what was wrong but not very good at telling you why it was wrong'[77] or what to do about it.

Not so Carmody. He says: 'There are 16 million white Australians and they're not going to go away, so the only bloody thing you can do is change the mentality that exists. Music is a small part of that. The more you work with people, the more you will understand them. If you don't understand something, nine times out of ten, you fear it. The only way we can understand each other as cultures is to get together.'[78]

ANOTHER singer–songwriter whose work lies somewhere on the border between folk and country is Pat Drummond. Although identified often as a country singer, Drummond was born in Sydney in 1952 and raised in the inner-city suburb of Macdonaldtown. He is probably best known to Sydney people from his eleven-year stand at The Rest hotel in Milsons Point near the northern end of Sydney Harbour Bridge. This ended in 1988, when the hotel was torn down.

After that, Drummond went on tour with the Bushwackers and then with John Schumann, and got plugged in to the Australiana style, with which he has been identified ever since. He now lives in the Blue Mountains, west of the city, and in April 1999 completed his 4000th professional gig. His songs, many written in partnership with his brother Geoff (also a member of his band), are stories about ordinary people going about their business.

Pat Drummond denies that he writes green songs, except once in a while by accident, preferring to think of himself as merely a reporter, a gatherer of other people's stories. He says that in fact some of his songs are just the opposite of green. 'I even have a song about a crop duster—his views would be anathema to most of the green movement,' he said in 1999.[79]

Drummond says he tries not to editorialise too much in his music, but rather present people with all their faults and foibles and opinions just as they are. He travels around Australia—and has done for years—performing in pubs and clubs and just listening to people from all walks of life. Whenever he feels that a particular anecdote has something universal about it, he writes a song. He writes of the joys of long-married couples, the troubles and fears of new migrants, about handicapped people, about parents and children.

Certain themes do appear in Drummond's work—even his choice of stories to present reveals much about his world view. He is a thoughtful, intelligent writer, though he hoots with laughter at a description of himself as 'a thinker'. He said: 'Oh, no. Don't tell people that! Do you know what they do to "thinkers" in the bush? They'd be veeery suspicious to say the least.'[80] But for all his jokes, Drummond is a proud Australian and identifies strongly with the people of this country, especially those in small places, and with the land.

In his song 'The Spirit of the Southern Shore', Drummond celebrates the beauties of the natural landscape and acknowledges its iconic role for our nation. But he says that the inner landscape—the values we stand for, the experiences that make us Australian—is something more important, 'beyond the brolgas and the kangaroo/More than the Olgas, more than Kakadu'.[81]

Drummond has a compassion for his fellow creatures, a real respect for the work of others, a wish to show their lives as filled with dignity and grace. With this goes an attitude of respectfulness and joy in all of creation. In the lovely 'Acceptable Losses', he laments the death of a young kangaroo, killed on the road by a speeding car. This song, dedicated partly to WIRES (the wildlife rescue service in New South Wales), shows the carelessness, the driver's absolute lack of awareness of the death he has caused. By extension, we can read this song as a protest against the speed and mechanisation of modern society, in which the helpless and the weak are often mowed down:

> The thing was huge she could not move
> Blinded by the light
> A rush of wind, a whimpering
> A final flash of pain
>
> A velvet face, the speed and grace
> So natural to her kind but
> A speed, in fact, that was no match
> For the thing that ran behind.[82]

'Shoalhaven Man' was written by Geoff Drummond and recorded by Pat Drummond and his band on the album *The Local Rag* (1990). The green message is very traditionalist, one which country people can understand and with which they can identify. It is a plea for the old ways of doing things, for respect for the land, for solid values and for beauty and grace. The story is told from the perspective of an old 'bushie', a timbercutter from the Shoalhaven region of the New South Wales South Coast. He laments the area being turned into a retirement village for city people, who have brought ugliness and stress with them and changed his home beyond recognition:

> They come for the stars at night
> They come for the peace and quiet

They come for the bushland that no man can claim
And they call themselves locals

They don't like the snakes so they flatten the greenery
They can't take the spiders so they Baygon the halls
And they bulldozed Culburra till it looked like Parramatta
I wonder why they ever went movin' at all.[83]

This plea for grace and beauty, for compassion and loveliness suffuses the music of Pat and Geoff Drummond. It is not activist or propagandist but rather gentle songs that show us who we are and what this nation can be, how we can treat each other gently and our land with love and respect.

A MORE political songwriter is Allan Caswell, a performer with a distinctly country bent. Caswell calls himself 'a bit of a leftie' and is deeply concerned about the Australian environment—both the actual physical environment and the cultural environment, which he sees as inevitably linked together. He says Australians have neglected both their natural and cultural heritage and that neglect is beginning to show.

'It's all about greed,' Caswell said in 1999. 'Money talks—and for enough bucks you can buy our environment. Everything in this country is for sale.'[84]

Caswell was born in Chester, England, and spent his early years in North Wales, arriving in Australia in 1966 at age 15. He has been a full-time musician since 1979, when his song 'On the Inside', the theme song for the television show 'Prisoner', became a huge hit. It has since become an Australian classic. Since then he has been busy and productive, 'touring and producing and writing and playing and singing', and has been to 'every State and every town in New South Wales'.

In 1999, Caswell had just finished an 18-month tour with his band the Self-Righteous Brothers. 'From time to time, I seem to disappear to the media, but I'm still here doing what I've always done—and I know where I am,' he said. Caswell noted that he had always been a dark horse in the music industry. 'Even when I won the Male Vocalist of the Year award in Tamworth in 1988, it was reported in the press as 'Slim Dusty doesn't win award'.'[85]

In 1984, Caswell wrote the song 'Moruroa', a humourous little ditty about French nuclear testing in Oceania. He sings:

Now I've met some nice French people
And I've drunk some nice French booze
And I love the way their football teams
Come over here and lose
While their leaders nibble Camembert
And sip their Beaujolais
Their celebrated arrogance is evident today
And when the fallout hits the fan
They'll be a thousand miles away
While it's raining radiation on Australia

Then the chorus, to the tune of Alouette:

> Moruroa, pretty Moruroa
> Moruroa, blow it all away![86]

In 1999, Caswell said he wrote the song out of 'anger—pure and total anger'. He said that about that time, he happened to give a lift to someone who was working for Greenpeace and they got talking about the French tests, as everyone was then, and this fellow had all the facts about how high the sea was rising, how many poisoned fish were being caught, and about all the cover-up. 'It made my blood boil at the arrogance, the fact that the French see the South Pacific as theirs to do what they like with,' Caswell said. 'Even now the statistics are frightening—the number of cancer victims there is way out of proportion to what it should be. It's a funny song, I hope, but that's often been my style as an artist—to make a serious point through humour. You can sometimes break down resistance to your ideas that way. People laugh, but it makes them think too, and maybe change their thinking.'[87]

There was quite an 'interesting' response to 'Moruroa', according to Caswell. 'I was banned from a number of RSL clubs in Sydney for a long time because of that song,' he recalled. 'And the Campbelltown Catholic Club wouldn't let me through the door. That ban is still in force, it seems, because we were booked there—the Self-Righteous Brothers—but when they found out that Allan Caswell was involved it all fell through—that's 15 years later.'[88]

Caswell has written a number of albums for children, an activity which remains close to his heart although his own children are now grown up. 'Writing for kids is wonderful,' he said. 'It is the most fulfilling and the most exciting thing I do. There are no limitations on you, because you are writing for the one section of the community which still has an imagination. With my Australian animal songs, for instance, I try to make kids think, to learn about animals and to think about them and appreciate them.'

For Caswell, songwriting is all about sharing insights. 'For me, the words are the most important part of a song,' he said. 'I've always been the "words" man and I've always wanted to be. When I was growing up and thinking about music as a career, it was the great lyricists I admired. I'd always wanted to be Ira Gershwin, not George. And even now, I'd sure as hell rather be Tim Rice than Andrew Lloyd Webber.'

Many of Caswell's kids songs have a strong green message. 'The Solar Powered, Practical, Combustible, Compactable, Responsibly Recyclable Machine', which is just as much fun as its title, introduces such concepts as solar energy, waste disposal and recycling. And 'Trees' discusses, well, trees, particularly their use as habitat for wildlife:

> I want to have a little conversation
> I want to talk about the birds and bees
> When it comes to accommodation
> Birds and bees, they live in trees.[89]

Caswell considers himself very much an environmentalist and, on the day we spoke, he had just resigned his long-time membership in the Australian Democrats over their 'cave-in to Howard' on the environment. He was twice a Democrat parliamentary candidate, but 'I'm disgusted now, I really am'.

Describing the Democrats' 'deal' which allowed the Howard Government to pass a goods and services tax, Caswell said: 'Prime Minister John Howard has given the absolute minimum compared to what he would have been quite willing to give [on environmental commitments which should have been pressed by the Democrats]. The Democrats have returned to their Coalition roots and I don't want any part of it. I'm just waiting for their acknowledgement of my resignation so that I can join the Labor Party.'

Caswell said uranium mining remained a big concern to him, 'although it keeps getting pushed onto the back-burner politically'. He said that was one reason why he remained pro-union and pro-worker, because 'when governments start pushing individuals around, the next thing you know they're carving up the national parks—it's a cycle'.

And he's concerned for Australia's beaches. 'I live two blocks from Maroubra Beach,' he said. 'Need I say more? And I love trees—I really do.'[90]

THE gentle, lilting folk ballads of Brenda Liddiard can be seen as a musical history of the Australasian environment movement. The songs, many of which are collected on the cassette album *Songs of Protest and Survival* (1991), have formed a sort of 'soundtrack' for the times—they were sung by protesters at the Franklin, at Terania Creek, and almost everywhere else that people gathered to express their outrage at environmental vandalism.

They have been used on several films about the period—such as *Earth First* and *Give Trees a Chance*, both by Jeni Kendell and Paul Tait (see Chapter Six).

Liddiard was born in England and came to Australia in 1974, she says, in search of adventure. After travelling around the country for a while, following the music festival trail, she ended up at Nimbin in northern New South Wales, where she became involved in the music scene. At that time, there was a great deal of antagonism between the original inhabitants of the area (dairy farmers mostly and country people) and the so-called new settlers. These were mostly young people ('hippies') who had come north in search of an alternative lifestyle—peace, love, lots of dope—and to be close to nature.

This culture clash soon led to open confrontations, as in the battle over Terania Creek. This was a patch of rainforest, which had long escaped logging but now was targetted by a local company. The new settlers objected and set up a peaceful protest, which turned into one of the pivotal battles of the environmental movement. It became the subject of Kendell/Tait's *Give Trees a Chance.*

Liddiard said in 1999 that the experience at Terania Creek marked her environmental and musical coming of age. She learned a lot of practical skills—surviving in the forest, how to deal with the media, and much about her musical craft—but, much more important, she discovered just how much she cared about that one small piece of rainforest and how much she wanted to protect it. 'When the chainsaws started, I hurt; when the trees fell, I cried,' she recalled.[91]

This passionate commitment radiates in Liddiard's music. The two Liddiard songs used in *Earth First* are important in recreating the atmosphere of intensity and commitment that all those who were there remember about the protests. The lovely, haunting voice of Lisa Yeates is heard in 'For the River', a song written at the time of the blockade of the Franklin River in protest against government plans to build a dam in south-west Tasmania. While the camera shows us images of the protesters in their little boats or taking on the bulldozers in the mud (the images include a shot of a sobbing Yeates) the song rings out. It is a moment of extreme power and emotion.

In another scene of great power in *Earth First*, the tune 'Save the Daintree' is sung by protesters seeking to prevent a road being built through the North Queensland rainforest, a road the State Government was determined to push through. In the pitch-dark night of the rainforest, the protesters are shown being hunted by police with attack dogs. A media blackout was in force so no journalist is there except for one lone radio reporter who sneaked behind the lines.

The film is dark, the pictures being taken surreptitiously, and creatures of the night are about, obviously disturbed by the noise of tramping feet and barking dogs. Then we see the protesters, with lighted candles, walking slowly along the beach singing 'Save the Daintree'. It is an unforgettable moment.

In the tradition of all good protest songs, 'Save the Daintree' is simple—easy to sing, easy to remember—with a straightforward message:

> Save the Daintree, hear our cry
> Don't let the wilderness die
> Save the Daintree, hear our plea
> It means the world to you and me.[92]

After the fight for Terania was won, Liddiard spent the next decade devoted to environmental causes. In 1986 she moved to New Zealand where she met the man who was to become her husband, Mark Laurent. He had been involved for many years in the gospel music scene, and through her growing relationship with him she rediscovered Christianity—what she refers to as the 'missing link' in her spiritual life.

Liddiard and Laurent, in the midst of a busy performing schedule in the cafes and clubs of New Zealand, have been trying to take the green message to church people for years, doing concerts and participating in services—anything to get the word out that the care of the earth ought to be the concern of Christians. Liddiard says she tries to encourage people to live simply, to live lightly upon the earth, so that others may have enough to meet their basic needs. She prefers to think of humanity not dominating the earth but rather that its care is a sacred trust. And she suggests that the fight to save this earth from destruction is as much a part of the good fight as any other mission.

According to influential commentator Susan Jarvis, environmental issues such as those enunciated by Liddiard are at the forefront of contemporary Australian folk music. 'Folk has always been political,' Jarvis said in 1999. 'Folk musicians have always commented on social issues, and hoped to change society. Someone like Eric Bogle has been a social conscience of music for decades now.'[93]

Jarvis, a music journalist, radio broadcaster with national exposure and a consultant with the Australian Key Centre for Cultural and Media Policy at Brisbane's Griffith University, said the growing interest in indigenous music was contributing to the growth of environmental consciousness in music and linking that with issues of social justice.

'Kev Carmody has had a huge influence on everyone,' Jarvis said. 'He was really the first to bridge that gap between black and white society and to introduce the Aboriginal way of understanding the land to a white folk audience.'

Critic Michael Hannan, an associate professor at Southern Cross University in Lismore and the head of its program in contemporary music, said protest songs had acted as a 'socially cohesive force' in Australia's environmental battles.

'These songs have become classics,' Hannan said. 'They're standards. The same songs that were written for Terania Creek were later sung at the Franklin, and now they're being sung at Jabiluka—and, of course, more are being created all the time.'[94]

'This is real community music-making, although many of the artists have developed a high degree of professionalism and sophistication. The protest songs are interesting, though, because they are truly of the people. They're not part of the mainstream culture that's imposed from outside.'

The songs of this period have become a worthy study for academics. Andy Parks, a student of Professor Hannan's at Southern Cross, has collected some of them as part of his degree requirement and put them on a CD called *Lock On! Songs to Save Australia's Rainforests*.[95] This is a terrific bunch of songs, all of which deserve to be much better known, and includes performances by Brenda Liddiard and Lisa Yeates.

Yeates, the singer of Brenda Liddiard's 'For the River', has been a long-time friend and associate of Liddiard. She still lives in the Nimbin area and is now a sort of unofficial archivist for the music of the period. Yeates has been singing all of her life—but always just for her own pleasure, and it wasn't until Terania Creek that she discovered the power of music to move people and to inspire them in a cause.

'Music helps people in so many ways,' she said in 1999. 'For one thing, it helps them to express pain and anger in a much better way than violence or hard words. I saw it time and time again—a confrontation would be brewing but people would start singing and then tension would be lifted. I myself have stood in front of a bulldozer singing 'Open Your Ears to the Sounds of the Forest'. I was angry—so angry—but it helped me to find my courage to stand there and to express that anger in an acceptable way.'[96]

'And music gives people a way to find hope when they're really in despair—it can really lighten your heart and make anything terrible easier to bear. We saw that in the Franklin protest. When people had been arrested and they were on the police launches waiting to be taken away, we would all sing to them—it helped. You could see them take heart again. And it brings people together in a group—gives them focus and makes them realise they're part of something bigger than themselves.'[97]

'Music is also a way to express horrible truths that have to be said but somehow in words they are too much to take in,' Yeates said from Nimbin in 1999. She told about hearing internationally renowned campaigner Dr Helen Caldicott on the issue of nuclear proliferation and being so horrified that she 'just closed down'. Yeates said: 'I couldn't listen any more—I couldn't bear it. But then some people started singing a song—'And the Winds Blow from Pennsylvania'—it was the same information, but somehow when it was in a song it could kind of sneak into your mind and not completely destroy you.'

Lisa Yeates has been leading workshops at Southern Cross University at Lismore since 1986 about music and the power of songs. When asked about 'Lurk Loiter Hide Secrete', the Saul Moony song which had helped to drum up public support for the Franklin protests, ridiculing as it did the 'instant laws' that were decreed to stifle protest, Yeates was amazed that people in 1999 still remembered it. 'I don't know if it had any effect,' she said. 'I didn't think it ever made it out of Tasmania—that anyone had heard it who wasn't there.' But she still knew the chorus:

> We're going to lurk, loiter, hide secrete
> I'm a green greenie and I don't care
> I walk through the forest and I've got long hair
> I haven't washed for a week or more
> But I strut right by
> With my japarah in the air.[98]

In 1999, Yeates was still singing, despite her claim to be a 'retired greenie' with a bad case of burnout. 'I know how hard we worked for such small gains,' she said, 'and now to see the Howard government undoing it all—it just makes me weep.' But she was in Sydney in November of the previous year while the New South Wales Parliament was debating the future of the forest agreements, singing in the Domain and in Macquarie Street outside Parliament House. She said Upper House Green MP Ian Cohen spoke to her to give her some ideas about future directions, where the movement would go from here and she replied: 'Ian, you got the wrong person. I'm just an old hippie singer now. Somebody else will have to do this.'

Maybe so. But we still have Liddiard's songs to inspire us and Yeates' lovely voice to give them life.

PERHAPS no singer conjures up the sight and smell and feel of the Australian landscape better than John Williamson. His music, which he has called 'gum leaf music' to differentiate it from the American country sounds of Nashville, crosses the genres of folk and country—with a decidedly Australian flavour. He has written songs about places he's travelled (Charters Towers, Broome, Lightning Ridge) and has immortalised towns that nobody ever thought worth singing about before.

Williamson is very much a son of the land. Born in 1945, he spent his boyhood in the farming community of Quambatook, in the Mallee country of Victoria, later moving to Croppa Creek, near Moree in northern New South Wales. He was a farmer for a few years until musical success sidelined that life. Although Williamson has lived in Sydney for many years, his songs have dealt with so many 'country' concerns that they must call forth a response from anyone who has ever lived in a small town or tried to farm in this unforgiving land. He sings about the colour of the soil, the patterns of the clouds, the backbreaking work, about horses and dogs and drought and isolation and the pull of the city.

For Williamson, love of the country and a concern for country people naturally extend to a love of the bush. 'I like romanticising or immortalising bush towns,' he said in 1986. 'I find mining towns fascinating, but even when I write a love song it seems to be in an Australian landscape setting.'[99]

He also says that it was his personal farming experience that formed his outlook on rural matters and conservation. In 1990, he told music reviewer Bruce Elder that, 'because I was involved in clearing I could see what we had done'. Williamson continued: 'I really know [Australia] from a farmer's point of view. I have seen the devastation that's happened.'[100]

Williamson has had so much recognition and success for his bush ballads that it is hard to remember that it wasn't always so—that Australians were not always so in love with the sound of their own accent, with songs about their country and their way of life.

After some very early and unexpected success with his song 'Old Man Emu', which shot to No. 1 in 1970 and won him a gold record, Williamson struggled for years, gradually building a following and producing several albums, before he found real recognition with his album *Mallee Boy* in 1986. *Mallee Boy* took off and had gone double platinum by the end of 1987. Suddenly, it was OK to be Australian—and Williamson along with other country artists began to reap the benefits. 'I think *Mallee Boy* came out just at that point when Australians finally accepted the full-on Australian thing,' Williamson told Susan Jarvis in 1995. 'They became comfortable with who they were and wanted to hear and know more about it.'[101]

Williamson became famous for his love of all things Australian. A longtime republican, a designer of flags and a booster of 'Buy Australian' campaigns, he communicates in his music a deep sense of what it means to him to belong to this country.

'There is an underlying theme throughout my music that I call the Spirit of the Land,' Williamson has written. 'On the surface, this is about my great love of the original nature of this country, but it goes much deeper. When you consider that the Olgas are over 500 million years old, there is a power out there that the Aborigines have tried to explain to us. The spirit of the land, I believe is still encapsulated in our old-growth forests that are being devastated by woodchip companies for short-term profit. It is about survival. We are still losing koala habitat to developers on the east coast, and we are destroying the life-blood of the country, our rivers. To destroy the spirit of the land is to destroy our own spirit . . . My dream is that one day we will resolve the differences between indigenous and non-indigenous people and all become Australians. The thing that will unite us is the spirit of the land, a respect for the nature of this country. It is well represented by the colour red ochre and that majestic rock out in the red heart.'[102]

In 1990, he told David Greason, writing in *Australian Playboy* magazine, that he had 'a genuine passion' for this land. 'Before Albert Namatjira we didn't see the colours out there,' Williamson told Greason. 'That's a fact as far as I'm concerned. We didn't realise the mountains were pink. The old painters came to Australia and painted the leaves on the trees. They couldn't see the leaves were hanging down.' Williamson said he had 'fallen in love' with the Australian landscape—'the twisted agony of the red gums, which were regarded in the past as ugly'. He said Australians had come to realise that 'all these things are beautiful'; he shared his spiritual concerns with Greason: 'The sand is beautiful and the coast is beautiful. And every time I see another bit of the bush being wiped out—and everything that lives in it goes with it—it hurts.'[103]

Williamson's passion for the bush has led to his involvement with a series of controversial issues over the years—most to do with the environment. He has tackled serious issues in his songs and has paid the price in terms of controversy and strife.

In 1981, his outspoken republican views—he said that the Diggers had an 'unnatural allegiance' to the Queen—got him banned from an RSL club at Inverell in northern New South Wales. Many angry fans vowed to shun his music forever after such disrespect.

And in 1989, the New South Wales Rugby League Grand Final decided to do a salute to the environment as its pre-game entertainment, thinking, one supposes, that they would put on a nice show about trees and fresh air and offend nobody. However, they invited Williamson to sing his new and controversial song 'Rip Rip Woodchip' and all hell broke loose:

> Rip rip woodchip, turn it into paper
> Throw it in the bin, no news today
>
> Chainsaw, eyesore, more decay[104]

The League was inundated by hundreds of phone calls and faxes objecting to the use of the song. The State Minister for Sport and Recreation at the time, Bob Rowland Smith, got into the debate on the side of the loggers. He called the song 'provocative' and warned that the league risked offending everyone, including the Canberra Raiders, one of the participating teams, which was sponsored by the timber industry. Rowland Smith said a football game was no place to get controversial and that overseas people knew better: 'They have all these lovely little girls kicking up their legs. Now why in the world would you want to bring into that atmosphere a question of whether it's right or wrong to log?'[105]

Williamson has said that he was taken aback by all the fuss about the song. But his 'blood boils at the thought of our old-growth forests still being whittled away for short-term, short-sighted economic gains.' Williamson has lent his name—and given freely of his time and money—to help the causes he believes in. A percentage of the revenues from 'Rip Rip Woodchip' has been donated to the Australian Conservation Foundation and profits (hundreds of thousands of dollars) from his song 'Goodbye Blinky Bill', about the loss of koala habitat, have gone towards the preservation of that unique marsupial. Countless times Williamson has appeared for free and paid all his own expenses to raise funds for a good cause. He is involved with LandCare Australia, he is a patron of WIRES and he has raised money for a fund to preserve the Mallee fowl.

He has a great love of birds, and says that he mentions more than 40 birds in his show. 'I notice the changes in birdlife when I'm in a new area,' he told Bruce Elder. 'If they've wiped out the native trees all you'll see are sparrows, starlings, a few crows and magpies, which can nest in the English pine trees. But none of the birds that live in hollows—and most of our native birds do. There's no hollows to live in if the bush has been destroyed. Aussies don't realise that once you wipe out all those old trees, the Australian birdlife is gone.'[106]

Williamson's stands on environmental matters have taken great courage because he has risked alienating his natural constituency: country people and those in the small towns who are the first to feel economic downturns and who have most to fear from change.

Williamson manages to be optimistic in his outlook on Australia's environment. He says that it would now be unthinkable to conduct British nuclear tests here. 'It's heart-rending to think what Menzies did, that he was keen for the British to bomb Maralinga,' he told *Australian Playboy*. 'That's almost unbelievable: that anyone could regard any part of Australia as being shit, you know. And the Aborigines weren't even consulted. As a group I think we're progressing. And that's one of the things about the green movement—the victories are hard-fought. No one could do something like Maralinga without having a fight on their hands.'[107]

A NEW generation of singer–songwriters has taken up the challenge of writing music to celebrate our Australian environment and to call for its protection. It is encouraging to know that the younger songwriters at the end of the millennium don't find the issues boring, or overwhelming, but as deeply relevant, disturbing and important.

One such songwriter is Ruth Apelt, a Brisbane-based folksinger who has written a number of green songs in a repertoire of politically and socially charged music. Apelt, who describes herself as thirtysomething but who looks and sounds much younger, has performed with Eric Bogle, Kev Carmody, Archie Roach—and just about everybody else on Australia's folk circuit, as well as such international performers as Janis Ian, Luka Bloom and LadySmith Black Mombazo. In 1999, she had a CD out—*Told You Twice* (1994), recorded at The Salad Bowl in Nashville—and another in the works.

In Apelt's songs, you sense a Generation X frustration that society is still messed up, despite the promises of the '60s. In '1969', for instance, she sings:

> Too many people have died
> Too many governments have lied
> Since the answer was Blowin' in the Wind
> Since the times were a changin'[108]

Apelt's Queensland roots show themselves in 'Tourist-Led Recovery', in which she decries the destruction of the environment and the social ambience of North Queensland as the local people learn to scramble for a buck:

> Hotels are going up
> And the locals are getting mean
> They've got overseas investors
> Moving in on their scene
> But green is dollars, baby
> Even more green
> Than the rainforest attractions
> That we show[109]

In her clear, lilting, yet somewhat whispery voice, Apelt sings about the evils of economic rationalism (she has a degree in Government and Economics) and environmental vandalism. In her song 'Mother Nature Blues', she demolishes the 'greens cost jobs' argument in a few poignant and clever verses:

> Imagine you're a miner
> Digging up the land
> Sweating in the sun
> Sifting through the sand
> Gold, nickel, iron ore, uranium too
> Oh you've got your private reasons
> Why you do the things you do
> But imagine you're the rock
> Standing for all time
> Blasted now to pieces

Though you're sacred and divine
Imagine you're the rock![110]

Apelt says she considers it the role of musicians to keep issues alive after the press and mainstream culture have decided to give them away. 'Remember a few years ago,' she said, 'when the environment was a really big issue? It was everywhere—in politics, on TV, in the papers. But now, they seem to have decided it isn't sexy any more. I think people stopped wanting to read about it. They'd hear about greenhouse gas and start feeling uncomfortable about driving their cars, or about all the waste they were generating. But with songwriters, well, it's our job to make people feel uncomfortable sometimes, to tell the truth about things. Even the unfashionable things.'[111]

Another teller of unfashionable truths is Penelope Swales. In what is perhaps the edgiest and most uncompromising folk music since the '60s, Swales takes on the evils of society. In a strong, rich contralto, she tackles tough issues—the spate of police shootings in Victoria, Aboriginal dispossession, drugs, human rights—even her more personal love songs are political and filled with trenchant social comment.

Swales, born in Melbourne in 1969, had a tough apprenticeship as a songwriter. She left home at 15 and spent the next few years living from hand to mouth in the streets of St Kilda and Kings Cross, busking to survive. The companions of her youth were drug addicts and prostitutes but she says now that an innate sense of survival kept her from being swept up in their destructive lifestyles. 'After busking I used to hang around in nightclubs snoozing until morning,' she said in 1996. 'During the day I would sleep in trains or in the park. I was terrified of losing myself and abstained while those around me spent their time trying to score or get stoned. The working girls saw that I was trying to make a go of it and loved me for it. There was one who would never walk by me without throwing coins in the guitar case.'[112]

Penelope Swales has travelled around Australia extensively during the '90s and is a regular on the folk music circuit. She has also travelled overseas, performing at the Edinburgh Festival in 1996 and throughout the UK and Europe.

In songs written during her time in Britain, through her homesickness and loneliness, a sense of her Australian identity emerges. In 'Sidmouth Song', she is surprised that she doesn't feel more at home in England and laments for the 'wide, brown [land] that's really in my soul'. And in 'Lost in London', ground down by the intensity of this huge city, she sings:

And I'm amazed at the persistence of life
Weeds growing out of windowsills
Flowers blooming on asbestos roofing
And pigeons nesting in factory grills
And how do children still skip and glow
With that strange pure luminosity
While they drink this hard water and breathe this foul air?[113]

In 'Bougainville', one of the radical, highly political songs that keeps Swales firmly out of the mainstream, she tells the story of the Rio Tinto zinc mine on that island and of the tremendous environmental and social destruction that it wrought. She sings of the poisoned fish, the billion tonnes of waste that were poured into the Jaba River, and the increasing poverty and despair of the islanders. She castigates Australia's role in the matter, that our government armed and financed the troops sent in to quash the villagers' small revolt. She writes:

> The company denied
> They were the cause of any of the damage
> But our fruit trees no longer bore fruit
> Their leaves were killed by acid rain
> There are no fish in the river
> The damage they did will be here forever[114]

The chorus of this song is a hymn to the importance of land in Bougainville society. Land is seen as 'our life'. It is food, it is sustenance, it is security, it is politics, it is status, it is marriage. At the end of the song, she calls for Australian solidarity with the struggles in Bougainville, not just for humanitarian reasons or because of any regional loyalty but because their struggle is our struggle, that we must fight to save our land, too:

> In fact, this is our only world
> Land is our life
> And if we destroy that land
> We're cutting out the very heart of our existence[115]

Swales wrote this song in England, where she was amazed to find that the situation in Bougainville was in the forefront of current events. 'People were really aware and indignant about it—it was at least as well known there as the problems of apartheid in South Africa,' she said in 1999, describing how 'Bougainville' was inspired by a pair of journalists who went to the island at immense personal risk and brought back stories of the devastation they found there. Swales says she finds this song inspires people because it's not just a story of destruction and ruin, but it's about grassroots action. 'The people there were able to organise and get the attention of the world to their problems and actually make a difference,' she said. 'I sing this song at Jabiluka rallies and demonstrations against toxic waste dumps. I find it really heartens people, it keeps them going, because it shows what a small, dedicated group of people can achieve.'[116]

Despite the dark intensity of so many of her songs, Swales is witty and upbeat in conversation and in concert. She lives in the small community of Gembrook in the Dandenongs about seventy kilometres from Melbourne, a place she describes as deeply rural—'I can't see another building from my house.' Here in the heart of the bush, Swales sees the effects of logging and is concerned that so much of it seems to happen in secret. 'I drive home from gigs in the city late at night and I see trucks one after the other in the middle of the night, carrying out big logs,' she said. 'Now, I suppose they'd say that was a safety issue—that they want to travel when there's less traffic—but there is also an element of not wanting it known that they're logging in the Melbourne water catchment area.'[117]

Swales lists among her concerns the creeping urbanisation of our bush, which she sees between her town and the city. 'It's terrible,' she said. 'They buy these quarter-acre blocks and then put up quarter-acre dream homes on them. You know the sort of places? Four or five bedrooms, family rooms, glassed-in patios, enormous laundry? They're huge—30, 40 squares. They'd be impossible to heat, they're not built to any kind of eco-design. They're just great temples of consumerism and they're having a huge impact on the land.' But she sees room for some optimism in that even amongst this urban sprawl, a kind of grassroots activism is getting a toehold. 'Up here not long ago, there was a huge outcry when they wanted to build a big Safeway store, and they managed to stop it,' she said. 'People are becoming aware and ordinary people are getting involved more and more—I see that as a good sign.'[118]

A PLETHORA of environmental concerns have suffused the works of Australia's leading poets over the past quarter-century, with Judith Wright leading the way. Wright, an activist for peace, disarmament, Aboriginal rights and an array of ecological matters, has written poems whose themes cover the broadest range—from dropping the atomic bomb on Japanese cities to Japanese coastal developments off Queensland.

In 1992, Wright received the Queen's Medal for Poetry in Canberra, becoming the first Australian to earn this prize, joining the likes of Poet Laureate Ted Hughes, W.H. Auden and Robert Graves.

At the start of the 1970s, an inner rage—what Wright would describe as 'almost pure fury'—was her driving motivation. She increasingly wrote about her ecological despair in poems such as 'Australia, 1970'. This and other verses were partly inspired by a drive through Queensland where she saw bulldozers rampantly flattening the earth. To Wright this was 'dreadful' and she felt the need to protest as loudly as she could in her poetry.

Wright's soul-sister in environmental protest was Kath Walker (later Oodgeroo Noonuccal), who by the mid-1970s told of how she had 'broken away completely' from her previous writings to concentrate more fully on ecological issues. Walker's poetry and other works like *Stradbroke Dreamtime* (see Chapter Three) were spurred by her demands for radical changes, such as the end of mining in fragile areas like her beloved Stradbroke Island. 'The immediate step is to STOP ALL MINING! IMMEDIATELY, STOP IT! . . . Australia at the moment is all set to mine herself out of existence,' she told *Meanjin* magazine in 1977. In that interview, Walker expressed her deepest misgivings about the prospects for human error in the mining of uranium, which she regarded as dangerous and unable to be properly controlled.[119]

The uranium issue continued to gnaw at Kath Walker's psyche, as she told a Canberra writers' symposium on Armistice Day 1986: 'There will come a time when our mother, whom the invaders so ruthlessly raped, by cutting down her trees, by slaughtering her creatures and mining her minerals, will retaliate. Are we so impudent as to expect that we can survive her backlash?'[120]

In a further eloquent statement of her identity, Walker changed her 'European' name to an Aboriginal one during Australia's Bicentennial celebrations in 1988, protesting against what she described as 'the murder and exploitation of our people' in the 200 years since the First Fleet, and because of the then Federal Government's refusal to recognise Aboriginal land rights.

As early as 1976, the poetry of Judith Wright, who began writing as a little girl, was concerned with many ecological contradictions, such as how whales self-destruct on beaches. She sees such mass suicide as symptomatic of a wider madness afflicting humanity. She discourses on these universal dramas in verses like 'Interface III':

> Whole peoples, countries, nations
> have died in the same way.
> Galaxies may be strewn
> with staring burned-out planets
> which took that path.[121]

The diverse poems in the collection *Fourth Quarter* (1976) reflect Wright's anger against a range of potent contemporary issues: militarism, a 'coca-cola culture' destroying Pacific countries, and the deadening conformity which has allowed the growth of a bizarre lifeless entity like Canberra.

Wright, born in 1915 and raised amid the eucalypt country of her parents' property near Armidale in northern New South Wales and who as an adult lived in Queensland and southern New South Wales, incorporates all of these influences in her work.

Anger combined with an added sense of desperation are evident in a later Wright collection, *Phantom Dwelling* (1985). 'We've brought on our own cancers,' she writes in 'Rockpool', a verse which is part of 'The Shadow of Fire (Ghazals)'.[122]

Wright continues to rejoice in nature and all its splendours but she is more concerned with human-caused disasters. She writes of decay and the aging process, and laments that harsh realities like drought and fire need to be more adequately addressed. For example, in 'Dust' Wright meshes these various themes, and in 'Patterns' there are even more ominous man-made blights:

> Round earth's circumference and atmosphere
> bombs and warheads crouch waiting this time.[123]

These concerns were an integral part of Wright's life and work, and in an interview with the *Canberra Times*' Marion Frith in 1994 she spoke, at age 78, of a 'despairing resignation to all that is impending'. The interview, at Braidwood in southern New South Wales, was summed up by Frith as a reflection of a 'resigned incredulity' afflicting the poet at the time: 'The hurt to the land, and the oceans and the skies, envelopes Wright in a perpetual grief, coupled with disbelief and outrage . . . we do not even realise what we have done and what we are continuing to do.'[124]

A life-long activist, in 1967 Wright joined the struggle to save the Great Barrier Reef, in 1971 she participated in a range of environmental and political protests, and in 1974 she addressed the National Press Club in Canberra on the pollution of the Reef.

Among all Australian poets, few have worked as assiduously as Wright to rid the world of destruction and to warn her readership of what needed to be done.

Adelaide critic and academic, Syd Harrex, has listed such themes as degradation and pollution among the 'intimate concerns' discussed by Australian poets. He believes Australia's leading poets have produced works of 'purpose and responsibility' and have had a significant influence. He described Les Murray as a writer who 'can evoke things so powerfully, and raises issues with which readers can identify'. To Harrex, Murray is an 'extrovert like Whitman', who sees himself as the voice of rural Australians, who have been neglected and misunderstood for too long. By contrast, Bruce Dawe is the powerful spokesman for urban and suburban who has been able to 'crystallise the important issues' of Australian cities. Harrex said of Murray and Dawe that they both write about anything they feel strongly about, and they did this with a 'common touch' and in an 'idiom' understood and loved by their readers.

The important contribution of 'popular' poets was also acknowledged by Harrex, who noted the 'intrepid and forceful' contribution of someone like Mark O'Connor who, for example, in poetry about marine life 'gets in there, dives in' and describes the underwater scene within the context of being 'concerned with the future'.

Above all, Harrex pointed to and praised the breathtaking contribution of Judith Wright, whom he described as 'in the forefront of issues throughout her writing life'. Despite Wright being born and raised in an environment which might have led her down a conservative path, she instead became 'consistently radical' and her vast body of work always revealed 'a commitment of one kind or another, either through traditional lyric forms or other poetic styles, and her writings always have a potent point'.[125]

The contribution of Judith Wright was also recognised by Canberra scholar and critic Adrian Caesar, who said she was an 'indisputably great poet' who had struggled to reconcile her writing with a 'sense of political despair' that she felt over environmental issues. Caesar described Wright's influence as transcending her immediate readership and her writing did much more than 'preach to the converted'.[126]

Caesar, a senior lecturer in the School of English at the University College of the University of New South Wales at the Australian Defence Force Academy, said Wright was among a range of important modern

Australian writers whose ecological concerns gelled with the pastoral and other traditions of our poetry. Caesar described how 'reader-friendly' poets, by using forms ranging from satire, the lyric mode and the romantic were able to 'speak to a great number of people' and touch them with their passion. For example, the poetry of Aboriginal writer Kevin Gilbert represented a 'collision between agitprop and the Anglo Saxon tradition' while 'the brave writing' of Dorothy Hewett had a strong 'political edge all the way through, one way or another' in describing her environmental reflections.

Discussing the seminal influence of Les Murray, Caesar said his work was 'contradictory' and often 'ultra right wing' but was important in 'forcing readers to consider their positions on issues . . . and only the truly great poets can do this'. Caesar said Murray was able to convey his nostalgia for the 'Australia of his own imagination, and he hates anything which despoils or impinges upon the Australia that he loves'.

THROUGHOUT a long career, Les Murray has expressed his environmental concerns in a broad range of works—everything from sonnets to short, clever rhyming lines, to novels, verse, film scripts, essays and various works of journalism. Les Murray was born in 1938 in the town of Bunyah, in the Great Lakes district of New South Wales, about sixty kilometres from Taree, and spent much of his early life on the family's dairy farm. He returned to Bunyah on the final day of 1985 to a 16-hectare property, after an absence of more than thirty years from the place of his ancestors.

Another writer who grew up in the district, Ian Moffitt (see Chapter Two), describes Murray as being imbued with 'an aching love for his "spirit country" between the Myall and Manning Rivers'—a love, and concern, that permeate his literary output.[127]

Poet and novelist Alan Gould embellished this theme, noting that Murray's work was based on this region 'as surely as, say, Thomas Hardy's was centred on Wessex, or [Eric] Rolls' is on The Pilliga'. And on a more universal scale, according to Gould, Murray's poetry shines a light on 'the character of Australian Civilisation and the nature of existence itself'. Gould also noted how Murray had redressed a 'notorious' neglect by other poets of 'the creations of technology'. In an analysis of *The People's Otherworld*, Gould said: 'In his celebrations of bulldozers, cranes, the space-shuttle launch-pad, geo-phones, projectors, Murray integrates into our literary tradition phenomena that are close to the actual experience and imaginings of our time.'[128]

American poet and critic William Logan has commented that Murray has shown a 'superb ear for dialect' and has revealed the same type of 'innocent cunning' of writers like Hardy and Robert Frost.

A man who arouses passion and controversy, Murray is by consensus Australia's most important and influential modern poet. He was described in 1998 by prominent critic James Tulip as a giant who 'squats or sprawls on the land of Australian poetry like the imposing monolith that is Uluru; and just as Uluru seems to be the guardian and sentinel of the great western deserts that lie behind it, so too does Murray seem to be the guardian of mystery, an object of confrontation and of vistas of wisdom'.[129]

Murray, who as the nation's pre-eminent poet and leading voice of rugged individualism was asked in 1999 by Prime Minister John Howard to write a preamble to the Australian Constitution, has dabbled in politics and social movements to get his ideas wider exposure. For example, in the early 1970s he was part of a group called the Australian Commonwealth Party, among whose stated aims, as enunciated by Murray, was 'a start on the work of eliminating all exploitative and exhaustive technologies from the Australian economy'. In this manifesto, which rejected any 'Luddite reversion to the past', Murray called for the replacement of 'bad, greed-motivated, typically violent technology' with an 'ecologically responsible' variety. Typical of the hard-headed attitude of a man of the land, Murray wrote in a tract well ahead of its time: 'We cannot and should not destroy the Machine—we must civilise it.'[130]

After the political involvement went by the wayside, this theme was developed in works like *The Peasant Mandarin* (1978). In this collection, Murray lauded many aspects of conservation but deplored how it had 'often been used as a stick to beat ordinary Australians' by arrogant zealots caught up in 'the radicalism of the Best People'; and a decade later in *The Vernacular Republic* (1988), in which he wrote 'Ecology? Sure,/But also husbandry'.[131]

On a broad, poetic canvas, Les Murray has regarded himself with typical assertiveness as 'one of the lightning conductors' of Australian attitudes and thinking, and he always sought to expound to as wide an audience as he could that 'the gum forest is always with me'.[132]

In the narrative, story-telling style of *The Boys Who Stole the Funeral* (1980), Murray laments the worsening fate of mankind. Murray observes how the earth's treasures—and its horribly abused inhabitants—are mere fodder, especially in the increasingly high-tech conflicts of the 20th century.

Side by side Murray uses words like 'dreams' and 'unhuman' to observe that morality has not advanced all that much since the trench-warfare killing fields of World War I. Not only the earth but also 'the cosmos' feel the devastation of what mankind reaps. The progress of civilisation is a medley of the 'inhuman, mild, indifferent/to chagrin or alarm'.[133]

In war and peace, writes Murray, a desensitised humanity crashes through with all its endeavours, without thought to longer-term consequences. As development proceeds pell-mell, there is nobody to champion nature against the encroachments of expanding populations. Every time a bulldozer roars, Murray feels the shudder:

> Sydney! It's a building site now. . .
> waiting to shovel. Ambition has gobbled this city
> in towers for companies wanting to build more towers.[134]

Such crash-through impetus is, naturally, often accompanied by breath-taking wastefulness, and Murray observes how less than half of the office space created in the burgeoning skyscrapers is occupied.

In this expansive work, Murray believes nothing less than that man's worst enemy is man when it comes to harming the ecology; the inventiveness and glorious achievements of civilisation are too frequently throttled by corporate greed:

> We fought with rabbits and with drought; we
> beat disease and flood
> but we couldn't fight the men in ties when
> they were out for blood.[135]

Sydney's urban blight is also examined in another well-received Murray collection, *The People's Otherworld* (1983), in which 'The Fishermen in South Head' features lyrical descriptions of the mouth of a glorious Sydney Harbour which has suddenly turned to 'floating gruel/of gull-blown effluent'. Murray grapples with the contradictions whereby South Head, a place of awe-inspiring majesty, is also the most notorious site for suicides. And, despite the various detritus of man-made follies, one still cannot escape the glorious vista 'atop the centuries/of strata . . .'[136]

In 'Anthropomorphics', Murray again marvels at what a piece of work is man—and what wonders are all living things—but humans are simultaneously noble and unfeeling and indiscriminate hunter-killers. And, in 'The Romantic Theme of Ruins' and 'The Forest Hit By Modern Use', the thread of contradiction is woven yet again, glorying in exquisite opposites: 'Earth itself as the great ruin: was she not the contracted wreck of Chaos?'[137]

Ruin is at the core of a later Murray collection, *Dog Fox Field* (1990), whose verses are filled with human and environmental decay, constantly referring to 'bareness', 'semi-desert' and 'algal water'.

Land and water are being poisoned, young minds are being poisoned, and the harvest is one of bitter fruit as in 'The Fall of Aphrodite Street':

> . . . the apples are stacked and juicy
> but some are death to eat.
>
> God help the millions that street killed
> and those it sickened too,
> when it was built past every house
> and often bulldozed through.[138]

The land was once pristine but the Europeans had turned it into an exploited slave. Murray angrily describes how the best and brightest in the nation are being belted by drugs, and as criminality takes over, bars are appearing on more windows—just as in 1788 the original white inhabitants were oppressed on land and sea by iron bars. Also, modern man is now moving much faster and more giddily than nature had decreed, with atrocious and unforeseen results.

Esteemed West Indian poet and dramatist Derek Walcott, describing Murray as 'the appointed bard of the Australian wastes', has likened some of his visions to those of the *Mad Max* films (see Chapter Six)—bleak, post-Armageddon and hopeless. Walcott, writing in the Washington-based weekly journal of opinion *The New Republic*, said in 1989 that Murray's poems 'quake on the page . . . They leave tracks like home-built engines. They stagger and they are staggering'.[139]

A tone of pessimism and outrage continues into the next phase of Murray's writings, as in the collection *Subhuman Redneck Poems* (1996), where his distaste for trends in modern society reaches a new peak of intensity.

One of the poems, 'On the Present Slaughter of Feral Animals', seems to enshrine Murray's sense of doom:

> It seems that merciless human rearrangement of the whole earth
> is to have no green ending . . .
>
> . . . this luxury massacre on landscapes draining of settlement smells
> of gas theory . . .[140]

Murray acknowledges the occasionally altruistic underpinnings of much of man's doings, but the shallowness of it all seems certain to engulf us all. 'Inside Ayers Rock', for example, shows how limitless bad taste and unthinking cultural cretinism can invade one of Australia's true natural wonders.[141]

The poems in this collection, despite the in-yer-face challenges they make, have been described as 'ecologically visionary' by James Tulip, who said they 'situate Murray's awareness in world terms but with local colour'.[142]

Murray has been labelled by some critics as overwhelmingly conservative in his thinking and in his artistic bent, so it is quite an achievement that, despite what commentator Andrew Taylor has described as his 'dislike of intellectual and social movements', he has 'dramatically and potently alerted his readers to the ecological peril facing the nation'.[143]

Leading Adelaide poet and reviewer Peter Goldsworthy, although conceding that he himself belonged to the 'class' so scorned by Murray, nevertheless wrote glowingly about his admiration for the Bard of Bunyah: 'Why do I love Murray so much? Because he is there, like his eponymous river, our one big river, a force of nature. Because he makes the goose-bumps erupt, makes the hair stand up on my neck—the Robert Graves Test of Poetry. Because he shows me things I didn't know I knew, or knew but had not yet articulated.' And, as for Murray's many controversial notions, Goldsworthy revealed how 'even at his most polemical, he shakes my views around, which is never a bad thing . . . he challenges complacency'.[144]

Goldsworthy went even further, likening Murray to Walt Whitman in that 'we see more things for the first time, and see more familiar things renewed in Murray's work than in the work of any other poet or painter or writer I can think of'. To Goldsworthy, Murray is 'an infinitely more interesting and diverse stylist than Whitman'.[145]

This view was echoed by Brisbane critic Rosemary Sorensen, who wrote in the *Courier-Mail* that for more than thirty years 'the disgruntled, often passionate, sometimes outrageous voice of the big bloke from Bunyah has carried the can for poetic standards'.[146]

The contradictory strands of Murray's works and personality have been commented on internationally, including by American critics. One of these, Professor William Logan of the University of Florida's Department of English, has described Murray as 'the most arresting poet to come out of Australia since A.D. Hope and Judith Wright' yet was a writer with a 'burly, somewhat shambling style, excursive and unmethodical, odd and clumsy and hard to categorise, like a platypus, but with an immense appetite for subject'. Logan, a prominent poet and essayist, has described Murray as being like Midas in that everything he touched turned to poetry, as someone who is 'not afraid of subject, and will tackle any old thing'.[147]

Throughout his career, Murray has been a paradox—sensitive yet seemingly arrogant; conservative yet exceptionally caring with a strong sense of community; deeply thoughtful yet irreverent—but these contradictions reflect accurately those in the Australian character and the challenges, environmental and social, that the nation has had to face in the past thirty years.

A KINDRED spirit to Les Murray, who shares many of his concerns, is Bruce Dawe, whose deeply compassionate verse bravely and steadfastly rails against senseless destruction and the unthinking basis of so much that alters the ecology forever. In his collection *Sometimes Gladness* (1997), which encompasses more than forty years of writing, Dawe dissects how the imaginations of both the rulers and the ruled are as arid as the depleted lands they massacre. The notion of 'Lucky Country' is to Dawe a vicious irony; and, human nature being what it is, he sees no hope of salvation.

Born in Geelong in 1930, Dawe settled with his family at Toowoomba in the late 1960s, and since then he and his poetry have become inextricably linked with southern Queensland. After a stint in the RAAF and as a secondary-school teacher, Dawe pursued a literary career and had his first major collection of poetry, *No Fixed Address*, published in 1962.

According to his literary biographer Peter Kuch, one of Dawe's major achievements has been to 'successfully challenge contemporary Australia to take account of itself'—and that would encompass everything from drab local suburbia to international outrages. Throughout his career, Dawe has examined 'a world perpetually being impoverished by routine, advertising, pollution, habit, stress and all the other etiolating forces of modern life'.[148]

Dawe sought to reach out to the widest possible readership and to exert the widest possible influence. He saw himself as a populist poet, seeking to touch the person in the street, rather than regarding poetry as an exclusive and inaccessible domain. 'I'd rather have six poems over six years in a newspaper than six poems in an issue of [the literary magazine] *Scripsi*,' he confessed in an annotation to his private papers. 'I don't have a snob aesthetic about this or that place of publication—I publish my own work in such things as newspapers, even *Retail News*—and I dream of the day when poems will be printed on paper bags!'[149]

In *Sometimes Gladness*, an unrelenting bleakness is at the core of poems like 'In a Time of Drought':

> it's too late now to look to the heavens for help;
>
> Everything is up for grabs.[150]

Dawe reviles the civilisations which have 'stripped away the living skin' and left the landscape a sordid 'bony frame'. Such are the images of 'When First the Land Was Ours' which documents the follies of untenable public policies and unjustified economic planning. Dawe describes how the earth's glories are being thrust aside:

> We ring-barked, burned and bulldozed
> the trees and scrub away.[151]

Dawe is outraged at the unsustainability of the way in which humanity treats its rare resources and decries how 'in a land we robbed of life . . . we could not hope to live'. According to Kuch, throughout *Sometimes Gladness* Dawe exposes the 'craven consumer' as being largely responsible for the outrages against nature. In possibly the most profound reflections on the environmental tragedies which are being wreaked, Dawe uses the full force of his extensive powers to depict the madness. He surveys a scene of degradation, one where 'we were not meant to pasture sand or harvest fields of stone'.[152]

His summary of 200 years of European settlement is equally a lament. 'Search and Destroy (a Bicentenary poem)' tells of bleakness experienced over several decades. The specific instances of often irreversible decline are many:

> Now fish float belly up downstream caught
> by chemicals too vague to be fought[153]

There are constant and relentless images of a careless throwaway society where nobody cares about what is happening—of detritus piled upon growing mounds of junk. The woes of society and inappropriate economic priorities create a climate of utter desolation and despair. The message in verses like 'Planning a Time Capsule' is that social justice must become the wellspring to heal the wider malaise afflicting us and the environment.

In 'You and Sarajevo', Dawe, like Murray a self-described Christian poet, examines the global implications of mass destruction whereby on every corner of the planet 'beautiful things and people are blasted apart all the day long'.[154]

Describing the build-up to the Commonwealth Games in Brisbane in 1982—the ultimate in bread and circuses during the Bjelke-Petersen era—Dawe provides a glorious precursor to the attitudes which would also infest preparations for the Sydney Olympic Games in the year 2000. 'The situation in Queensland created by the promotion of the Commonwealth Games . . . was absolutely crazy,' Dawe said in a 1991 interview. 'It's that sort of false optimism, that blind boosterism, that I am very much opposed to. I don't accept it in any of its forms. I object to it in Australian Babbitts, as I do in American ones.'[155]

Another Dawe literary biographer, Ken Goodwin, has noted that although Dawe was sympathetic to many conservative causes, he had nothing but disdain and contempt for the agenda of Sir Joh, both on environmental and human-rights issues. For example, in 1986, Dawe wrote a letter to the editor of the *Courier-Mail* in which he noted that Bjelke-Petersen's attitude to coastline development in Queensland gave a new meaning to the words 'coastal strip'. Elsewhere, he similarly blamed Sir Joh for 'ravaging the environment, crushing opposition and erecting ugly structures'.[156]

Goodwin has noted that Dawe dissected the ills of Toowoomba as a microcosm of all that was evil in Queensland, and the Sunshine State in turn reflected the ills of the wider world.

Dawe saw Australians as victims of manipulation and a large part of the problem was that they were a 'transplanted people'. He described a national suffering based on the fact that 'we have not been able to bring much of our original soil with us', and this has had devastating consequences for our environment.[157]

A SIMILARLY wide scope of subjects is embraced in the work of Dorothy Hewett, who in poetry spanning almost fifty years reflects thoughts also expressed in her many plays (see Chapter Six). Throughout her poetic works, Hewett deals equally with the plight of living in a huge metropolis like Sydney and coping with land degradation and other blights throughout arid areas in Western Australia.

Concerns about the big city and the degraded Outback cover the spectrum in her later poetry embracing *Alice in Wormland* (1987) and *Peninsula* (1994).

Alice in Wormland vividly portrays a T.S. Eliot-like cityscape of destroyed lives amid detritus and despair in the slums of Sydney's Darlinghurst:

> . . . a roomful of ash
> cans books and butts piled up
> . . . Sydney's burning this tragic city . . .[158]

In the remote, degraded areas of the vast continent as depicted in *Peninsula*, the scene is just as bleak, depressing and hopeless:

> . . . in the middle of nowhere
> on the same road
> bitumenised now
> littered with busted tyres
> and the corpses of animals[159]

Similarly, 'Pylons' has images of forests being trampled savagely underfoot as human invaders create 'the blood-soaked highways stinking of sweat and speed'. In tragic contrast, 'The Frogs' is a rhapsody for the splendours and majesty of those creatures which have always defined nature at its most benign and tranquil.[160]

'South-West' is a paean for Georgiana Molloy—also lauded by William J. Lines (see Chapter Three)—who studied the plants of Western Australia during its first years of settlement. In her joyous discovery of these new and strange plants, Molloy infused them and the land with a kind of spirituality, far removed from the attitude of exploitation and utilitarianism that surrounded her in the new colony. Hewett makes clear just what glories are still being constantly destroyed by exploiters.[161]

Also transcending Hewett's poetry is her enduring fear and abhorrence of all things nuclear, worries which had emerged during her years as a member of the Communist Party when she was keenly aware of how Cold War brinkmanship and the arms race threatened civilisation.

ENVIRONMENTAL themes are also the concern of younger poets like Anthony Lawrence. Born in 1957, Lawrence has lived in several quite different places across Australia and travelled extensively overseas; and everywhere he has observed environmental outrages first-hand. After many years on the West Australian coast, he settled in the New South Wales Blue Mountains, where in the late 1990s he became a teacher of English and Drama at Blaxland High School. In 1997, he moved to Hobart where he taught at the University of Tasmania.

Lawrence's collection *The Darkwood Aquarium* (1993) decries society's cavalier attitudes towards environmental disasters. The spread of industry in New England is evoked in 'Mountview', which describes fields littered with 'yellow glow', 'metal cross-bars', 'hornet's nests' and 'rotting trees'. The locals seem powerless to do anything about it:

> The Tamworth electricity commission ripped
> a passage through the Moonbi Ranges
> then planted steel towers . . .
> . . . feeding miles of cable inland.[162]

Another strong theme throughout this collection is a disgust at war, weapons and the militaristic mentality. Spanning many conflicts across many geographical trouble-spots, Lawrence tells of how the culture of warlike materialism is inconsistent with living a life of quality and meaning.

Similarly, 'The Deadline' derides the geopolitical notion that life is expendable and oil is priceless. In this poem, the Gulf War is defined as 'something cruel and most appallingly human'.[163]

In 'The Fan', there are graphic descriptions of whale slaughter, which in turn is a metaphor for the senseless violence that can be found whenever and wherever humans gather.[164]

THE nation's best known—and most controversial—'environmental' poet is Mark O'Connor, whose *The Eating Tree* (1980) shows how nature has the capacity to resist even the most diabolical of human incursions upon it. O'Connor regards the laws of nature mostly as a hallowed thing, whereas man's laws and activities (although often enlightened) are a travesty by comparison.

O'Connor sees man's destructiveness as not only unjustified and lacking in logic but also as having a manic quality about it: the waters are 'fished out' and much of nature is 'choked, cleaved'. The totality of the incursions is dramatised in verses like 'The Rainbow Serpent—ix: Man':

> Tree-killer, all killer,
> Ignorant promiser,
> not knowing why
> or whence or whither,
> I breed, I eat, I excrete, and die . . .[165]

Many poems in O'Connor's other collections are also suffused with the incompatibility between the glory of nature and the base way humans treat it, especially in the Third World. O'Connor describes oppressed people fighting for survival being forced to grind and plunder the earth.

O'Connor is in awe of nature but does not idealise it; instead he bends to its force and potency. He explains how cyclones and ocean currents have no consciences or cares about how they affect those who live upon the planet.

Humanity's incursions upon nature permeate O'Connor's verses in collections such as *Fire-stick Farming* (1990), whose individual poems cover many topics. Urbanisation is at the heart of 'In the Gardiner Valley', which laments how 'Today I watched the last trace of country going/forever out of the suburb'.[166]

O'Connor treats these wide-ranging concerns with both gravity and lightness as he dissects global catastrophes such as over-population, a subject on which he has written widely and controversially.

As a poet, O'Connor fell silent through most of the early 1990s as he adopted a high public profile in movements—including Writers for an Environmentally Sustainable Population (WESP)—aimed at controlling Australia's immigration intake and other measures towards population control. O'Connor's feelings about population control largely stemmed from his own observations while travelling through Europe, as he described to the *Canberra Times*' Christopher Deere:

> In Europe . . . I saw what could be Australia's future if we don't watch out . . . I can remember going to a beach in Yugoslavia which was described as one of the wonderful beauty spots of the Dalmatian coast. It was a beauty spot because there was actually some sand on the beach, in amongst the beach umbrellas and the bouncing soccer balls, and you could see the sewerage outlets from the town and the various luxury hotels running out into the green, eutrophic bay. And that was what was passing in Europe for an experience of nature and beauty. I don't want that to happen to Australia.'[167]

By the late 1990s he was again in full poetic flight, consistent with the person who fled an academic career and Victoria to journey in 1972 to North Queensland where the islands of the Great Barrier Reef 'turned me into a poet . . . I hadn't been one previously'. In the Reef he discovered 'a muse a thousand miles long'.[168]

From the reef to the rainforests to the Blue Mountains to almost every nook and cranny of the continent Mark O'Connor has travelled and told of his delights and concerns.

From his Canberra home and on his travels throughout Australia, O'Connor regularly rails against a 'human chauvinism' which he believes is inconsistent with a sustainable ecology. 'We are only one among five million species on this planet,' O'Connor wrote in 1986. He said Australian history had been traumatised since the First Fleet because the earliest settlers 'saw the bush as of value only for what they could make out of it when it was chopped down'.[169]

Such views have won O'Connor widespread favour and recognition among green groups. In 1992, the vice-president of the Cairns and Far North Environment Centre, Jon Metcalfe, wrote that O'Connor's poetry and proclamations represented 'insightfulness and reverence for the natural environment not often achieved since the Dreaming'. Commenting on his role as a poetic conscience, O'Connor told Metcalfe that inner feelings needed to be put into words to counter the 'dollars-and-cents arguments for destroying rainforest'.[170]

Despite the fierceness and stridency of many of his views, especially on the need to control population, Mark O'Connor is a warm and thoughtful poet who engenders appreciation of the glories which he seeks to preserve and protect.

ANOTHER Canberra poet who strongly expresses a range of environmental concerns is R.F. Brissenden, who stands out because of a keen ability to describe everyday situations. As in his novels (see Chapter Two), Brissenden's poetry is full of a true parish-pump, kitchen-sink realism with which all readers can identify.

Brissenden believes each and every one of us shares responsibility for the state of our surroundings, and different verses of his collection *The*

Whale in Darkness (1980) are testament to this. 'By the Shore' is typical:

> The weekend rubbish of our dirty tribe
> Lies on the grass—cans, plastic, shattered glass:
> At least I bashed and burned and buried ours.[171]

Brissenden tells how we have all 'fouled the nest'. He angrily condemns Australians for the manner in which they have spoiled their surroundings and created a mess which is 'filthier than most'.

The verses in this collection include searing laments for a world diminished and depleted. Brissenden presents in his poetry the no-nonsense hardness of the tough-guy sleuths who inhabit his crime fiction. Brissenden rails against senseless cruelty to animals, the idiocy of hunting and the thoughtless non-relationship between humans and other living things. He is as concerned with the dispassionate tinkering with microbes under a microscope as with the callous harpooning of whales. Brissenden sees these are equally de-humanising acts which undermine any pretence of civilisation.

To Brissenden, the Third World is a metaphor for the most extreme cases of environmental degradation, as in 'Leaving Jakarta', in which he describes the Indonesian capital as a place of 'corrugated iron, cardboard, plastic'—all indistinguishable among the 'animals and people'. Typically indifferent tourists are oblivious to the overwhelming decay around them:

> We cannot smell the excrement and filth
> Rotting in the clogged canal.[172]

Brissenden—who wrote his first collection of poetry, *Winter Matins* (1971), at age 43—displays a maturity and depth of understanding gathered throughout a career of writing and academe. Writing until his death in 1991, Brissenden espoused many genres and many issues, but foremost was a 'distinctive ecological vision'.[173]

POIGNANT and harrowing accounts of Aboriginal realities are the essence of the poems of Kevin Gilbert, a major black Australian voice whose *People Are Legends* (1978) shows the degradation of his people as total and the savagery of their oppressors as complete. Gilbert, who died in 1993 aged 60, had spent many years in jail, where he contracted the emphysema which is believed to have eventually killed him. At age 37 he was sentenced to life imprisonment for murder after killing his wife while drunk. During the 14½ years he was locked up he educated himself and began a path which would make him one of Aboriginal Australia's leading poets and playwrights. A funeral service was held for him at the Aboriginal tent embassy site outside the old Parliament House in Canberra.

In 'The Black Drunkard', Gilbert shows the unfitness of Europeans to be masters of their conquered acres:

> [The whites] claim that this land is God's Country
> If so, then it's vacant, to let

He ain't lived here for 200 years
Ever since the whites came, I bet![174]

Gilbert has written that such attitudes are possible because of acceptance of the long-established concept of Terra Nullius, which was 'disinformation' and a 'government lie' based on the ludicrous premise that 'this land was a wasteland and unoccupied' before 1788.[175]

The devastation of the Australian ecology also features heavily in Gilbert's anthology of contemporary Aboriginal verse, *Inside Black Australia* (1988), in which both well-known and obscure poets proclaim that there was nothing much to celebrate in the nation's Bicentenary. Anger and outrage suffuse the poems in this collection.

Prominent among the contributors is W. Les Russell, who in 'The Developers' tells of how Aboriginal traditions and aspirations were thrust aside in the European conquest of an ancient land:

Like a spear thrust deep within my heart
The drill turns deep within the earth.[176]

To reaffirm that Aborigines were victims of the nuclear age, Russell dedicates 'The Nuclear Winter' to those forced to endure the horrors of British testing:

The agony of Nuclear,
The land turned to glass,
Then came long night, cold drear,
And years bitter pass.

No future but slow cruel death . . .[177]

The nuclear issue and the way in which a nation brushed aside the birthright and wishes of Aborigines in remote areas are heart-rendingly chronicled by Gilbert. In 'The New True Anthem', he suggests there is nothing for white Australia to feel proud of or patriotic about in the sunburnt country they have irradiated and trashed:

you pollute all the rivers
and litter every road
your barbaric graffiti
cut scars where tall trees grow . . .[178]

In 'Celebrators '88', Gilbert furiously repeats that Australia has very little to rejoice at after 200 years of a history which has seen 'the river now replaced by sheep torn gullies and muddy silt'.[179] Everybody is complicit in such evil deeds, says Gilbert, who urges his countrymen to take responsibility for past wrongs and ensure that the future is more humane and fair.

Poems such as Gilbert's and other passionate black Australians have raised consciousness on environmental issues in tandem with the other big issues of the past thirty years affecting indigenous peoples.

The essence of poetry and songs was defined even more poignantly by the *Canberra Times*' literary editor Robert Hefner in an obituary of Bob Brissenden. Hefner recalled Brissenden reading his poetry at the first Writers Against Nuclear Arms Symposium in Canberra in 1987 and described in 1991 the aching yet inspiring scene. 'Struggling mightily with the Parkinson's disease that racked his body, Bob stood with grace and dignity and read his poem "The Old Track". I was moved almost to tears then by the essential humanness of its expression, and last week, reading it again, I knew, perhaps more vividly than I have ever known, how poetry can transcend death. The great spirit that had departed Bob Brissenden's physical body touched me deeply again as I read [the] lines.'[180]

Bob Brissenden, like Mark O'Connor, like Les Murray, like all real poets, took his verse directly to the people, to touch them directly. So it is with all worthy balladeers who have taken their songs directly to audiences and to the barricades.

Peter Garrett has said that as soon as a band like Midnight Oil stops performing 'live', the band becomes dead. But if the band can continue to touch and inspire each person it plays to, there is a power far greater than the utterances or actions of politicians and officials.

8

CHILDREN
Hope of the future

A HOST of wonderful books for children have taken up environmental themes in recent years. They have discussed many of the same issues that appear in literature for adults—species extinction, water and air pollution, preserving the rainforests—all with a view to inspiring children to care about the natural world. In this chapter, we look at books aimed at the youngest preschoolers to those for teenagers and young adults. Many of the books have been written by specialist children's authors, but some have been produced by writers who have made their names in other genres. A great deal of talent has gone into the production of these fine books—the best writers, painters and illustrators, photographers and scientists have all thought them important and a worthwhile forum for their skills.

Writers are involved with children's books because they are convinced that such books are an important means of communication—and that children are the world's hope for the future. As songwriter Allan Caswell has said, 'children are the one section of the community which still has an imagination'—and these wonderful books are all working to capture that imagination and fire it up to change the world.

One leading expert in Australian children's literature, Margaret Aitken, sees a continuing theme of 'What is our relationship to the land?' and since the 1970s a growing environmental consciousness in the genre. She says many of the prominent authors of books for children and young adults have increasingly tackled ecological themes which have been integrated into other specific issues like the pain of adolescence and youthful relationships. Aitken, a sessional lecturer at the School of Early Childhood Studies at the University of Melbourne, said in 1997: 'Since the environmentalist movement gathered momentum, more and more children's books have honestly tackled important issues, and these will prosper culturally.'[1]

Two other experts, Stella Lees and Pam Macintyre, while specifically discussing Christobel Mattingley, laud many of the writers for children and young adults as confronting 'moral and social issues with courage, humour and resourcefulness'.[2]

A prominent publisher of, and one-time lecturer in, children's literature, Walter McVitty, believes the genre has adapted to fill the cultural needs of young people on issues like the environment. According to McVitty, the significant Australian children's authors have moved on from the 'yarns of blissful, sunny childhood' and are now producing works which 'look unflinchingly into the minds of genuine characters and to describe the sometimes stressful realities and shadows of the world in which they live'.[3]

McVitty, who has published such major authors as Colin Thiele from his base at Montville in Queensland, told in 1981 of a 'growing willingness' for writers in the genre to discuss vital issues like the 'despoliation of the environment' within a context of 'an integrity which entails benign respect and responsibility'.[4]

Furthermore, McVitty said the importance and significance of children's literature had not been properly recognised in Australia. The genre is 'part of the mainstream of literature and not some minor branch . . . given the right priorities, any society ought to regard the writing of literature for the young as one of its most important priorities'.[5]

The importance and significance of the genre are also seen in a study by the Australia Council, published in 1990, which found that of all books bought in the previous year 30 per cent were aimed at the young market, and of these 30 per cent fell under the category 'Australian'.[6]

From the 1950s onwards, Australian writers of children's fiction adopted specific issues such as the environment. Furthermore, according to literary critic Brenda Niall, 'the question of finding the right relationship with the land is linked with that of reconciliation with the Aborigines'.[7]

According to a range of critics, including *Grass Roots* magazine's correspondent Marion Boetje, Australian authors are 'world-class' when it comes to picture books, and an increasing number of titles are published every year.[8]

Many Aboriginal writers have sought to speak directly to the younger generation—white and black—and in the forefront is poet Kath Walker (Oodgeroo Noonuccal), whose *Father Sky and Mother Earth* (1981) is a classic touchstone for the youngest readers. Walker's drawings are stunning Dreamtime motifs—in vivid colours and patterns—from the mastery of sunrise to the reflected splendours of sunset. Earth, water, clouds and vegetation all have their distinct and proud persona.

The purity and marvel of Aboriginal culture and an appreciation of the interlinked majesty of all living things are part of Walker's heart-felt storytelling which would heighten the environmental consciousness of any pre-schooler.

Kath Walker writes of a pristine time destroyed by man, a time when the earth 'lived in peace and happiness. And there was beauty everywhere':

> Until . . .
>
> Until . . .
>
> One time a strange animal came among them. This animal was not like any animal they had ever seen and this animal was called Human.[9]

Here are poignant and powerful yet simply-told lessons on how despoliation killed off all quality of life. The book contains sufficient anger to attract the attention of the youngest readers, yet enough love and empathy so as not to depress them or scare them off. To capture the imaginations of the young readers, there are greatly exaggerated illustrations of a jet belching fumes, causing plants and all life beneath its flight path to wither. *Father Sky and Mother Earth* is a clarion call to youth for effective and committed green activism.

Strong emotional strands are woven through this book, as they are in all the classics for young readers as tears, outrage and anger often combine to touch them and make them aware of what is happening to their environment.

CLASSIC novels abound in spreading green and enlightened messages to a young readership. One of the most legendary and enduring writers of children's fiction is Colin Thiele, most famous for his early works set in South Australia like *Storm Boy* (1963). Thiele's environmental consciousness has permeated all his books, most of which are set in the Coorong country where he grew up and the South Australian landscape dominates his prolific writings.

In his autobiography *With Dew on My Boots* (1997), Thiele describes how he is singularly well-placed to present a view of the world and the threats to it. 'Most of the children who have to live in the vast urban communities of today can never roam the hills as I did, never run up grassy uplands in a world pristine and free, never experience breathtaking moments of discovery there as part of their daily lives,' Thiele writes.[10]

In Thiele's fond look back at the forces which guided his formative years, he exults in the 'effect of the natural environment, and the freedom to explore it in my own way'. He describes and defines the touchstones that have guided his prolific output of warm-hearted and caring writings: 'The tender intricacy of a bird's feather, the texture of pebbles in a creekbed, the skitter of a rabbit or the lope of a fox, the majesty of an eagle, the fragility of a dragonfly's wing—these and a hundred other revelations that were part of everyday life for me are largely beyond [urban children's] reach. Yet how else can anyone learn to stand in awe of the miraculous diversity of the universe?'[11]

The Thiele philosophy is evident in *Pinquo* (1983), set in a small and quiet township in the Coorong whose citizens realise rudely and suddenly that an instant and unprovoked act of nature can change everything—that humans have nothing to feel complacent or arrogant about.

It is a charming story whose main characters are penguins, especially one named Pinquo who is befriended by ten-year-old Kirsty Kemp and her nine-year-old brother Tim. Lovingness, nurturing and care dominate the prose as Thiele chronicles how Kirsty, Tim and a compassionate scientist, Dr Piper, tend the badly injured creature and learn of its habits. The endearing storyline reflects Thiele's philosophy as he defines Pinquo's struggle to recovery: 'The whole world is wonderful. Everything is made to fit into its place.'[12]

Before any complacency can set in around such heart-warming scenes,

Thiele quickly reveals that human avarice and disrespect seem certain to invade the intricate and charming life cycles of the growing penguin community in the Southern Ocean off Coorong:

> By now the story of Dr Piper's penguins was known far and wide. Tourists stopped and photographers wanted pictures.
>
> Dr Piper joked about it to Kirsty and Tim. 'At this rate,' he said, 'the penguins will be more important than the people. Pinquo will be the most famous citizen of Sickle Bay.'
>
> Dr Piper didn't know how true his prophecy was going to be.[13]

Thiele, born at Eudunda north of the Barossa Valley wine-growing district of South Australia, continues the theme so powerfully visited in *Storm Boy* and in his non-fiction writings about the Coorong which describe the marauding and destructive manner in which hoon-like thrill-seekers discover the region in increasing numbers:

> . . . more and more campers and visitors and tourists were coming to Sickle Bay, littering the beaches and disturbing the penguins. They poked sticks into the rocky hollows and broke down the roofs of the burrows with their big feet. Whether it happened purposely or accidentally didn't matter—the damage was done all the same. As there were always some penguins in the colony, even in the off season, they were being molested more and more.[14]

Simultaneously the community learns that, when it comes to human survival amid natural catastrophes like earthquakes or tidal waves, Pinquo and his penguin pals have a lot to teach the arrogant and supposedly superior species. It is up to the penguins to lead the people to the safe higher ground as the tidal wave approaches. The penguins are much more capable of adapting to the new destructive reality: 'There was a great deal of noise. The penguins were still yapping in a frenzy and the people were worse.'[15]

As if to punctuate the short-sightedness and self-centredness of mankind, Thiele then describes how these natural disasters are matched by the man-made ones against which no adequate provision has ever been made. Thiele dissects the cruel aftermath of an oil tanker dumping its load. Those who suffer the most, naturally, are the penguins—those penguins who had done so much to save the humans.

Only the bravest and most intensive of interventions by man is able to save the lives of hundreds of the birds which had such a short time ago saved them. But for Pinquo, choked by the oil which the stricken tanker had belched into the wild ocean, it is a slow and painful death.

Thiele submits that caring people must learn from such disasters and be able to prevent such horrors in the future.

The inter-relationship of boy and creature is also warmly examined in *Brahminy* (1995), whose plot is centred on a young boy, Luke Knott, nursing a weak fledgling kite from near death to robustness. Luke's father is Alexander Knott, a Northern Territory park ranger in Arnhem Land, who early in the book takes up the cudgels of Thiele's core philosophy: 'Wouldn't it be magical if all living creatures on earth could live beside human beings . . .

trusting and natural and unafraid? But people spread fear and violence across the world. Wherever human beings walk, terror and cruelty follow.'[16]

The loving and nurturing relationship that develops between Luke and the increasingly frisky Brahminy contrasts markedly with the behaviour of the hordes who trample across the Top End:

> June brought a flood of tourists, campers, fishermen and shooters. The rangers had to travel far and wide 'with eyes in the back of their heads', as Luke's father said, pouncing on poachers, checking for nets and firearms, protecting the breeding birds, culling feral animals, guiding scientists and international visitors . . .[17]

As Luke's winged protégé thrives and becomes the talk of the local community, the routine activities of visitors continue to be a bane:

> Poachers and mindless travellers were an endless problem. They fished illegally in the estuaries, destroying the breeding cycle of the barramundi, and they shot protected birds and animals—with lead shot, which made things even worse. Most of the litter they left was dangerous to wildlife. Plastic bags, pieces of fishnet, bottles, ring-pulls, broken glass, rusting cans, discarded rope or twine—all these were things that could trap the heads, throats, wings or claws of waterbirds, or entangle the feet of animals. When this happened, they died a slow death from starvation . . .[18]

Luke and his family behave quite differently. They represent a culture that is warm and loving. Here is a way of life in which humans and creatures, as in *Pinquo*, recognise and can repay all their debts to each other. Thiele describes an indelible bond between Brahminy and Luke—based on friendship, mutual understanding and a code of existence missing from so much of humanity. Inherent throughout the novel is a plea that the Arnhem Land wilderness be allowed to retain its uniqueness and that many of the values as enshrined in the Knotts be allowed to prevail.

High Valley (1996), set in Queensland, depicts the anger and solidarity of a small, remote community as it responds to government plans to expropriate its land and flood its valley. Jamie Judd and his primary-school classmates must put the diversions of youth aside to help their parents face the crisis. These are families who love nature and are nurtured by it, who wish to preserve their bucolic way of life—for whom a tree is the very fountainhead of existence; for whom the destruction of trees spells catastrophe.

Against this handful of local inhabitants is the demographic imperative of such things as the supply of water and power being demanded by Queensland's rapidly growing population. Thiele describes to his young readers the powerlessness of reason, logic and human concerns against a government determined to fulfil a development scheme: 'The listeners [at a public meeting] felt defeated. They were sure that the engineers and builders were already preparing to move in with bulldozers and jackhammers, and that nothing in the world could stop them. The sense of despair affected everyone . . .'[19]

This onrush of damming, diversion and hydro-electricity is a death-knell for old-fashioned values, as defined by Jamie's mum:

> 'I don't know why people can't make do with rainwater from a tank,' she said bitterly. 'Here in the valley everyone has lived like that for a hundred years. Why do they have to build reservoirs all over the place?'[20]

And Jamie's father is equally seething with rage:

> 'I didn't think the valley would ever be destroyed by water . . . I've always thought water was a blessing, but not any more. Now it's a death sentence. Doomwater.'
>
> The word echoed in Jamie's mind, like the tolling of a bell. '*Doom*,' it said. '*Doom, doom, doom.*'[21]

In a gentle and uncluttered, yet evocatively erudite manner well understood by young readers, Thiele shows how such a community can be utterly crushed, losing its sense of purpose and direction. Thiele is forthright and unsparing in describing what is at stake when a community is evicted: 'The dam was not just going to force the gentle community of High Valley to move away. It was going to blow it apart for ever.'[22]

Thiele tackles the personal dimensions of such dislocations, especially as they affect schoolchildren like Jamie:

> [Jamie] was appalled. Rows of beautiful trees were being bulldozed and cut up—trees that Mr Schulz had grown so carefully . . .
>
> As the driver of the bulldozer uprooted the trees he laid them down neatly for the sawyers, side by side, like bodies laid out in rows after a plague—High Valley corpses, lined up to be burnt.[23]

Among the many strands of this environmental plight is that of one very strong-willed adult fighting back, and Thiele eloquently describes how this influences the next generation like Jamie.

Another issue confronted by Thiele is that of the dangers created by Australia's introduced animals, sharing with a young readership the concerns of writers like Eric Rolls. In *Little Tom Little* (1981), Thiele describes what happens when buffalo and bantengs, introduced from Asia, are allowed to run wild: 'Now they were ruining the wetlands, plunging about and uprooting the plants, churning the beautiful feeding grounds of the birds into ugly black mudholes, and destroying the whole delicate balance of nature.'[24]

The story is bolstered by the photographs of David Simpson, who illustrates the animals and locales of the Top End—and shows the destruction of the introduced species. Thiele describes the extent of damage caused just by pigs and bantengs: 'not a stem or a plant remained in the once beautiful places where there had been reeds and water plants and bright flowering lilies. Now there was nothing but a horrible mess of churned-up mud.'[25]

Throughout this charming tale of the adventures of a park ranger and his son, Thiele's overwhelming conservationist ethic and a call for commonsense and care shine through.

These philosophies and concerns were evident in Thiele's writings more than twenty years earlier in *Albatross Two* (1974), reissued in 1986 under its US title *Fight Against Albatross Two*. Set around the Coorong country, it tells of how nervous townsfolk in a small community await the arrival of an oil rig off the South Australian coast. Some of the locals see the gigantic rig as good for the economy of the quiet little town of Ripple Bay while others see it as a 'Frankenstein monster'.

Some marvel at this magnificent achievement of mankind's technology, such as respected boat-builder David Banks who exclaims: 'Men are incredible creatures—to create a thing like that.' But as the rig Explorer King 'moved relentlessly shorewards' to Ripple Bay many others are extremely worried.[26]

What nobody disputes is that the entire character and way of life at Ripple Bay will be changed forever, and that a period of traumatic adjustment is ahead. The immensity of what is happening cannot be ignored amid the feelings of apprehension and suggestions of an impending disaster:

> The rig was too big; its effects too great. From the moment the spirit of the little town which had nestled there on the southern coast of Australia for over a hundred years was torn to pieces, and when the crisis was over, a different town had taken its place. A harder place. A place more likely to survive in the new world of steel and oil, than the soft gentleness of an old fishing village. And so it was.[27]

In one of the most eloquent descriptions of how youths must confront revolutionary and sudden change, Thiele tells of how Dave Banks' twelve-year-old daughter Tina articulates the way children of her age form impressions about the coming of such a mammoth rig:

> She was trying to work out the reason for her feelings about it. Not that it was ugly, exactly. Even from this distance the booms of the cranes and the jutting helicopter deck gave it a modern angular look. Like a signalling station semaphoring with arms of steel lacework.
>
> No, it was what it stood for that was the important thing. It was a sign, a symbol of everything that was foreign to Ripple Bay, of freeways and steel jungles and jet power and smog—and war. It just didn't go with white water, or crayfish that were crawling about in rocky grottoes, or ancient untouched coast lines, or clean winds over the headlands, or bight sand in little inlets. It didn't fit in. It was black where the breakers were white; it was greasy where things were clean; it was brash where life was gentle; it was fire where the world was water.[28]

Just as in *Pinquo*, a fairy penguin called Piglet (who had already survived a series of misadventures) is potentially at fatal risk from the coming of the rig. Quite apart from what the oil-drilling activities can do, the children of Ripple Bay learn how Piglet is also at risk from the routine activities of people going about their business. Tina confronts one thoughtless kid among a group of youthful interlopers:

> . . . you'll do as much damage as you can while you're here. You're like all the other tourists—if you're not disturbing the fairy penguins here, you're frightening off the pelicans on Pelican Island, or the swans on the Coorong, or the seals on Kangaroo Island. Till there'll be no wild life left in South Australia at all and we'll be forced to go and look at *people* in cages instead.[29]

Before the drilling platform is engulfed in catastrophe, Tina's fourteen-year-old brother Lincoln gets a job on the rig and is agog at its awesome dimensions. Through Link's experiences, Thiele teaches his young readers what life is like on a rig, and how it is a marvel of engineering, logistics, organisation and efficiency. But young Link also learns from the hardened rig veterans that sometimes things can go wrong.

As the ominous plot unfolds, Thiele concedes that all of civilisation's recent progress is fuelled by oil and gas, but he warns that alternative energy sources must be found and adopted immediately. Meanwhile, out on the rig, Link philosophises on the clash of incompatible attitudes into which he has been thrust:

> Things were getting clearer all the time, now. There were two worlds, that was as plain as a mountain. Ripple Bay and Explorer King—two empires, two island-states separated by a narrow strait. And they were in a conflict that could only get worse. Irritation, resentment, anger, violence. It happened so often, even between one person and another . . . if the rig struck oil . . . there would be pipelines ashore, and big tanks, and oil spills and an endless fight with the townspeople . . .[30]

Despite such realisations of what a lucrative offshore gush might bring, Link cannot in all conscience hope for the failure of the rig crew to strike a productive payload, having been part of a vigorous and noble effort by good and solid men. But he realises increasingly that in oil drilling there is always a litany of things that can go wrong—despite the advance of knowledge, know-how and technology.

The forces of nature cannot be treated lightly. Here is a fast-paced and suspenseful adventure set against the backdrop of nature's omnipresent wrath at those who abuse it.

What the million-dollar enterprise of skilled and sophisticated men has wrought in this instance is a spreading oil slick which devastates all the life forms in its path.

Fight Against Albatross Two is among Thiele's longest books—and possibly over-long in testing the endurance and patience of a young adult reader—and it is stylised in a way his later books are not. For example, the clichéd American accents and the knee-jerk behaviour of the townspeople come across as somewhat hackneyed. As critic Rosemary Wighton put it, 'the didacticism is overt, the characters stereotyped, the transcribed accents simply tiresome'.[31]

But on the other hand *Albatross Two* provides invaluable in-depth technical information and covers every aspect of environmental concern not usually available in books directed at such a young audience.

This array of works by Thiele displays what Wighton has described as a mixture of the writer's 'commonsense practical nature [and a] strong combination of irony, realism, and old-fashioned morality, that rejects the facile assumptions of much writing for children, especially of novels of action and suspense'.[32]

Regardless of their setting, Thiele has a consistent theme in his books whereby he passionately believes humankind has the potential to destroy itself utterly and that every effort must be made to raise the consciousness of the next generation.

MANY of the much-loved and enduring writers of children's literature have conveyed their strong environmental concerns. The authors embrace a wide range of philosophical concerns and writing styles but share a deep passion for conservation and ecological values. The writers include Patricia Wrightson, Hesba Brinsmead, Ruth Park and Christobel Mattingley—all of whom have captivated young readers for several generations.

The shattered tranquillity of a rural region is at the heart of Patricia Wrightson's classic *The Nargun and the Stars* (1973, 1988), set in the New South Wales Hunter Valley where human incursions have disrupted spiritual harmony. This novel's young hero is Simon Brent, at the crossroads in a timeless land that is being utterly transformed. No one—especially not Simon—has any control as machines roar across Wongadilla, which is being altered beyond recognition. Simon is perplexed at how such devastation can happen considering that Aboriginal folklore warns that the death of even a frog can cause a storm of fury from nature.

Young readers are dramatically reminded of the effects of an axe being swung among ancient trees. Instead of 'bird calls; wind among leaves' there was a 'thunk! thunk! thunk!' as precious belongings were lost. Wrightson vividly describes the rude interruption: 'Then would come a flat, solid *boom!* that Simon guessed was blasting; another tree would begin its long slow crash and then the bulldozer would start again.'[33] Simon's observations leave nothing to the imagination as 'already the bulldozer had wrecked a whole slice of forest'. His shock and bewilderment are dramatically conveyed to young readers: 'Slaughtered trees lay where they had fallen, or where they had been snigged aside . . . The yellow bulldozer worried at the edge of the forest like some great blunt beast, roaring above the howl of the chainsaw . . .'[34]

Wrightson's character, although impressed by the magical and metaphysical aspects of a forest, is nevertheless in many ways still a callow youth equally impressed by the state-of-the-art machinery hacking away at the foliage. Wrightson makes the point that the myths and idealism which guide youth must often co-exist with more worldly realities.

Although *Nargun* contains a lot of realism, it is recognised as Wrightson's first serious work of fantasy, a genre from which her widest fame grew. It is simultaneously a novel of faith and optimism, whose frequent absence in children's literature Wrightson has described as 'criminal'. Her feelings on this score are quite adamant: 'If you stop offering children hope, you're doing an enormous damage to the whole human race,' she commented in 1994.[35]

Wrightson, born in Lismore in 1921, spent part of her childhood in the township of Bonalbo in the Clarence Valley which is the setting for her early novels. In *The Nargun and the Stars*, according to Brenda Niall, Wrightson was among the earliest modern authors to write fantasy, which until then had been 'awkwardly handled'. Niall likens this keynote novel as the crossing of a cultural threshold in the writing for young people.[36] Other critics, like Stephen Matthews, have noted how Wrightson's works often transcend their genre, by having a universality as well as 'a particularly Australian kind of fantasy'.[37]

Wrightson's first novel, *The Crooked Snake* (1956), dealt with conservation and ecology before children—and, indeed, adults—were aware of these concepts. This book, in which kids rally in defence of a local wildlife sanctuary being threatened by shooters, reflected the concerns of a writer described by Walter McVitty as 'Australia's most talented, accomplished and distinguished children's author'.[38] Similarly, *The Bunyip Hole* (1957) is about a family's struggle against the destructive forces in the bush.

Afterwards, Wrightson's novels would grow in sophistication and complexity. *An Older Kind of Magic* (1972) delved deep into Aboriginal folklore and motifs; it had a wider framework of fantasy to describe mythical creatures who unite to fight off plans to turn part of Sydney Botanical Gardens into a huge commercial car-park.

Ecological concerns are also at the core of a later Wrightson novel, *Moon-Dark* (1987), which instils in young readers a keen sense of the balance of nature and the empathy and integrity among creatures. Wrightson constructs clever and witty conversations between Australia's beloved wild creatures on the best ways they can survive.

The setting of *Moon-Dark* is based on the land, foliage and animals around her cottage on the Clarence River near the town of Maclean, where she has lived in later life with her son Peter. In the novel, a wallaby surveys the landscape and despairs at the European invasion. 'In this known, familiar country there was nowhere to go,' the wallaby laments. 'A new race had taken the wallaby's old range, the grass was eaten at the roots and dying, its own clan were fugitives down the river.' Similarly, the onset of the white man was a barbaric invasion, not answerable to any of the codes as in the Bandicoot War as described by the wildlife: 'They knew the war was not funny, since nothing could happen to one of their kind without happening to them all; they might fight, compete, even eat each other, but they must care for and respect each other. It was the law.'[39]

There follows a lament from deep in the wilderness that 'there are too many men' who seem to destroy everything.

Throughout *Moon-Dark*, Wrightson describes the wild creatures growing concerned about how their habitat is being diminished, about how there simply is not even room for all the creatures great and small to co-exist.

Both *The Nargun and the Stars* and *Moon-Dark* have what another prominent children's author, Lilith Norman, has described as a 'toughness' and a constantly clear intention to never 'dodge hard decisions'. Above all, Norman hails the 'compassion with which Wrightson treats creatures like the Nargun'.[40]

Like other fine authors of children's fiction, Wrightson, who since the 1960s has been studying the relationship between young people and the land, manages a humorous mix to her more serious observations, and crusty locals and forest spirits all have a free rein.

In an interview at Maclean on 4 June 1998, Wrightson said the themes in her books arose naturally, and the plots often revolved around concerns she faced in real life both in the Clarence River region and elsewhere—such as powerful developers seeking to change the face of the riverfront or a caravan park which was dumping effluent into the river, largely because the septic system was inadequate.

'There can't be any more developments along the Clarence without polluting the river,' she said. 'No structures can go up without chopping down mangroves.'

She lamented the insatiable demands of 'progress' all over the world, whereby 'rivers were being diverted from their natural course'.

In an ironic twist, Wrightson lambasted those who dismissed as 'fantasy' many international problems such as global warming.

At the same time, Wrightson fondly referred to the abundance of local natural splendours at Maclean. She told of how her property was surrounded by greenery and of how she constantly sought to grow more trees. She explained there were now two agreements with the New South Wales Parks and Wildlife Service to preserve her property and the adjoining one.

Another feisty voice for conservation is Hesba Brinsmead, who has captivated children of advanced understanding with books such as the remarkable *Echo in the Wilderness* (1972), in which this much-loved and internationally celebrated author reveals the nastiness of Tasmanian politics and the environmental horrors wreaked in the Apple Isle.

The novel begins as Clippie Nancarrow, a youth from the New South Wales Western Plains district, suddenly finds himself in Tasmania after he learns of inherited money. Clippie—a character loosely based on the ill-fated Tasmanian pilot Max Price, who perished with Lake Pedder campaigner Brenda Hean after their light plane was sabotaged—very quickly learns about the harsh realities of the island State, including the extensive environmental degradation which has become the norm. Clippie is appalled at what he sees: 'a steeply tilted landscape, a mountainside of ruined forest . . . Huge, blackened monoliths of dead trees stood clawing at them, with decayed, skeletal arms; some were grey, petrified in death; some were prone upon the mountainside. It was a ravaged land.[41]

Clippie learns even more about the extent of the ravages from the adults around him:

> 'The mountain was covered in rainforest,' said Ted. 'Huge mountain ash, grey gums, tree-ferns. Took hundreds of years to grow—if not thousands. Someone drops a match—they're gone . . . Civilisation—progress—they're taking the world away. The Tasmanian tiger—almost gone—and now it's Tara.'[42]

Lake Tara is about to be flooded, and Clippie is told of the impact this will have, such as the destruction of rare flora and fauna. The significance of the flooding is explained to Clippie by one of the locals, the Prof, who laments that too many things have been destroyed supposedly 'in the interests of science'. The Prof tells Clippie: 'One tiger, in captivity—even a pair—could only die. They could never breed. Simply die . . . this wild creature—so very wild—so primal. It represents the soul of the wilderness country. It's to do with the life force of the earth.'[43]

Here we have an enlightened older generation of Tasmanians explaining to the younger generation how everything in their benighted State was established purely on the basis of resource exploitation—for example, the hacking down of the majestic Huon pines. Clippie's new friends explain to him how only a lack of interest by a large resource company is keeping their community intact—for the moment.

With poignancy and piquancy, the flooding of Tara is shown to be something which will affect not just the landscape but also humans, most of whom have numbingly become resigned to mindless destruction. Brinsmead describes the increasing endangerment of wildlife in a way caring younger readers can relate to:

> 'What killed the mother [of a potoroo]?' asked Bev.
>
> 'A bullet.'
>
> 'A bullet! Someone—shot—her! But surely they don't let people *shoot* the creatures of the wilderness. Isn't it—well—declared a refuge—or something?'
>
> 'Some of it's national park. But not around Tara . . . up until now it's been perfectly safe for wildlife, simply because of its remoteness. The only people with the guts to get there wouldn't be the despoilers. But now they're building the dam, well, they've made the access road . . .'[44]

Echo in the Wilderness also teaches children that they must not build within themselves a generalised and simplistic rage, especially towards people's behaviour that they cannot fully understand:

> 'People! . . . "If it moves, shoot it." '
>
> 'Still—they are what they are. They build, too, remember.'

Hesba Brinsmead laments the insane actions of those who have despoiled Tasmania yet she hopes for a broader understanding by the younger generation, and that they will take a more caring and sympathetic approach to their natural surroundings.

Australians' inability to properly manage the environment is at the heart of another Brinsmead novel, *The Sand Forest* (1985), in which Clippie also comes to the rescue. The problems of how short-sightedness and inadequate preparation and research have led to land degradation are eloquently noted:

> She's a thirsty land, from the Indian Ocean, here, clean across to Adelaide . . . Yet we get a good enough rainfall. But it all comes at once, you see; in the early winter. Down it all comes—and disappears, into the sand. Then not a drop for nine months. The sun hammers down; hammers the country flat.[45]

These problems are caused by man, who without much ado is able to fell a karri tree which has flourished for more than 300 years. Man in his insatiable quest for development-at-all-costs has an in-built urge towards self-destruction, and the sense of doom is reflected upon by Uncle Hamish: '[Mankind is] right now well set up to destroy himself, quick smart, along with his gimcrack stressed concrete.'[46]

Man in all his manifestations, wherever he chooses to tread, has a propensity to damage. Mankind—'sightseers and plain looters'—thoughtlessly puts both forest and reef at risk by hooning through in four-wheel-drives and motorcycles.[47]

The cultural underpinnings which result in children growing up to be hoons and in turn destroyers is discussed in a Brinsmead novel released in the following year, *Someplace Beautiful* (1986). Brinsmead's heroes this time are a group of children (of whom Clippie would be proud) who gang together to fight the might and power of city hall over a proposed development. These kids are thrust into fighting a culture which decrees that a betting shop is more important than a book store built on knowledge and dreams. Brinsmead eloquently explains to young readers how the redevelopment process occurs and how all citizens must fight and organise to preserve what is precious:

> 'They want the shop, you know. We only lease it. Now they want it back. Redevelopment, it's called. Oh, they'll get us out. I dare say they'll win.'
>
> 'Redevelopment? How?'
>
> 'The Council wants to see this little bit of land redeveloped. In other words, they want to bulldoze away the Flying Trunk, and put up a lovely new betting shop instead.'
>
> 'A betting shop! Bulldoze away the Flying Trunk, and put up a betting shop?'
>
> 'That's right. They'll get us out,' said Tandy. 'Joy Dove leased this place when it was a mess. It's her life, this shop. Well, mine too. We've both worked. Look at it.'[48]

Tandy, king of the kids in the manner of Clippie, snorts at the philistinism of such attitudes: 'For a betting shop . . . they'll bulldoze away someplace beautiful.'[49]

Amid the outrage of such happenings, Brinsmead, in a very humorous and homely way, gives tips to kids about how they can change things through means such as petitioning a council and making their views most effectively heard.

Brinsmead was born in Bilpin in the New South Wales Blue Mountains in 1922, and has spent much of her life in various parts of Australia—including Tasmania, Victoria and at Terranora in northern New South

Wales where she has retired—all of which have been included in her varied works.

Her environmental concerns emerged as early as *Beat of the City* (1966), whose characters reveal a deep concern about what is happening to the ecology.

Brinsmead's novels have what Walter McVitty describes as 'positive didacticism' which, although often humorous and absurdist, gives her adolescent readers 'not what they want but, rather, what she feels they need'. In this vein, McVitty believes that *Echo in the Wilderness* can be regarded simultaneously as 'a fast-moving adventure-romance as well as a propagandist plea for conservation of the natural environment'.[50]

Some of the realism in Brinsmead's books stems from her journalistic training, enabling her to provide accurate reports of significant events. This training was put to best use in the non-fiction work for adults, *I Will Not Say the Day is Done* (see Chapter Three).

If *Echo* was the result of her concerns about the Franklin dam controversy, many of her other themes stem from her philosophy of 'being an ecologist'. As she describes this: 'I believe that Genesis 1 defines the purpose of Man: his role is to care for the Earth. And that Revelation defines the self-destruction of Man: "The destroyers of the Earth shall be destroyed." So one must align oneself on the side of Creation, not Destruction.'[51]

Another classic and much-loved Australian children's author to have tackled environmental issues is Ruth Park. In *My Sister Sif* (1986), Park presents young readers with a frightening and devastating scenario which they must contemplate and confront. In a very mature and reflective plot, *My Sister Sif* makes young adults realise that when cultures and sets of values clash, the end result is not always a happy one.

Ruth Park presents vivid imagery as young readers are told of humanity's depredations and how this will affect the world they will inherit: 'the ocean swelled into inexplicable turbulence, rumbling like a boiling pot, loosing gas bubbles that wobbled for a second and burst. The stench was suffocating.'[52]

At the start of the novel, Sif, 17, and her sister Erika, 14, learn of the powerlessness of individuals against overwhelming international forces seeking to destroy the world they love. They are as one with their friend Stig, who wants more than anything 'a man to man confrontation with those who were poisoning the ocean'.[53]

The sisters were brought up on Polynesian islands where they learned a love and respect for ocean creatures, and later they discovered that the city life of Sydney was often too burdensome. 'The city was a torment to Sif,' writes Erika. 'She was stupefied by its grinding noise, the tainted air, the people—millions of people rushing, crushing, not looking at one another or the sky. Sif could not adapt to the city because she never would be of the city.'[54]

Such urban blight is only one of many concerns for the cosmopolitan Erika—Rikoriko to her island people—who had already learned about nuclear testing in the Pacific, heavy-metal pollution and the dumping of poison wastes into the ocean.[55]

Ruth Park's portrait of the sisters reveals their affinity to the sea and its creatures. Sif and Erika preserve a natural link that mechanised, technologised and industrialised man has not yet been able to destroy. Their empathy with their natural environment is heart-warmingly described by a supreme writer: 'For they are angels, the whales, holy creatures that during their long majestic lives do nothing except what their Creator planned for them. Their world is the Garden of Eden unless man enters it.'[56]

Similarly, the narrator Erika describes the Pacific as being battered by civilisation's thoughtlessness:

> I saw a coral reef, dead as a doornail after a thousand years of gorgeous life, poisoned by herbicides and fertilisers leached from farm soil and sent down rivers into the sea. I saw penguins so sludged with chemical waste that they were no more than blind suffering bundles of congealed tar. And seals and dolphins and whales ill and dying, bearing freakish young, because they are on the end of the food chain and the fish or krill they must eat are polluted with lead, mercury, and other destroyers.[57]

My Sister Sif also describes exotic sea cities whose values are so remarkably different to the destructiveness and thoughtlessness elsewhere. Erika remarks on how so-called modern civilisation has a tendency to 'pull to pieces anything rare and strange'. The inability of ordinary people to properly defend what is there utterly appalls Erika, whose anger rises against all grown-ups who 'hang around like wet dishcloths and let dreadful things happen to our planet'.[58]

Walter McVitty has noted that *My Sister Sif* urges young readers to 'realise the peril the world is in and to do something about it'. Similarly, *Canberra Times* reviewer Eleanor Stodart has recognised 'the deep feeling that it evokes of the need for care of our planet'.[59] Part-fantasy and yet often gnawingly realistic, *My Sister Sif* is a futuristic look at how people's treatment of the planet will rebound on them.

The perceptiveness of teenagers is also at the heart of a popular work by another prominent writer of children's fiction, Christobel Mattingley, who in *The Battle of the Galah Trees* (1973) presents another heart-warming chronicle of a group of committed youngsters ganging together to save something precious—this time gum trees near their playing field. Our hero, Matt, develops a close and affectionate relationship with his pet galah and learns the importance of the bird's gum-tree habitat. This book teaches kids the mechanics of how to influence local councils and other authorities to preserve something precious, but it also concedes, as does Colin Thiele's *High Valley*, that sudden and brutal violence can result despite the noblest and most hard-fought of battles.

As the bulldozers move in against Matt's beloved gum trees, the enormity of it is conveyed to all:

> Matt stopped, shocked at the scene . . . somehow even his worst fears and nightmares had never prepared him for the stark irreversible reality which confronted him now . . . [A tree's bough] lay on the ground quivering like a wounded creature, then gradually all movement ceased . . .[60]

The consequences of such destruction are dramatically shown as Matt observes how 'the tree, reduced to its trunk and four main branches, had been mutilated almost beyond recognition, a stump of a thing shorn of its symmetry and grace'. But at the same time Matt displays a poignant optimism and a fighting, defiant and brave spirit:

> But [what was left of the tree] was alive. Its life-giving leaves had been hacked away but down in the dark of the earth its great roots were still entrenched. For a season or two with rain the roots could keep the tree alive, drawing its food from the soil, the weathered particles of thousands of years, until the wounded tissues healed and the tree pushed out new growth to hide its scars . . .[61]

Like Thiele, Mattingley explains the utter devastation of how a campaigner feels when a cause is lost. She is perhaps the most poetic and descriptive of Australia's impressive band of children's novelists and she paints the most poignant and harrowing of word portraits:

> . . . [Matt] was almost numb with shock at the way things had happened and mesmerised by the steady, inexorable pattern of destruction. For the present at least he was incapable of any action whatever . . . The great grey-green gum tree, survivor of a thousand storms, a hundred droughts, stood defenceless now before man-made metal . . .[62]

At the time of the book's release, Mattingley, a teacher/librarian, and her teacher husband David were the parents of three young children and lived in an old stone farmhouse at Stoneyfell, an Adelaide suburb. Both keen conservationists, the Mattingleys built a bird sanctuary in their garden by retaining the natural vegetation and beautiful wattles in the creek at the edge of their property.

Christobel Mattingley is the author of more than 35 books for children as well as film scripts, short stories, articles and poetry. In 1982, Mattingley, who was born in 1931, described many of her writings as 'ahead of their time' because of such themes as conservation, Aboriginal rights and nuclear issues. In that year she and her husband, history master and director of studies at Prince Alfred College in Adelaide, protested against government policies on Aboriginal land rights.[63]

A NEW wave of fine writers has produced many environmentally conscious novels aimed at the later teens. Among the most influential is Nadia Wheatley, whose *The Blooding* (1987) is a highly entertaining and witty novel. *The Blooding* personalises the struggles to conserve a typical forest—as seen from one family's and one community's perspective. The narrator is an intelligent, good-humoured lad, Colum Ferris.

Colum is a teenager of Higher School Certificate age who is having to cope with the conflicting influences of an environmentally enlightened education and the attitudes of his family, whose income for generations has been derived from logging. The plot revolves around greenie activists converging on the town of Cornwall, where 100 loggers' jobs are in jeopardy as a State government, dependent on green preferences at an upcoming by-election, is about to pull the plug on logging activities in the area.

Young Col describes the opposing attitudes to the logging debate with the straightforward and quite undiplomatic zest of youth. Col discovers that one cannot be indifferent or neutral when one is confronted with an environmental issue; when a major ecological battle looms, there is no escape. Nadia Wheatley relies not only on Colum to tell the tale; newspaper clippings reporting the unfolding events reveal to youthful readers the Machiavellian nature of green balance-of-power politics.

Nadia Wheatley, born in Sydney in 1949, incorporates into this and her other works (mainly for much younger readers), an array of experiences gained overseas, in the inner-Sydney suburb of Newtown and at Blackheath in the New South Wales Blue Mountains.

With great skill and verve, *The Blooding* shows the pain and confusion that individuals go through during sudden and dramatic change. It describes how all lives are altered when logging operations in a proud and close-knit community are suddenly shut down. The manner in which the young people of Cornwall adapt makes for a heart-warming and humorous tale.[64]

Margaret Aitken describes Wheatley as among the few authors able to examine for her young audience the complexity of such issues as jobs versus the destruction of natural resources—a vision further developed by showing the unfolding events through the eyes of a young man.[65]

As for the novel's descriptions of the conflict between protesters and loggers, Wheatley herself stresses the need to give an even-handed appraisal of what makes people behave as they do when confronted by opposing points of view. 'It wasn't difficult to put both sides because, though it may sound fey, I *was* Col while I was writing,' she told critic Agnes Nieuwenhuizen in 1991. '*He* was confused. *He* could see both sides. *He* was caught and absolutely split. *He* was jumping back and forth. And so was I.' In another interview, Wheatley discussed the 'shades and subtleties' which she sought to examine in the conflict between working-class timber workers and the middle-class conservationists seeking to save forests. 'I tried to put both sides of the debate, and not to be dogmatic,' she told Toss Gascoigne in 1988. 'I can see the point of the loggers. I have sympathy with individual loggers, and I believe the Government has a responsibility to find them alternative employment.'[66]

As in non-fiction works like Ian Watson's *Fighting Over the Forests* (see Chapter Three), which sought to heighten understanding of the attitudes of timber workers and their proud traditions, Wheatley said she sought to counter ingrained and clichéd prejudices—attitudes which usually were inculcated into young minds regardless of which side of the argument they were brought up on. When *The Blooding* was released, she commented: 'The kids from the logging areas should enjoy the book, and I desperately want them to read it. The real test . . . will be if those kids like it.'[67]

In the same interview, Wheatley told of how she aimed for a universality of issue and place, and of how her prevailing and contemporary concerns made their way into *The Blooding*. For example, although originally inspired by anti-Vietnam protests, she latched on to the more contemporary fight for the Franklin River. 'The environment issue intervened,' Gascoigne wrote. '[Wheatley] followed the Franklin Dam issue closely on television, and thought how interesting it would be to live on the Tasmanian west coast and side with the greenies. Then came Farmhouse Creek, another Tasmanian battleground, with graphic television shots of protesters being swept aside by the loggers' machinery while the police were limited to a passive watching role.'[68]

Similarly strong feelings about one's own immediate environment permeate Gillian Barnett's *The Inside Hedge Story* (1981), in which a group of local children await the developers about to move in on their precious bit of turf in suburbia, an estate known as Hedgerow. Like Wheatley, Barnett draws a highly dramatic and emotion-charged picture: 'Abby had a vision of the bulldozers moving in, and smashing flat all the sheltered places and gentle views. She felt that the house was like an injured whale, ready to be torn apart by sharks.'[69]

The Inside Hedge Story teaches the harsh realities of what is at stake when defending one's own patch, about how hired toughs and a police presence usually accompany any controversial development. Young readers learn the realpolitik of preserving something of value, and how local councils and the National Trust need to be approached—and how several parallel campaigns are often a prerequisite for effective political action. The action strategy in which the neighbourhood kids are embroiled is about as thorough and sophisticated as anything found in *Greenpeace Australia News*:

> You've got to keep a petition short . . . People won't sign what they haven't read thoroughly. And they won't read it if it's too long. You need a few short sentences to outline the situation. The describing and persuading will have to be done aloud . . .[70]

Barnett's novel also sadly shows how, in the end, ordinary citizens are always hard-pressed to have a voice in matters which directly affect them. This is graphically defined by the emotional and physical destruction by heavy machinery in the process of demolishing a neighbourhood.

The idea for the novel flowed from Gillian Barnett's own experiences as a child. In the Melbourne suburb of Balwyn where she saw a beautiful house surrounded by about four hectares of gardens subdivided for the mass-produced homes being built in the 1950s. 'This had an effect on me; there was a real sense of loss,' Barnett said in 1999. 'Right opposite my place, divided only by a hedge, was a grand Italianate mansion complete with stables and marble staircases. This was all demolished at a time when there was no sense of heritage.'[71]

Barnett, who in 1999 described herself as a 'low-key environmentalist', said her consciousness also arose from accompanying her mother on trips to the countryside where 'my attention was drawn to insects, what kind of cows I was seeing and a knowledge and appreciation of nature'.

The writing of *The Inside Hedge Story* came about when, while a lecturer in children's literature at Monash University, she was asked to organise a seminar attended by some of the world's leading writers for children at that time. 'After meeting these famous writers, I discovered that, apart from being impressed with them, they were just ordinary people and I felt that "Maybe I can do this too,"' Barnett said. 'I got rid of any mystique I had about writers being a breed apart, so I set about writing my first book.'[72]

In 1999, Barnett stressed that the subject matter of *The Inside Hedge Story* was much more contentious at the start of the 1980s than it would be at the end of the next decade because 'the notion then of campaigning to save a building is not like it is today . . . the book reflects a time when the media were not interested in such issues and the public were fairly apathetic.'[73]

Barnett's first children's book was followed in 1989 by *Gumboots and Other Risks*, about a young big-city girl dealing with the culture shock of being forcibly transferred to the country, as well as two picture books for younger readers. In 1999, she was working on her first novel for adults.

Barnett described all her writing as 'making sense of the world around me' and she tried to create characters who had complexity and depth and reflected the real problems of the time. She admitted to 'getting a terrific buzz' if any of her books struck a chord with children. Of *The Inside Hedge Story*, Barnett said that if she were able to exert even the smallest influence on a child to grow up to be an adult who stopped an undesirable demolition or redevelopment she would be extremely satisfied.[74]

CHILDREN'S literature embraces a range of genres, among them science fiction, whose voyages into fantasy and the imagination captivate young readers. One very potent and futuristic tale is Caroline Macdonald's *The Lake at the End of the World* (1988), which chronicles a distant generation coping with the debris and moral squalor of its ancestors. This novel can be seen as a version for young readers of *Feral City*, by fellow New Zealand-born writer Rosie Scott (see Chapter Two).

Caroline Macdonald has been described by the *Australian* as 'one of Australia's most enigmatic and tantalising writers of children's literature'. Her plots are 'complex, surprising and full of implied threat and tension' and at the same time 'not afraid to explore different styles, different genres and different states of mind'.[75]

The Lake at the End of the World describes such ecological blights as widespread land clearing, the total disappearance of wildlife habitats, and a nation being turned into a quarry—all for the sake of foreign money.

Caroline Macdonald, born in 1948, does not spare her young readers any of the horrors of the post-industrial age as she tells of nuclear accidents, toxic spills and the wanton use of herbicides. She warns that such actions cannot go on forever if dire consequences are to be avoided:

> . . . the gradual accumulation of all these things [continued] until the combined effect was just too much for the earth to absorb any more. But then the panic got so great there was no time for theories . . . Evacuation was the main thing. Get the survivors away before everyone was lost.[76]

Macdonald, who died of cancer in 1997 aged 48, also poses the question in *The Lake at the End of the World* of how severe a depletion of their lives would people be prepared to tolerate. She wants young readers to consider how serious the ravaging of their environment would have to get before they were moved to fight back. Looking to a distant and horrible future, Macdonald asks what could have been done to prevent things getting to such a horrendous stage.

The Lake at the End of the World was described in 1989 by prominent reviewer of children's literature Sally McInerney as one of the most significant novels to warn adolescents of ecological woes. She sees Macdonald's book as 'an ingenious, enjoyable forerunner of all the greenhouse effect stories which are undoubtedly advancing in their hordes'.[77]

Another novel with a similar theme—and similarly titled—is Lee Harding's *Waiting for the End of the World* (1983), which also depicts an authoritarian central government in tight control after the collapse of everything. Timeless agricultural skills had been forgotten in the mass migrations to and from cities. A central authority even exterminated all pets; feral cats, which managed to retain many of their most primal instincts, were among the few animals to survive.

Here was an age when anonymous Patrols kept the population under complete thrall. Here was a citizenry racked by disease, pain and no prospects of anything better. This citizenry is so volatile and under such overwhelming pressures that the State is forced to control all aspects of life:

> [The central government has] been getting away with it for years. Imagine! Pumping the water full of all kinds of drugs—drugs to make you calm when you're not, drugs to make you feel well when you'd rather feel rotten . . . they also dose the water with contraceptives, to make sure the population remains stable. Don't look so surprised! How else do you think they keep everything under control? By keeping people drugged, *that's* how . . .[78]

One of the rare thinkers in this barbaric society of the future is Manfred, who has almost mystic contemplations about how humanity can best survive, and how nature must one day reassert itself to ensure the continuation of all species. But, for the moment, it is all pessimism. 'Our lives don't matter to the Patrol,' muses Manfred. 'To them we're only vermin, to exterminate when necessary.'[79]

Lee Harding, born in the Victorian town of Colac in 1937, said he put himself among writers who are concerned with humankind as it approached the millennium. Harding, whose early career was as a professional photographer, said in 1991 that the times of fear and uncertainty were probably responsible for 'a resurgence of fantasy in literature'. He sought to tap into 'all sorts of ideas up in the air' and that his was part of 'the collective unconscious [which] must be experiencing a turbulent time'.[80]

Harding, while presenting such a bleak futuristic outlook, also probes the problems of teenagers relating to an increasingly complex world—and indeed relating with each other. On another level, this novel is an extremely realistic portrayal of relationships at both the best and the worst of times.

Discussing the sci-fi genre aimed at young readers, Stella Lees and Pam Macintyre, the editors of *The Oxford Companion to Australian Children's Literature*, see it as dealing with many of the 'pressing problems' faced by children and teenagers, ranging from 'the likelihood of a nuclear holocaust or environmental disaster on the one hand, to the alienation from society experienced by many people on the other'. Lees and Macintyre see sci-fi for kids as preparing them for some of the harsher realities which they are likely to encounter.[81]

A VAST array of imaginative picture books have raised the consciousness of the youngest of readers towards their environment. Some are bedtime stories and others are non-fiction, usually including sumptuous photographs or artwork, depicting the glories of nature and the creatures within it. Some of the nation's most creative minds have applied their talents to explaining ecological themes for the youngest readership.

One of the most serious and powerful picture books is by a survivor of the atomic bomb blast at Hiroshima, Junko Morimoto, who has conveyed to young readers the utter horror of nuclear atrocity in *My Hiroshima* (1987). Morimoto, born in Hiroshima in 1932, presents everyday and recognisable family scenes which coalesce with the utter destruction and carnage. Tots are not spared a single horror as Morimoto seeks to explain the shame and futility of military technology gone mad.

Morimoto presents the events at Hiroshima in ways the young reader can understand, creating scenes with which they can easily relate. 'Every school became a hospital for the badly injured,' Morimoto writes. 'I heard people screaming and moaning in pain, and there was a horrible smell of burnt skin . . . Many people died, one after another. Their bodies were taken to the school's playing field and burnt.'[82]

In Japan Morimoto was an illustrator and teacher, as well as Art Director of the Children's Art Studio in Osaka. She gives a philosophical underpinning to her childhood experiences of an event that so many did not survive:

War, the atomic bomb . . . only human beings have produced them. They are the crimes of adults who forget the precious value of life. I believe it is the duty and the responsibility of adults to teach our children the importance of not repeating these mistakes and to give them the heart to care for and value all life on earth.[83]

My Hiroshima was the seventh picture book released by Morimoto after she arrived in Australia in 1982, settling in the Sydney harbourside suburb of Mosman, and she has been published regularly ever since. But *My Hiroshima* held a special place for her, as she explained in 1987: 'My most terrible memory is of the atomic bomb attack on Hiroshima. I was lucky that I survived and so did my family, but all the students in my school were killed by the bomb. It is still an unforgettable sad memory that I collected the remains of the bones of my school friends and teachers about a year after the bomb. The last half of my [time] in primary school was badly influenced [by World War II], especially by the militaristic education and by the lack of food and so many other things necessary to life.'[84]

Through impressionist drawing and a double-page actual view of the total desolation which followed the blast, Morimoto leaves her young readers appalled at the madness of the nuclear option that laid waste Hiroshima.

Just as serious but with a different emphasis is Colin Thiele's *The Sknuks* (1977), in which the novelist indulges in some quite serious and portentous whimsy. *The Sknuks*, deftly illustrated by Mary Milton, shows a perfect, idyllic place called Htrae, in its pristine state the loveliest planet in the universe. Thiele's word pictures and Milton's comic and captivating illustrations tell of the most ideal of creations: 'There were forests of tall trees, and beautiful blue mountains, and rivers of sparkling water, and seas like a million mirrors . . . The Sknuks were very lucky.'[85]

But these Sknuks had a very human-like quality of never being content with their lot, of always wanting more and more:

> They growled and grumbled, and pillaged and plundered, and squabbled and squandered. They eyed and they envied, grew frantic and frenzied, they coveted and craved and threatened and raved. They even stole from their neighbours and spied on their friends.[86]

The Sknuks are depicted as uncaring ferals who ravaged and destroyed their glorious environment and the stunning variety of creatures among whom they lived—and only the Sknuks were able to spread and multiply across the rapidly depleted land. The greed of these automaton-like predators knew no bounds:

> They knocked down the trees and plundered the soil and blocked up the rivers and streams. They dug out the coal and used up the oil and squandered the copper and iron. They built up their burrows till there were burrows on burrows, and burrows on burrows on burrows, in clusters and clumps that were higher than mountains and wider than plains.
>
> They made millions of useless things to cram and clutter their burrows. They made things to keep for a week or a day, and then they threw them away . . .[87]

And all this was just the beginning—Sknuks showed they could be far more ruthless and hideous than this. As their industrial development and weaponry spread, 'before long the rivers were putrid and the sparkling waters were ribbons of sludge. The fish gasped and died'.[88]

In *The Sknuks*, young readers are shown how such an upside down and inside out way of living is not sustainable on Htrae, and their conscience is stirred to make things better and more sensible in their own increasingly worrisome surroundings.

Important environmental issues are discussed in *Oil Spill* (1996), in which Tricia Oktober describes to young readers how the delicate balance of nature in the oceans can be shattered—with seals, gulls and penguins equally affected. Oktober is a writer and illustrator whose many picture books have been in the forefront of voicing environmental concerns—dealing in turn with such ecological topics in *Bushfire* (1995), *Drought* (1998) and *Rainforest* (1998). As she put it in 1999, writing a book for children is her way of tying herself to a tree if someone is trying to chop the tree down.

The enormity of a tanker discharging its load is frighteningly shown in *Oil Spill*, and an illustration of a distant ship being pounded by gigantic waves and lightning flashes is accompanied by equally disquieting text:

> Hundreds of tonnes of greasy black fuel oil are pouring from the ship's damaged tanks. It spreads quickly, driven by strong winds and carried by the currents towards the shore. In just a few hours, the tide washes a thick blanket of oil over the beach, leaving it black and ugly. In time, the storm passes, but the oil has been spread far and wide.[89]

Oktober draws and describes penguins soaked in filthy sludge and tells of how 'oil destroys the natural waterproofing of birds and animals, causing them to drown or die of cold, so they must be helped swiftly'.[90]

Oil Spill is a litany of the many species grievously affected by a spill. It also reveals how an average clean-up operation can take many months.

Release of this book was timely in that a tanker was spilling its load off Tasmania at the time, putting penguins at risk. Similarly, *Bushfire* was released just as huge fires were surrounding Sydney in the summer of 1995. And drought is an ever-present harsh reality for people on the land.

The ecological conscience which suffuses the works of Tricia Oktober emerged from her earliest girlhood. She was born in the Bellinger Valley of northern New South Wales in 1940, and during her infancy her family moved to Eden and Bermagui on the New South Wales South Coast where she saw fairy penguins, whales spouting in the ocean and a profusion of wildlife in the lands around her and in the rainforests where she regularly went exploring. She went to school in Wollongong, which was then a comparatively undeveloped place where water birds frolicked offshore and a wondrous array of creatures congregated.

As a young adult Oktober became a professional painter—honing her skills in France, Spain and Quebec—and her love of nature shone through as she captured the minutiae of insects (such as the wings of cicadas) and the patterns within flowers, mosses and plants. Back in Australia, seeking a picture book for her infant daughter, Oktober discovered that Australian fauna and flora were largely neglected. So after some prodding from friends, she showed her drawings to a publisher, and with the release of *Bush Party* (1986) began her brilliant career.

The environmental essence that drives Oktober on is evident in everything she produces. 'I felt guilty that I wasn't out there chained to a tree; I always felt there weren't enough of us supporting those who were chained to trees,' Oktober said in 1999. 'There are still pristine areas in Australia, but I'd like to know where they are. Animals far from where environmental outrages are happening have DDT. You can't isolate one happening from another.'[91]

In her quest to win over the hearts and minds of her young readers, Oktober felt that 'I was given a voice and a platform, and I decided to use it . . . I wanted to do my best'. As her books were released, she talked to primary and high-school kids to further spread the message.

Since 1993, Oktober has been living at Katoomba in the Blue Mountains, where she enjoys a glorious garden and peaceful surroundings, but she is gnawed by the environmental ills that plague that world. 'The state of the world's oceans is my biggest concern,' she said. 'I wake up at night worrying about it. Fishing is a problem everywhere, as is the destruction of all coral reefs.'[92]

Oktober believes every ecological disaster has a series of repercussions, and that an oil spill such as that described in her book has knock-on effects everywhere. In *Oil Spill*, she warns that Australia has to be ready—through up-to-date legislation and the most effective technology—to deal immediately and effectively should a catastrophic accident occur.

The types who would vandalise the planet are seen in Jiri Tibor Novak's *Birdman* (1994), in which an angry man—depicted as a supposed pillar of the conservative community—looks at a glorious singing bird and sees it merely as a pest. Here is a man without a sense of beauty or without any tinge of care for the natural world. As the bird delightfully and harmlessly warbles, the man 'built a factory with many chimneys, which made poisonous brown smoke'. This is a forceful creation by Novak—an illustrator, painter, print-maker, graphic designer and writer—who since arriving in Australia in 1970 has written several picture books for children as well as a range of other works. His paintings are regularly exhibited in private galleries as well as at La Trobe University and the Australian National University.

Novak was educated in Fine Arts at the Hollar School in Prague and at the Ontario College of Art and has a Diploma in Fine Arts from the Phillip Institute in Melbourne. He brings to books like *Birdman* the grittiness and self-belief which allowed him to escape the then Czechoslovakia and travel to Australia via Egypt. 'All I knew about Australia when I fled was tennis and kangaroos, because the information we received in Czechoslovakia was very bad,' he noted in 1999.

The bird in *Birdman* somehow manages to survive every one of the man's attempts to discover fouler and fouler pollutants and destroyers. But Novak asks his young readers to imagine a world without a singing bird, and he regards the book as 'a parable for humanity . . . and I hope its message is much bigger than the book looks'.

Novak got the inspiration for *Birdman* after a visit to Prague, where he was born in 1947: 'I was horrified by the pollution . . . it made me realise what pollution really is'. In 1999, Novak described the Czech capital as a city where 'you can't breathe on the streets sometimes, and it made me depressed that kids were brought up in such an environment'.[93]

Living in the tranquil seaside town of Aireys Inlet with his partner, writer and editor Nan McNab, and their five-year-old son Tibor, Novak in 1999 preferred to stay away from big cities like Melbourne where he is appalled by the constant car fumes and a public transport system which is not sufficiently used.

Birdman is a wonderful, witty and optimistic affirmation of how the more positive aspects of the human character can sometimes prevail.[94]

A similar glow permeates Ted Greenwood's *Joseph and Lulu and the Prindiville House Pigeons* (1972), in which infants gain an insight into what happens when a low-rise CBD landmark comes under threat of developers. This habitat of 28 humans and many other denizens—and their ways of life—will be sacrificed to progress. But where there is inventiveness and courage it is possible to create a better future with many elements of the bygone age. Greenwood has created a heart-warming and charming interlude.[95]

The book is typical of the work of Greenwood, born in 1930, a writer and illustrator who produced most of his output in a timber cabin in the Victorian settlement of Ferny Creek. Greenwood produced 27 children's books during a long career which at times coincided with his job as a primary-school teacher and a lecturer in art at teachers college. Greenwood, who died in 1999 after fighting Parkinson's disease for thirteen years, was described as a pioneer of new ways to present picture books. In an obituary, fellow children's author Paul Jennings wrote in Melbourne's *Age*: 'The circle of life and the inevitability of change was one of his major themes.'[96]

Just like Greenwood's Prindiville pigeons, a leprechaun-like lot called the Scribblies seek to preserve their habitat in Nancy Cato's *Nin and the Scribblies* (1976). Cato, a popular novelist and environmental campaigner (see Chapter Four), tells the tale of a young girl Ninette Martin who encounters the Scribblies, keen creatures who write on all the trees beseeching humanity to preserve them.

For Nin, it is a time of growth and greater understanding:

> 'But why are you writing, "Help save our homes"?' [Nin asked.]
>
> 'Ah!' The little man stopped writing and leaned on his style. 'It's a p'tishun, you see. We're writing it on all the trees, and then perhaps the humin bings won't cut them down.'[97]

This fortuitous meeting of minds at Granite Bay unites the Scribblies and their human friends, led by Nin, to force the local council to save their precious beachside trees.

The plight of animals (specifically, lizards awaiting an uncertain fate) is described in *Goanna* (1989), in which Jenny Wagner describes goannas in a hollow log awaiting the arrival of bulldozers which will trample over everything.

Wagner arrived at Eumundi on the Sunshine Coast via Essendon in Victoria and a career writing scripts for the long-running TV soap 'Bellbird'. She delivers a delicious sense of irony as a sign atop the creature's favourite mountain reads 'This Is The Site For A Complex Of Shops' yet the goanna believes it reads 'Beware of the Goanna'. The truth was far sadder. As the bulldozers advanced, the goanna could only look as 'his mountains and valleys [were] swept aside and his treasures buried beneath them'.[98]

The utter helplessness of the goanna is dramatically shown in text and drawings as its environment has been totally covered in concrete. All we see is a vast expanse of nothing. The trees so admired by the goanna are now in concrete tubs on the edge of the car-park leading to the supermarket and junk-food outlets which in turn were responsible for moving the lizard on.

Another Queensland-based writer and illustrator, Narelle Oliver, ponders the plight of an endangered species—ospreys—in *High Above the Sea* (1991). Through her range of stunning linocut prints Oliver shows the balance of nature, and the birds are seen in their exuberant glory. Minutely drawn curlews poke their beaks towards their meal of worms and crabs, and two pied oyster-catchers peck away at pipi shells. Above all this, the drama unfolds as ospreys circle majestically on high.

Amid the glory, Oliver warns of the risk to ospreys posed by DDT which is killing off the fish they consume. Her message to young readers is powerful and to the point:

> . . . many of the osprey's traditional nesting sites are being taken over by development of coastal areas for housing, recreation and industry.
>
> Hopefully, there is still time to halt the decline of the Australian osprey population. The banning of DDT and the protection of traditional nesting sites in some countries have allowed declining osprey populations to recover significantly.[99]

These themes were evident in an earlier picture book, *Leaf Tail* (1989), in which Oliver shares the struggle of all creatures, especially lizards, in a rainforest under siege. A struggle it is, but Oliver maintains an exultant note as her joyous page spreads show the flora and fauna under the canopies.[100]

The exultant mood is a feature of Oliver's many picture books, and *The*

Best Beak in Boonaroo Bay (1993) is a typical explosion of linocuts hand-coloured with water-colours, ink and pastel. These illustrations show tropical vegetation and life at its most extravagant.

Oliver suggests to her young readers that when they hear squawking from the waterways and forests, it may be coming from glorious creatures like spoonbills, darters, cormorants, curlews, pelicans and oyster-catchers in fierce disputation over who indeed has the best beak. This fanciful notion is at the core of a witty and exuberant celebration of the senses.[101]

THE broader threat to the planet and the many issues affecting the ecology are examined for young readers by Margaret Dunkle and Robert Ingpen in *Conservation* (1987), which, as its subtitle indicates, is intended as 'A thoughtful way of explaining conservation to children'. Dunkle and Ingpen warn: 'Conservation is one way of reminding ourselves that if we change our world thoughtlessly we can cause serious damage, perhaps forever.'[102]

Conservation includes simple drawings by Ingpen of the planet, of a nesting bird—as well as of the smokestacks and high-tech industries which are changing things forever. 'If we make changes that are sound, safe and sensible we can even improve our world,' the authors show.[103]

Amid an illustration of children at play—a most potent vision of the happiness which existence should provide—Dunkle and Ingpen inspire their readers with the notion that 'conservation means using our imagination to plan for what we hope the future will be like, so that those who follow us may inherit the best possible world'.[104]

The authors suggest simple solutions which are far from being simplistic. They warn against dogmatic attitudes and show that politically correct or clichéd responses are not always the best solution. The treatment of the environment, like nature itself, is a matter of delicate balance: 'Conservation helps us to see that nothing in our world is that simple. It means deciding what is BEST for something, balancing the good and the bad, advantages and disadvantages, for everybody today and tomorrow.'[105]

Conservation is a collaboration between Ingpen, a long-respected illustrator whose credits include the 1974 edition of Colin Thiele's *Storm Boy*, and Dunkle, a recognised expert in Australian children's literature. Dunkle said in 1999 that the project was the result of a long-held concern for environmental issues which began during her upbringing in California, where she was born in 1922.

Dunkle's conservationist ethic was what brought her to Australia in her 30s, following the establishment of California's Silicon Valley which she described as the last straw. She told of rebelling against how a beautiful orchard area fertile with Mediterranean fruits was turned over to the computer industry. Her husband Robert Dunkle, a scientist at the University of California, could no longer continue research into solar energy as funds were diverted to military-related projects. As a Quaker, as part of the organic agriculture movement and as an anti-nuclear advocate, the Dunkles quit the USA.

With her husband, who became a senior principal research officer with the Department of Engineering at the CSIRO, Margaret Dunkle established an eight-hectare property at Gembrook in Victoria's Dandenongs region. The property was managed on permaculture principles and contained flora and fauna reserves for such creatures as wombats.

In 1999, Dunkle told of her empathy for Aboriginal concerns about caring for the land. 'We are temporary custodians of the land,' she said, 'and I want to be as good a custodian of it as I can be in the time that's been allotted to me.'[106]

A CREATIVE feast for young readers are the books of Jeannie Baker, who has combined her skills as writer and illustrator to present dazzling montages of environments at risk. Margaret Aitken has described Baker's works as non-fiction classics of the picture-book genre. Aitken lauds Baker as a cultural force who writes from conviction after having undertaken pains-taking research. Baker's books, to Aitken, are 'didactic, but strongly felt'.[107]

An early celebrated book by Baker, *Where the Forest Meets the Sea* (1989), introduces a magical quality, inviting young readers to exult in the wonders which surround them. Baker shows half a rainbow in the distance to where a father and son are heading. In the foreground of her complex montage are exotic shells and ancient sea life, with lush vegetation on the opposite shore.

The destination of the father and son has a promise of wondrous bounty, and young readers are not yet exposed to the threats to the natural splendours. At this stage all is magical and immortal: 'My father says there has been a forest here for over a hundred million years.'[108]

There is a stupendous, awesome quality to the forest before our eyes. The dense tropical growth is lush and luxuriant, and the gulls epitomise the freedom and joy of nature at its most profuse. The deeper Baker takes her young charges into the forest, the more stupendous are the discoveries. Firm trunks and branches represent thousands of years of evolution and splendour.

In 1998, after having sold 100,000 copies worldwide, *Where the Forest Meets the Sea* was re-released, and its message was deemed to be just as irresistible. In that year, a group calling itself the Daintree Rescue Program asked Baker to reproduce images of her artwork on signs along the Daintree walking tracks. 'We can't possibly do a better job than Jeannie has already done,' program manager John Lane was quoted as saying.[109]

The preparation for the book was fired up by Baker's first-hand exposure to how development could wreck the area. 'Even while I was visiting the Daintree in 1985 to generate the visual impressions and feelings for the artwork and the book, there was a large bulldozer preparing lots for sub-division and totally clearing the sites of trees,' Baker told reporter Lucy Clark in 1998. 'It struck me as a great absurdity when the one thing that makes the area so special is the tropical rainforest itself. We are strange creatures when we go to remote, unique and wonderful environments to explore and enjoy their unique qualities, and yet we have to make our mark, build our buildings, transplant our lawns and ultimately make this faraway place the same as the place we're getting away from.'[110]

Behind the childlike innocence and feelings of supreme beauty in Baker's work are worrisome ecological realities such as the pestilent spread of the imported weed rosy dock, as dramatically documented in *The Story of Rosy Dock* (1995). 'Without their normal predators, some non-native plants and animals multiply so quickly they change whole landscapes and push many native plants and animals to extinction,' Jeannie Baker explains.[111]

The collages vividly show how an introduced plant species can utterly submerge and dominate all else, transforming the original landscape and ecology.

Children's book reviewer Meg Sorensen said *Rosy Dock* did much more than issue a timely warning to children. Sorensen said it celebrated 'the life of the desert and the weed's beauty . . . this is not a heavy cautionary tale of human destruction'.[112]

Jeannie Baker, born in 1950, conveys through children's books an attitude which permeates all aspects of her life, as evidenced by her arrest, with about five hundred others, during a protest in May 1989 against the logging of old-growth forests in south-eastern New South Wales. Several celebrities from the arts community, including fellow painter John Coburn, faced charges of trespassing in a prohibited area of the National Estate-listed Egans Peak Forest. Amid the popularity of her books and her art, Baker regularly receives collages and letters from children all over the world.

Baker's wordless book *Window* (1991) is a stunning collection of montages yet closes with an 'Author's Note' which defines her environmental concerns and philosophies. 'We are changing the face of our world at an alarming and an increasing pace,' Baker writes, warning ominously that by the year 2020 'no wilderness will remain on our planet outside that protected in national parks and reserves'.[113]

Baker similarly estimates that, unless there is a dramatic change in humanity's set of values and aspirations, one-quarter of plant and animal species that existed at the start of the last decade of the twentieth century will be lost within thirty years. 'Already, at least two species become extinct each hour,' Baker laments.[114]

Nevertheless, this and Jeannie Baker's other pictorial tours de force are a groundswell of optimism and a hope that the planet's glories can somehow be saved.

AS YOUNG people grow up to take over a battered world, they will have to be prepared. Schools are doing their bit to teach environmental responsibility, respect for nature, sustainability, but the spiritual understanding of our earth—love for its beauty, its freshness and all the creatures who dwell upon it—comes from the books they read, the films they see and the ideas that are in the air as they mature. It is grounds for hope for the future of our world that there is such a rich heritage of books to nurture environmental wisdom.

The significance of picture books capturing the imagination of young children has been highlighted by Junko Morimoto who believes they 'will have reference to their lives in the future'. Morimoto has told of being 'very excited by the possibility that the messages in my picture books, which I can produce in the few years of my life left to me, may live forever through children's minds and the minds of their children'.[115]

Tricia Oktober has eloquently described the role of writing and illustrating to spread enlightened ideas to the nation's young: 'Every little Australian who takes notice of what I write will be a big Australian one day. And if only one out of a hundred of the little Australians decides to go into politics and seeks to change things, there may be some hope for us.'[116]

And Walter McVitty best expressed the significance of literature for children and young adults in raising environmental consciousness. Discussing Ruth Park and *My Sister Sif*, McVitty observed that 'whereas moralising in children's books once used to be aimed at saving the *souls* of the young [it was now] aimed at saving their *lives*—and the life of the whole planet—and so its impassioned message is desperately urgent'.[117]

ENDNOTES

AG: *Australian Geographic magazine*
CML: *Courier-Mail*, Brisbane
CT: *Canberra Times*
EG: *Earth Garden magazine*
GR: *Grass Roots* magazine
SL: *Simply Living* magazine
SMH: *Sydney Morning Herald*

Chapter 2 Novels

1 Webby, Elizabeth: telephone interview with Pollak, 28 June 1999.
2 Webby interview.
3 Bennett, Bruce: telephone interview with Pollak, 12 July 1999.
4 Bennett interview.
5 Davidson, Jim: *Sideways from the Page (the Meanjin Interviews)*. Fontana/Collins. Melbourne, 1983. p106. (From an interview in *Meanjin*, 6 March 1978.)
6 Herbert, Xavier: *Poor Fellow My Country*. Fontana/Collins. Sydney, 1983. (First published William Collins, 1975; first published Fontana, 1976.) p430.
7 Herbert, *Poor Fellow My Country*, p24.
8 Herbert, *Poor Fellow My Country*, p25.
9 Herbert, *Poor Fellow My Country*, pp1302–1303.
10 Herbert, *Poor Fellow My Country*, p97.
11 Herbert, *Poor Fellow My Country*, p97.
12 Herbert, *Poor Fellow My Country*, p137.
13 Goodwin, Ken: *A History of Australian Literature*. (Macmillan History of Literature series.) Macmillan Publishers Ltd. London, 1986. p109.
14 Herbert, *Poor Fellow My Country*, p1451.
15 Herbert, *Poor Fellow My Country*, p1454.
16 Harrison-Ford, Carl: 'Compelling power', *SMH*, 13 September 1975.
17 Clancy, Laurie: *A Reader's Guide to Australian Fiction*. Oxford University Press Australia. Melbourne, 1992. pp8–9; 153.
18 Richards, Dave: 'The last words of Xavier Herbert', *The National Times*, 18–24 January 1985.
19 Kendell, Jeni and Eddie Buivids: *Earth First (the struggle to save Australia's rainforest)*. ABC Enterprises. Sydney, 1987. p91.
20 Marr, David: *Patrick White: A Life*. Random House Australia. Milsons Point, NSW, 1991. p240.
21 Marr, *Patrick White*, p256.
22 Marr, *Patrick White*, p290.
23 Marr, *Patrick White*, p350.
24 Marr, *Patrick White*, p350.
25 Marr, *Patrick White*, p480
26 White, Patrick: *The Eye of the Storm*. Vintage (Random House) London, 1973. p63.
27 White, *The Eye of the Storm*, p176.
28 White, *The Eye of the Storm*, p177.
29 White, *The Eye of the Storm*, p252.
30 White, *The Eye of the Storm*, p334.
31 White, *The Eye of the Storm*, p346.
32 White, *The Eye of the Storm*, p347.
33 Marr, *Patrick White*, p509.
34 White, Patrick: *Patrick White Speaks*. (Edited by Paul Brennan and Christine Flynn.) Primavera Press. Leichhardt, NSW, 1989. p28.
35 White, *The Eye of the Storm*, p352.
36 White, Patrick: *A Fringe of Leaves*. Vintage (Random House). London, 1977. (First published by Jonathan Cape Ltd, London, 1976.) p248.
37 White, *A Fringe of Leaves*, p256.
38 Marr, *Patrick White*, p550.
39 Marr, *Patrick White*, p601.
40 Clancy, Laurie: 'Pessimism and anger', *Australian Book Review*, No 48 February–March 1983, p27.
41 Houghton, Greg, Untitled, *World Literature Written in English*, Vol 22, No 1, Spring 1983, pp223-225.
42 Ellison, Jennifer: *Rooms of Their Own*. Penguin Books. Ringwood, Vic, 1986. p61. (Interview conducted at Cairns, North Queensland, October 1985.)
43 Astley, Thea: *An Item from the Late News*. University of Queensland Press. St Lucia, Qld, 1982. pp56, 62.
44 Astley, Thea: *It's Raining in Mango*. Viking (Penguin Books Australia Ltd). Ringwood, Vic, 1987. p32.
45 Astley, Thea: *Vanishing Points*. William Heinemann Australia. Port Melbourne, Vic, 1992. p102.
46 Astley, *Vanishing Points*, p52.
47 Astley, *Vanishing Points*, pp23, 32.

48 Loane, Sally: 'Point of return', *SMH Spectrum*, 15 August 1992, p41.
49 Pierce, Peter: 'Blunt attack', *The Bulletin*, 17 November 1992, p91.
50 Astley, *Vanishing Points*, pp221, 227.
51 Astley, *Vanishing Points*, p168.
52 Geason, Susan, 'Take paradise—put up a parking lot', *The Sun-Herald* Books Page, 13 September 1992.
53 Baker, Candida: *Yacker (Australian Writers Talk About Their Work)*. Pan Books (Australia) Pty Ltd (Picador). Sydney, 1986. pp41, 48.
54 Baker, *Yacker*, p35.
55 Hefner, Robert: 'Recognition for Thea Astley in novelist award', *CT*, 18 November 1989, p4.
56 Willbanks, Ray: *Speaking Volumes (Australian writers and their work)*. Penguin Books Australia Ltd. Ringwood, Vic, 1991. p41.
57 Willbanks, *Speaking Volumes*, p27.
58 Loane, *SMH*, 15 August 1992, p41.
59 Willbanks, *Speaking Volumes*, p36.
60 Brown, Jenny: 'Write man for the job', *The Sun* (Sydney), 31 May 1982.
61 McQueen, James: *Hook's Mountain*. The Macmillan Company of Australia Ltd. Crows Nest, NSW, 1982. p149.
62 McQueen, *Hook's Mountain*, p34.
63 Bennett, Bruce, 'Perceptions of Australia, 1965–1988', pp433–453 in Laurie Hergenhan (general editor): *The Penguin New Literary History of Australia*. Penguin Books Australia Ltd. Ringwood, Vic, 1988. p448.
64 McQueen, James: 'City of two times—Hobart', *The Bulletin*, Australia Day, 26 January 1988—Special Bicentennial Edition, pp200–202; pp201, 202.
65 McQueen, James: 'Hook's country', *Overland*, No 98 April 1985, pp35–38; p35.
66 McQueen, *Overland* No 98, p36.
67 Hassall, Anthony J.: *Dancing on Hot Macadam (Peter Carey's fiction)*. University of Queensland Press. St Lucia, Qld, 1994. pp2, 5, 8.
68 Carey, Peter: *Bliss*. University of Queensland Press. St Lucia, Qld, 1982 (first published 1981), p135.
69 Carey, *Bliss*, p14.
70 Carey, *Bliss*, p79.
71 Carey, *Bliss*, p137.
72 Willbanks, *Speaking Volumes*, pp47, 51-52.
73 Lamb, Karen: *Peter Carey: The Genesis of Fame*. Angus & Robertson. Pymble, NSW, 1992. p21.
74 Carey, Peter: *The Fat Man In History*. The University of Queensland Press. St Lucia, Qld, 1974. p69.
75 Carey, *The Fat Man In History*, pp91–93.
76 Carey, *The Fat Man In History*, p97.
77 Carey, *The Fat Man In History*, p102.
78 Carey, Peter: *War Crimes*. University of Queensland Press. St Lucia, Qld, 1981. (First published 1979.) p10.
79 Carey, *War Crimes*, pp75–76.
80 Carey, *War Crimes*, p263.
81 Hassall, *Dancing on Hot Macadam*, p31.
82 Carey, Peter: *The Tax Inspector*. University of Queensland Press. St Lucia, Qld, 1991. pp116–117.
83 Carey, *The Tax Inspector*, pp163-164.
84 Carey, *The Tax Inspector*, p167.
85 Hassall, *Dancing on Hot Macadam*, p147.
86 Hassall, *Dancing on Hot Macadam*, p5.
87 D'Alpuget, Blanche: *White Eye*. Viking (Penguin Books Australia Ltd). Ringwood, Vic, 1993. pp57–58.
88 Barrow, Graeme: 'Much more than just a thriller', *CT*, 6 November 1993, p C10; Elliott, Helen: 'Eagle-eyed prose hits the heights', *Sunday Age* Agenda, 12 December 1993, p9.
89 Elliott, *Sunday Age* Agenda, p9.
90 D'Alpuget, Blanche: *Turtle Beach*. Viking (Penguin Books Australia Ltd). Ringwood, Vic, 1981. pp69–70.
91 Veitch, Kate: 'A question of placing women in books', *SMH* Summer Agenda, 24 January 1989, p14.
92 Johnston, Dorothy: *Maralinga My Love*. McPhee Gribble/Penguin Books. Fitzroy, Vic, 1988. p24.
93 Johnston, *Maralinga My Love*, p94.
94 Johnston, *Maralinga My Love*, pp172-173.
95 Johnston, Dorothy: *One for the Master*. Wakefield Press. Kent Town, SA, 1997. p23.
96 Johnston, *One for the Master*, p30.
97 Johnston, *One for the Master*, pp216, 218.
98 Winton, Tim: *Shallows*. George Allen & Unwin. North Sydney, NSW, 1984. p6.
99 Eldridge, Marian: '*Shallows* complex, disturbing', *CT*, 9 March 1985, p18.
100 Sen, Veronica: '*Shallows*, a sobering, challenging novel', *CT*, 10 November 1985, p8.
101 Myers, David: 'Stark work of brooding loners', *Age* Saturday Extra, 1 December 1984, p18.
102 Ikin, Van: telephone interview with Pollak, 9 July 1999.

103 Cranny-Francis, Anne: telephone interview with Pollak, 15 July 1999.
104 Cranny-Francis interview.
105 McMullen, Sean: *Call to the Edge*. Aphelion Publications. North Adelaide, 1992. p61.
106 Dowling, Terry: 'Stricken Sydney remains commercial', *Australian*, 22 October 1990.
107 Lord, Gabrielle: *Salt*. McPhee Gribble. Ringwood, Vic, 1990. p3.
108 Lord, *Salt*, p5.
109 Lord, *Salt*, p9.
110 Lord, *Salt*, pp14–15.
111 Lord, *Salt*, p19.
112 Lord, Gabrielle: *Jumbo*. The Bodley Head Ltd. London, 1986. pp7–8.
113 Lord, *Jumbo*, p9.
114 Lord, *Jumbo*, p71.
115 Scott, Rosie: *Feral City*. Minerva. Melbourne, 1993, p3. (First published by William Heinemann Australia, Melbourne, 1992.)
116 Scott, *Feral City*, pp1, 8.
117 Scott, *Feral City*, p4.
118 Scott, *Feral City*, p79.
119 Baker, Candida: 'Living a childhood dream', *Age*, 6 May 1992, p19.
120 Hutchings, Peter: 'NZ becomes land of the long dark cloud', *SMH*, 18 April 1992.
121 Mills, Claire: 'Rogernomics run rampant', *Australian Bookseller and Publisher*, Vol 71 No 1023 February 1992, p27.
122 Turner, George: *The Sea and Summer*. Faber & Faber Ltd. London, 1987. p297.
123 Turner, *The Sea and Summer*, p24.
124 Stone, Deborah: 'A fervent prayer for the future', *Australian*, 17 October 1988.
125 Stone, *Australian*, 17 October 1988.
126 Stone, *Australian*, 17 October 1988.
127 Steele, Colin: 'All washed up in bleak tomorrows', *CT* Saturday Magazine, 20 February 1988, p B2.
128 Stone, *Australian*, 17 October 1988.
129 Turner, *The Sea and Summer*, p19.
130 Turner, *The Sea and Summer*, p19.
131 Turner, *The Sea and Summer*, p21.
132 Hanrahan, John: 'A novel to read before high tide', *Age* Saturday Extra, 16 January 1988, p10.
133 Turner, *The Sea and Summer*, p39.
134 Turner, *The Sea and Summer*, p304.
135 Ikin, Van: 'A Sweet life on Swill Street, *SMH* Spectrum, 10 September 1988, p76.
136 Turner, George: *Brain Child*. William Morrow and Company Inc. New York, 1991. p11.
137 Turner, *Brain Child*, pp138–139.
138 Turner, *Brain Child*, p190.
139 Turner, George: 'Flowering Mandrake', in Peter McNamara and Margaret Winch (editors), *Alien Shores, (an anthology of Australian science fiction)*. Aphelion Publications. North Adelaide, SA, 1994. p41.
140 McMullen, *Call to the Edge*, pp33, 54, 55.
141 McMullen, *Call to the Edge*, p36.
142 McMullen, *Call to the Edge*, p234.
143 McMullen, *Call to the Edge*, p229.
144 'An interview with Sean McMullen', on Sean McMullen web page, originally published in *Eidolon* magazine, July 1993, pp16–26.
145 Dowling, Terry: *Rynosseros*. Aphelion Publications. North Adelaide, 1990. pp15, 24.
146 Dowling, *Rynosseros*, p53.
147 Dowling, Terry: *Twilight Beach*. Aphelion Publications. North Adelaide, SA, 1993. p18.
148 *SMH* Agenda, 4 February 1993, p12.
149 Reid, John: interview with authors at his home in Wyong, NSW, on 27 March 1998.
150 Coupe, Stuart: interview with authors at his home in Balmain, NSW, on 29 June 1998.
151 Coupe, Stuart: 'Walking the detectives', *SMH* Good Weekend, 4 April 1992, pp8–17; p11.
152 Fuller, Peter: 'Crime writer: old-fashioned heroes in hard-boiled Sydney', *CT*, 1 February 1986, p B1.
153 Pierce, Peter: 'Bleak times for a Sydney sleuth', *Bulletin*, 11 February 1997, p95.
154 Pierce, Peter: 'Cliff Hardy, back in the old milieu', *Age* Saturday Extra, 8 November 1986, p12; 'A private eye gives up on jaded Sydney', *National Times*, 16–22 May 1986, p35.
155 Reid interview.
156 Fuller, *CT*, 1 February 1986, p B1.
157 White, Judith: 'Canberra story "close to the bone" ', *Sun-Herald*, 17 March 1991
158 Corris, Peter: *The Japanese Job*. Angus & Robertson. Pymble, NSW, 1992. p95.
159 Corris, Peter: *The Washington Club*. A Bantam Book (Transworld Publishers). Moorebank, NSW, 1997. p128.

160 Corris, *The Washington Club*, p129.
161 Corris, *The Washington Club*, p133.
162 Corris, *The Washington Club*, pp137–138.
163 Corris, Peter: *The Reward*. A Bantam Book. Moorebank, NSW, 1997. p20.
164 Corris, *The Reward*, p21.
165 Corris, *The Reward*, p21.
166 Corris, Peter: *Wet Graves*. A Bantam Book. Moorebank, NSW, 1991. p12.
167 Corris, *Wet Graves*, pp22, 170.
168 Bedford, Jean: *To Make a Killing*. Angus & Robertson. Pymble, NSW, 1992. p7.
169 Bedford, *To Make a Killing*, pp10, 11.
170 Bedford, *To Make a Killing*, p23.
171 Bedford, *To Make a Killing*, p94.
172 Bedford, *To Make a Killing*, p89.
173 Ellison, *Rooms of Their Own*, pp88–89. (Interview conducted in Sydney, June 1985.)
174 Moffitt, Ian: *Gilt Edge*. Pan Australia. Sydney, 1991. pp5, 6.
175 Moffitt, Ian: 'A Good Rexona Town' in Geoffrey Dutton (editor): *Country Childhoods*. University of Queensland Press. St Lucia, Qld, 1992. pp88–105; pp101, p94.
176 Moffitt, 'A Good Rexona Town', *Country Childhoods*, p96.
177 Wright, Steve: *A Drop in the Ocean*. Pan Macmillan Publishers Australia. Chippendale, NSW, 1991. pp10–11.
178 Wright, *A Drop in the Ocean*, pp27–28.
179 Wright, *A Drop in the Ocean*, p29.
180 Wright, *A Drop in the Ocean*, p70.
181 Wright, Steve: interview with Pollak at his home in Balmain, Sydney, on 9 July 1999.
182 Wright interview.
183 Nowra, Louis: *Red Nights*. Picador/ Pan Macmillan. Sydney, 1997. p11.
184 Nowra, *Red Nights*, pp157, 202.
185 Nowra, *Red Nights*, p9.
186 Brissenden, R.F. (Robert Francis): *Poor Boy*. Allen & Unwin Australia Pty Ltd. North Sydney, NSW, 1987. p66.
187 Brissenden, *Poor Boy*, p66.
188 Brissenden, *Poor Boy*, p218.
189 Brissenden, *Poor Boy*, p90.
190 Brissenden, R.F.: *Wildcat*. Allen & Unwin Australia Pty Ltd. North Sydney, NSW, 1991. p114.
191 Brissenden, *Wildcat*, p162.
192 White, Judith, 'Canberra story "close to the bone" ', *The Sun-Herald*, 17 March 1991.
193 'Collective call', Letter to the Editor, *SMH*, 17 July 1989, p12.
194 Wongar, B.: *Walg (A novel of Australia)*. Macmillan. South Melbourne, Vic, 1986 (first published 1983). pp50–51.
195 Drewe, Robert: 'An uneasy journey over the color line', *Australian*, 3 March 1984.
196 Wongar, *Walg*, p35.
197 Wongar, *Walg*, pp167, 170.
198 Wongar, B.: *Karan (A novel of the Australian hinterland)*. Dodd, Mead & Co Inc. New York, 1985. ppxiv, xv-xvi.
199 Standing, Sue: 'Aborigines enter the nuclear age', *Boston Herald*, 8 December 1985, p110.
200 Moss, Chuck: 'Atomic dream time: tortuous novel pulls reader into non-logic of Aboriginal Australia', *Kansas City Star*, 12 January 1986.
201 Wongar, B.: *Gabo Djara*. Macmillan Company of Australia Pty Ltd. South Melbourne, Vic, 1988 (first published by Dodd Mead & Co, New York, 1987). p24.
202 Wongar, *Gabo Djara*, pp99, 166.
203 Simons, Margaret: *The Ruthless Garden*. Minerva Australia. Port Melbourne, Vic, 1993 (first published Angus & Robertson Bookworld, 1993). p41.
204 Houbein, Lolo: *Walk a Barefoot Road*. ABC Enterprises. Crows Nest, NSW, 1988. pp17–20.
205 Houbein, *Walk a Barefoot Road*, p139.
206 Houbein, *Walk a Barefoot Road*, p234.
207 Houbein, Lolo: *Wrong Face in the Mirror (an autobiography of race and identity)*. University of Queensland Press. St Lucia, Qld, 1990 (1988). pp245–246.
208 Houbein, *Wrong Face in the Mirror*, pp258–259.
209 Clancy, Laurie: telephone interview with Pollak, 16 July 1999.
210 Elliott, Helen: 'Eagle-eyed prose hits the heights', *Sunday Age* Agenda, 12 December 1993, p9.

Chapter 3 Non-Fiction

1 Hay, Peter: telephone interview with Pollak, 13 August 1999.
2 Hay interview.
3 Haigh, David: telephone interview with Pollak, 14 August 1999.
4 Haigh interview.
5 Timothy Fridtjof: *The Future Eaters (An*

ecological history of the Australasian lands and people). Reed Books. Chatswood, NSW, 1994. p101.
6 Flannery, *The Future Eaters*, p129.
7 Flannery, *The Future Eaters*, p185.
8 *The Future Eaters*, pp178–179.
9 Flannery, *The Future Eaters*, p84.
10 Flannery, *The Future Eaters*, pp95-96.
11 Flannery, *The Future Eaters*, p354.
12 Flannery, *The Future Eaters*, p271.
13 'Swap your dog for a quoll', *Herald Sun*, 5 June 1997, p11.
14 Nankin, Harry: 'Killing of Gondwana', *Age* Saturday Extra, 17 December 1994, p8.
15 Williams, Sue: ' "Dr Doom" and his radical solution', *Australian*, 6 December 1994.
16 Delvecchio, Julie: 'Citizens band together in defence of bandicoot colony', *SMH*, 2 July 1997, p5.
17 Benson, Simon: 'Rape of Sydney—property "exploiters" destroying our city', *Daily Telegraph*, 20 August 1998, p21.
18 Whittaker, Mark: 'They ate the future—stirring the mega-possum', *Australian* magazine, 8 August 1998, p14.
19 Smith, David: *Continent in Crisis (A natural history of Australia)*. Penguin Books. Ringwood, Vic, 1990. p68.
20 Smith, *Continent in Crisis*, p134.
21 Smith, *Continent in Crisis*, pp132–134.
22 Smith, *Continent in Crisis*, p68.
23 Smith, *Continent in Crisis*, p4.
24 Smith, *Continent in Crisis*, p66.
25 Smith, David: *Saving a Continent*. (Expanded and updated colour edition of *Continent in Crisis*.) University of NSW Press. Sydney, 1994. page x.
26 Smith, David: telephone interview with Pollak, 7 August 1999.
27 Gare, Deborah F.: 'A journey in Australian history: Geoffrey Bolton and the middle way', *Limina*, Vol 4 1998, pp85–89; pp87, 88.
28 Bolton, Geoffrey: *Spoils and Spoilers (a history of Australians shaping their environment)*. 2nd edition. Allen & Unwin. North Sydney, NSW, 1992. (First edition 1981.) p40.
29 Bolton, *Spoils and Spoilers*, pp15, 107.
30 Gare, Deborah F.: 'A journey in Australian history: Geoffrey Bolton and the middle way', Limina, Vol 4 1998, p88.
31 Lines, William J.: *Taming the Great South Land (a history of the conquest of nature in Australia)*. Allen & Unwin Pty Ltd. North Sydney, NSW, 1991. pp26, 40.
32 Lines, *Taming the Great South Land*, p41.
33 Lines, *Taming the Great South Land*, pp249, 264.
34 White, Judith: 'Significant work of history', *Sun-Herald*, 1 September 1991, p107.
35 Lines, *Taming the Great South Land*, p220.
36 Lines, William J.: *An All-consuming Passion (Origins, modernity, and the Australian life of Georgiana Molloy)*. Allen & Unwin. St Leonards, NSW, 1994. pp56–57.
37 Lines, *An All-consuming Passion*, pp102–103.
38 Lines, *An All-consuming Passion*, p103.
39 Lines, *An All-consuming Passion*, p116.
40 Lines, William J.: *False Economy (Australia in the 20th century)*. Fremantle Arts Centre Press. South Fremantle, WA, 1998. p50.
41 Lines, William J.: *A Long Walk in the Australian Bush*. University of NSW Press Ltd. Sydney, 1998. p106.
42 Lines, *A Long Walk in the Australian Bush*, p3.
43 Lines, *A Long Walk in the Australian Bush*, p20.
44 Rolls, Eric: *A Million Wild Acres (200 years of man and an Australian forest)*. Thomas Nelson Australia. Melbourne, 1981. pp10–11.
45 Rolls, *A Million Wild Acres*, pp10-11.
46 Rolls, *A Million Wild Acres*, pp10-11.
47 Rolls, *A Million Wild Acres*, p399.
48 Rolls, *A Million Wild Acres*, p292.
49 Rolls, *A Million Wild Acres*, p328.
50 Rolls, *A Million Wild Acres*, p399.
51 Rolls, *A Million Wild Acres*, p408.
52 Synnott, John: 'The key issues', *Sun-Herald*, 5 March 1995, p18; Rolls, Eric: 'A state of degradation', *Sun-Herald*, 19 March 1995.
53 Rolls, Eric: *They All Ran Wild (the animals and plants that plague Australia)*. Angus & Robertson. Sydney, 1984 (first published A&R, 1968). pxiv.
54 Rolls, *They All Ran Wild*, p484.
55 Rolls, *They All Ran Wild*, p507.
56 Rolls, *They All Ran Wild*, p174.
57 Robinson, Harry: 'The loves of Eric Rolls', *Weekend Australian* magazine, 26–27 June 1993, pp10-12, 15; p15.
58 Trezise, Percy: *Last Days of a Wilderness*. William Collins (Aust) Ltd. Sydney, 1973. p151.

59 Trezise, *Last Days of a Wilderness*, pp159–160.
60 Fowler, Andy: 'Why the rat-bags make Percy angry', *Australian*, 10 May 1980.
61 Rothwell, Nicholas: 'Tropical meltdown', *Australian*, 28 February 1998, p22.
62 Watson, Ian: *Fighting Over the Forests.* Allen & Unwin. North Sydney, 1990. ppxvi–xvii.
63 Watson, Ian: telephone interview with Pollak, 7 July 1999.
64 Ian Watson interview.
65 Ian Watson, *Fighting Over the Forests*, pxx.
66 Ian Watson, *Fighting Over the Forests*, p46.
67 Ian Watson, *Fighting Over the Forests*, pp98, 108.
68 'Recent Releases', *Grass Roots*, No 82, December 1990, p69.
69 Ian Watson interview.
70 Noonuccal, Oodgeroo (previously Kath Walker): *Stradbroke Dreamtime.* Angus & Robertson. Pymble, NSW, 1992 (first published 1972). p3.
71 Oodgeroo, *Stradbroke Dreamtime*, p4.
72 Oodgeroo, *Stradbroke Dreamtime*, p4.
73 Oodgeroo, *Stradbroke Dreamtime*, p27.
74 Baker, Candida: *Yacker 2.* Pan Books (Picador). Sydney, 1987. p288.
75 Baker, *Yacker 2*, pp290–291.
76 Davidson, Jim: *Sideways from the Page (The* Meanjin *interviews).* Fontana/ Collins. Melbourne, 1983. p54. (Interview in *Meanjin*, 8 July 1977.)
77 Turcotte, Gerry, 'Recording the cries of the people—an interview with Oodgeroo (Kath Walker)' in Anna Rutherford (editor), *Aboriginal Culture Today.* Dangaroo Press. Sydney, 1988. pp16–30; pp29–30.
78 Hogan, Christine: 'Kath Walker—an extraordinary life', *Portfolio*, Vol 3 No 9, July 1987, pp14–16; p16.
79 Hocking, Susan: 'Kath Walker's Dreamtime school', *Australian Women's Weekly*, April 1987, pp85–86; p86.
80 Walker, Kath: 'My day', *Woman's Day*, 17 July 1978 (as told to Frank Robson and Drena Parrington), p85.
81 Van Oudtshoorn, Nic: 'Poet and painter—poet Kath Walker takes Aboriginal art back to its roots', *People*, 1 July 1981, pp48–49; 49.
82 Thompson, Peter: *Bob Brown of the Franklin River.* George Allen & Unwin Pty Ltd. North Sydney, NSW, 1984. p63.
83 Thompson, *Bob Brown of the Franklin River*, p38.
84 Thompson, *Bob Brown of the Franklin River*, p51.
85 Thompson, *Bob Brown of the Franklin River*, p95.
86 Thompson, *Bob Brown of the Franklin River*, p96.
87 Collins, Paul: *God's Earth (religion as if matter really mattered).* A Dove publication (an imprint of Harper/ Collins Religious). North Blackburn, Vic, 1995. p1.
88 Collins, *God's Earth*, pp1–2.
89 Collins, *God's Earth*, p4.
90 Collins, *God's Earth*, p120.
91 Green, Shane: 'Portrait of a heretic', *Age* News Extra, 14 February 1998, p7.
92 Lindsay, Elaine: 'Heart of the matter', *Australian*, 3 June 1995.
93 Birch, Charles: *On Purpose.* NSW University Press Ltd. Kensington, NSW, 1990 (1991). p17.
94 Birch, *On Purpose*, p112.
95 Birch, Charles: *Regaining Compassion (for humanity and nature).* NSW University Press, Kensington, NSW, 1993. p174.
96 Knuckey, Marie: 'Will the grass roots people one day run the world?', *SMH*, 10 July 1980.
97 Birch, *Regaining Compassion*, pp110–111.
98 Birch, *Regaining Compassion*, p111.
99 Leser, David: 'Against the flow', *SMH* Good Weekend, 15 July 1995, p48.
100 Brinsmead-Hungerford, Hesba: *I Will Not Say the Day Is Done.* Alternative Publishing Co-operative Ltd. Chippendale, NSW, 1983. p7.
101 Brinsmead-Hungerford, *I Will Not Say the Day Is Done*, pp34, 46.
102 Brinsmead-Hungerford, *I Will Not Say the Day Is Done*, pp39–40, 138.
103 Brinsmead-Hungerford, *I Will Not Say the Day is Done*, p153.
104 Milliken, Robert: *No Conceivable Injury (the story of British and Australia's atomic cover-up).* Penguin Books Australia Ltd. Ringwood, Vic, 1986. p81.
105 Milliken, *No Conceivable Injury*, pp94–95.
106 Milliken, *No Conceivable Injury*, p135.
107 Milliken, Robert: telephone interview with Pollak, 9 April 1999.
108 Milliken interview.
109 Tame, Adrian and F.P.J. Robotham:

Maralinga (British A-Bomb Australian Legacy). Fontana/Collins. Melbourne, 1982. pp139, 140.

110 Tame/Robotham, *Maralinga*, p113.

111 Robotham, F.P.J.: telephone interview with Pollak, 17 June 1999.

112 Robotham interview.

113 Robotham interview.

114 Robotham interview.

115 Robotham interview.

116 Tame, Adrian: telephone interview with Pollak, 22 June 1999.

117 Tame interview.

118 Blakeway, Denys and Sue Lloyd-Roberts: *Fields of Thunder (testing Britain's bomb)*. George Allen and Unwin Australia Pty Ltd. North Sydney, NSW, 1985. p8.

119 Blakeway/Lloyd-Roberts, *Fields of Thunder*, p10.

120 Lloyd-Roberts interview.

121 Beder, Sharon: *Toxic Fish and Sewer Surfing (how deceit and collusion are destroying our great beaches)*. Allen & Unwin. North Sydney, NSW, 1989. pp32–33.

122 Beder, *Toxic Fish and Sewer Surfing*, p98.

123 Beder, *Toxic Fish and Sewer Surfing*, p98.

124 Beder, *Toxic Fish and Sewer Surfing*, pp154–155.

125 Beder, Sharon: 'Sydney's toxic timebomb', *SMH* Spectrum, 4 November 1989, p80.

126 Cawcutt, Linda and Catherine Watson: *Pesticides: The New Plague*. Friends of the Earth. Collingwood, Vic, 1984. pp16, 18.

127 Cawcutt/Watson, *Pesticides: The New Plague*, pp17–18, 19.

128 Cawcutt/Watson, *Pesticides: The New Plague*, pp8–9.

129 Cawcutt/Watson, *Pesticides: The New Plague*, p72.

130 Gordon (Cawcutt), Linda: telephone interview with Pollak, 6 May 1999.

131 Gordon interview.

132 Watson, Catherine: telephone interview with Pollak, 7 May 1999.

133 Murray-Smith, Stephen: *Sitting on Penguins (Australia and the Antarctic)*. Mandarin Australia. Port Melbourne, Vic, 1991 (first published by Hutchinson Australia, 1988). pp50–51.

134 Murray-Smith, *Sitting on Penguins*, p51.

135 Murray-Smith, *Sitting on Penguins*, p52.

136 McLaren, John: Obituary. *Notes & Furphies*, No 21, October 1988, p19.

137 Beder, Sharon: *Global Spin (the corporate assault on environmentalism)*. Scribe Publications Pty Ltd. Carlton North, Vic, 1997. p15.

138 Beder, *Global Spin*, p233.

139 Beder, *Global Spin*, p233.

140 Beder, *Global Spin*, p234.

141 Grant, Peter: 'A half-open door', *Island*, No 53, Summer 1992, pp30–34.

142 McLaren, John: Obituary. *Notes & Furphies*, No 21, October 1988, p19.

Chapter 4 Illustrated Books

1 Milne, Christine: telephone interview with Pollak, 24 August 1999.

2 Rankin, Robert: interview with Pollak, October 1997.

3 Chester, Quentin: telephone interview with Pollak, 7 January 1999.

4 Chester, Quentin and Alasdair McGregor: *Australia's Wild Islands*. Hodder & Stoughton. Rydalmere, NSW, 1997. pp1–2.

5 Chester/McGregor, *Australia's Wild Islands*, pp20–21.

6 Chester/McGregor, *Australia's Wild Islands*, p149.

7 Chester/McGregor, *Australia's Wild Islands*, p2.

8 Chester/McGregor, *Australia's Wild Islands*, p7.

9 Chester/McGregor, *Australia's Wild Islands*, p15.

10 Chester/McGregor, *Australia's Wild Islands*, p15.

11 Chester/McGregor, *Australia's Wild Islands*, p41.

12 Chester/McGregor, *Australia's Wild Islands*, p72.

13 Chester/McGregor, *Australia's Wild Islands*, pp2–3.

14 Chester/McGregor, *Australia's Wild Islands*, pp102–104, 108.

15 Chester/McGregor, *Australia's Wild Islands*, pp124,126.

16 Chester/McGregor, *Australia's Wild Islands*, pp132–134.

17 Chester, Quentin and Alasdair McGregor: *The Kimberley: Horizons of Stone*. (Photographs by Rob Jung, Alasdair McGregor and Quentin Chester.) Hodder & Stoughton (Australia) Pty Ltd. Rydalmere, NSW, 1992. p198.

18 Chester/McGregor, *The Kimberley*, pp73, 90, 198.

19 Chester/McGregor, *The Kimberley*, p90.

20 Chester/McGregor, *The Kimberley*, p93.

21 Chester/McGregor, *The Kimberley*, p153.
22 Neilson, David: telephone interview with Pollak, 29 August 1999.
23 Neilson, David: *South West Tasmania (A land of the wild)*. Rigby Ltd. Adelaide, 1975. p15.
24 Neilson, *South West Tasmania*, pp34–35.
25 Neilson, *South West Tasmania*, p156.
26 Neilson, *South West Tasmania*, p156.
27 Gee, Helen and Janet Fenton (editors): *The South-West Book (A Tasmanian wilderness)*. Collins/Australian Conservation Foundation. Sydney/ Melbourne, 1978 (3rd impression, 1983). p212.
28 Gee, Helen: telephone interview with Pollak, 11 July 1999.
29 Gee interview.
30 Fenton, Janet: telephone interview with Pollak, 22 July 1999.
31 Fenton interview.
32 Fenton interview.
33 Brown, Bob: *Wild Rivers (Franklin/ Denison/Gordon)*. Photographs by Peter Dombrovskis; drawings by Peter Jackson; Peter Dombrovskis Pty Ltd. Sandy Bay, Tas, 1983. p20.
34 Brown, *Wild Rivers*, pp21–22.
35 Brown, *Wild Rivers*, p29.
36 Brown, *Wild Rivers*, p31.
37 Cockburn, Milton: 'Senator farewells apple isle for ACT', *SMH*, 1 April 1989.
38 Bruer, Mark: 'Abuse and travel too much for green Tasmanian senator', *Age*, 1 April 1989.
39 Sanders, Norm: *A Time to Care (Tasmania's endangered wilderness)*. (Photographs: Chris Bell; foreword: Bob Brown.) Chris Bell. Blackmans Bay, Tas, 1980. p23.
40 Sanders, *A Time to Care*, pp20–22.
41 Sanders, *A Time to Care*, p20.
42 Bell, Chris: *Beyond the Reach (Cradle Mountain–Lake St Clair National Park)*. Laurel Press. Sandy Bay, Tas, 1990. p18.
43 Bell, *Beyond the Reach*, p6.
44 Bell, *Beyond the Reach*, p30.
45 Bell, Chris: telephone interview with Pollak, 18 August 1999.
46 Bell interview.
47 Bell interview.
48 Hawes, Martin: telephone interview with Pollak, 25 August 1999.
49 Hawes, Martin (text and photographs): *Above Me Only Sky (a portrait of the Tasmanian wilderness)*. Drinkwater Publishing. Hobart, 1981. pp54–56.
50 Hawes, *Above Me Only Sky*, p54.
51 Hawes, *Above Me Only Sky*, p56.
52 Hawes, *Above Me Only Sky*, pp80–84.
53 Hawes, *Above Me Only Sky*, p56.
54 Hawes interview.
55 Russell, Rupert: *Daintree (Where the rainforest meets the reef)*. Photography by Leo Meier and by Cliff and Dawn Frith. Published jointly by Kevin Weldon and Associates Pty Ltd and the Australian Conservation Foundation. McMahons Point, NSW, 1985. p13.
56 Russell et al, *Daintree*, p195.
57 Russell et al, *Daintree*, p184.
58 Frith, Dawn and Clifford: *Cape York Peninsula (A natural history)*. (Wilderness photography by Kerry Trapnell.) Reed Books. Chatswood, NSW, 1995. p222.
59 Friths, *Cape York Peninsula*, p225.
60 Friths, *Cape York Peninsula*, p228.
61 Cato, Nancy: *The Noosa Story (A study in unplanned development)*. Jacaranda Press. Milton, Qld, 1979. p56.
62 Cato, *The Noosa Story*, p86.
63 Cato, *The Noosa Story*, p5.
64 Cato, *The Noosa Story*, pp8–9.
65 Cato, *The Noosa Story*, p106.
66 Cato, *The Noosa Story*, p103.
67 Green, Glenis: 'Building war looms', *CML*, 24 November 1998, p6.
68 Andrews, Malcolm: 'At the heart of the novelist, a burning passion', *Weekend Australian*, 21–22 April 1990, p wkd 2.
69 Nankin, Harry: *Victoria's Alps (An Australian endangered heritage)*. Australian Conservation Foundation/William Collins Pty Ltd. Sydney/Melbourne., 1983. p56.
70 Nankin, *Victoria's Alps*, p10.
71 Nankin, *Victoria's Alps*, p7.
72 Nankin, *Victoria's Alps*, p40.
73 Nankin, Harry: telephone interview with Pollak, 21 August 1999.
74 Nankin interview.
75 Vandenbeld, John: *Nature of Australia (A portrait of the island continent)*. William Collins Pty Ltd and ABC Enterprises. Sydney, 1988. p35.
76 Vandenbeld, *Nature of Australia*, p65.
77 Vandenbeld, *Nature of Australia*, p243.
78 Vandenbeld, *Nature of Australia*, p260.
79 Vandenbeld, Margaret: telephone interview with Pollak, 17 June 1999.
80 Margaret Vandenbeld interview.
81 Vandenbeld, *Nature of Australia*, p266.
82 Kennedy, Michael (general editor):

Australia's Endangered Species (The extinction dilemma). Simon & Schuster Australia. Brookvale, NSW, 1990. p17.
83 Kennedy, *Australia's Endangered Species*, p17.
84 Kennedy, Michael: telephone interview with Pollak, 27 August 1999.
85 Kennedy, *Australia's Endangered Species* p18.
86 Kennedy, *Australia's Endangered Species* p18.
87 Jarver, Peter (author and photographer): *Kakadu Country*. Thunderhead Publishing. Darwin, 1988. p4.
88 Jarver, Peter: telephone interview with Pollak, 24 August 1999.
89 Jarver interview.
90 Jarver interview.
91 Milliken, Robert: interview with Pollak, 9 April 1999.
92 Milliken, Robert: *On the Edge (The changing world of Australia's farmers).* (Photographs by Lorrie Graham.) Simon & Schuster Australia. East Roseville, NSW, 1992. p133.
93 Milliken interview, 1999.
94 Kendell, Jeni and Eddie Buivids: *Earth First (The struggle to save Australia's rainforest).* ABC Enterprises. Sydney, 1987. p46.
95 Kendell/Buivids, *Earth First*, p54.
96 Frith, Dawn and Clifford: *A Walk in the Rainforest (An introduction to the environment).* (Illustrated by David Stacey.) Tropical Australia Graphics. Townsville, Qld, 1985. pp8, 10.
97 Friths, *A Walk in the Rainforest*, pp8, 10.
98 Rankin interview.
99 Andrews, Malcolm: 'At the heart of the novelist, a burning passion', *Weekend Australian*, 21–22 April 1990, p wkd 2.

Chapter 5 Magazines

1 *Grass Roots* (*GR*), No. 1, April–June 1973. p2. (Then edited by David and Meg Miller, Shepparton, Vic; in 1999, published by Grass Roots Publishing Pty Ltd, Seymour, Vic, edited by Megg Miller and Mary Horsfall.)
2 Power, Steven: 'Building with natural materials', *GR*, No. 1, p10.
3 Redwood, Jill: 'Weeds—food and medicine for free', *GR*, No. 37, June 1983, p18.
4 Meehan, Elaine: 'Feedback from far out' Letter, *GR*, No. 6, 1974.
5 Avent, Roz: 'Marjoram and oregano', *GR*, No. 5, 1974, p52; Miller, Meg: 'Chamomile—the herb of humility', No. 9, 1976, p45; Miller, 'Parsley—the herb of rejoicing', No. 10, 1977, p48.
6 Miller, Meg: 'Gumnut Gossip', *GR*, No. 26, August 1981, p98.
7 Miller, Meg: 'Gumnut Gossip', *GR*, No. 34, November 1982, p106.
8 Finkel, Maurice: 'Whole foods for better health', *GR*, No. 19, 1979, p44; Finkel, 'Avoiding the tragedy of modern medicine', No. 24, 1980, p60.
9 Miller, Megg: 'Gumnut Gossip', *GR*, No. 60, April 1987, p98.
10 Robinson, Jose: 'Helping nature', *GR*, No. 132, April/May 1999, p60.
11 Lynch, Bev: 'Practical agroforestry, seeing the benefits', *GR*, No. 132, April/May 1999, p61.
12 Hirsch, Martin: 'Self-sufficiency on seven acres', *GR*, No. 132, April/May 1999, p6.
13 *Earth Garden* (*EG*), No. 1, 1972, edited and published by Keith Vincent Smith and Irene Smith, Balmain, NSW. (In 1999, edited by Alan T. Gray, Balmain, NSW.) First pages.
14 *EG*, Nos 3, 6.
15 Editorial, *EG*, No. 13, November 1975, p5.
16 *EG*, No. 14, February 1976; No. 15, July 1976.
17 *EG*, No. 24, September/November 1978.
18 Massy, Charles: 'Death in the bush', *EG*, No. 32, November 1981–January 1982, pp32–34; p32.
19 Robinson, Jose: 'A thatched grass shelter for a contemplative spot', *EG*, No. 95, March–May 1996, pp57–59.
20 Greenland, Hall: 'The two of us: Keith and Irene Smith', *SMH Good Weekend*, 27 February 1999, p12.
21 Gray, Alan T: 'Woodchips, shares and senators', *GR*, No. 98, December 1996–February 1997, pp74–75.
22 Gray, Alan T: 'Lumbering the children of the forest', *EG*, No. 101, September–November 1997, pp48–50.
23 Gray, Alan T: 'Solar, not nuclear for Kakadu', *EG*, No. 104, June–August 1998, pp70–73; Gray, 'Diary to stop Jabiluka', No. 105, September–November 1998, pp70–73.
24 Gray, Alan T: 'The book the woodchippers wanted to ban', *EG*, No. 108, June–August 1999, pp70–72.

25 Stewart, David: Editorial, *Simply Living* (*SL*), Vol. 1 No. 4, 1977, pp3–4. (First issue Vol. 1 No. 1, 1975; founding editors David Stewart and John Stewart; Shree Media Productions Pty Ltd, Mona Vale, NSW; in the late 1990s, published by Otter Publications with Phillip Keir as publisher and Matthew W. McPherson as editor.)
26 Daemion, Jonathan (words and photographs): 'Pilgrimage to the breath of whales', *SL*, Vol. 1 No. 4, 1977, pp9–13.
27 Shiels, Rosa: Editorial, *SL*, No. 17, 1981, p5.
28 'Simply stated', editorial, *SL*, Vol. 2 No. 1, 1983, p15.
29 Jones, Richard: 'Is the earth dying?', *SL*, Vol. 2 No. 13, 1987, pp36–41; p41.
30 Brouwer, Stephen: 'The vanishing landscape', *SL*, Vol. 4 No. 4, pp34–41; pp35,37. Burnett, Jolyon and Simon Leake: 'Healing the earth—a step by step guide to saving our soil', pp42–47; p45.
31 McAllister, Peter: 'Fraser Island's last stand', *SL*, Vol. 5 No. 6, pp24–29.
32 Kennedy, Jean: 'Urban Sprawl—the death of our river system', *SL*, Vol. 7 No. 1, p25.
33 Baxter, Chris: Editorial, *Wild*. (originally *Australian Wild*.) First issue July/Aug/Sep (Winter) 1981 Vol. 1 No. 1. Wild Publications Pty Ltd, PO Box 415, Prahran Vic 3181. Editor and publisher: Chris Baxter. p3.
34 Schirmer, Geoff: 'Wild Interview—Geoff Mosley', *Wild*, Vol. 1 No. 1, July/Aug/Sep (Winter) 1981, pp37–45; Schirmer, 'Wild Interview—Milo Dunphy', Vol. 2 No. 1, Jan/Feb/Mar (Summer) 1982, pp39–41; Mitchell, Rob: 'Wild Interview—Bob Brown', Vol. 2 No. 4, Oct/Nov/Dec (Spring) 1982, pp56–61.
35 Heckinger, Bruce: 'The Franklin: threatened river', *Wild*, No. 4, April/May/June (Autumn) 1982, pp24–27; p26.
36 Baxter, Chris: 'The wilderness that was', Editorial, *Wild*, Vol. 2 No. 4, Oct/Nov/Dec (Spring) 1982, p3.
37 '*Wild* Information', *Wild*, Vol. 9 No. 2, Apr/May/June (Autumn) 1989, pp13–21.
38 'Green Pages', *Wild*, Vol. 14 No. 2, Autumn 1994, p23.
39 'Green Pages', *Wild*, Vol. 63, Summer 1997, pp21–25.
40 Baxter, Chris: 'Giving back' Editorial, *Wild*, Vol. 63, Summer 1997, p3; 'Green Pages', p21.
41 Bechervaise, John: 'The Overland Trail' (reprint from *Walkabout* magazine, 1952), *Wild*, Vol. 63, Summer 1997, pp44–46; Plowman, Cathie: Breakout story, pp47–49.
42 Wilde, John: 'Of islands and albatrosses', *Wild*, Vol. 63, Summer 1997, pp60–62.
43 Curtain, Stephen: 'Far-off places', *Wild*, Vol. 63, Summer 1997, pp66–71.
44 Chester, Quentin: 'Eyes in the wild', *Wild*, Vol. 63, Summer 1997, pp31–35.
45 Baxter, Chris: 'The Wild person', Editorial, *Wild*, Vol. 65, Winter 1997, p3.
46 Baxter, Chris: 'Jobs for the boys . . . and damn the rest of the world', Editorial, *Wild*, Vol. 67, Summer 1998, p3.
47 Baxter, Chris: 'Unnatural resources, the Victorian Government turns feral in national parks', Editorial, *Wild*, Vol. 65, Winter 1997, p3.
48 Baxter, Chris: 'Changin' times', Editorial, *Wild*, Vol. 70, 1998, p3.
49 Baxter, Chris: 'Ethiopia' photo essay, *Wild*, Vol. 71, Summer 1999.
50 'Wongungarra: can Victoria's last major unprotected wilderness be saved?', *Wild*, Vol. 71, Summer 1999, pp49–51; Patrick Moore and Geoff Law, 'To log or not to log', *Wild*, Vol. 71, Summer 1999; 'Jabiluka, an overview of the issues', Vol. 72, Autumn 1999, pp56–57.
51 Law, Geoff: 'Bob Brown', *Wild*, Vol. 72, Autumn 1999; Humann, Doug: 'Charles McCubbin', Vol. 71, Summer 1999, pp54–58.
52 Barnier, Cheryl: Editorial, *Geo* (First published by Paul Hamlyn Pty Ltd Magazine Division, Dee Why West, NSW; since 1993, published by GEO Productions, Chatswood, NSW.) Vol. 1 No. 1, 1979, p5.
53 Barnier, Cheryl: Editorial, *Geo*, Vol. 1 No. 2, 1979, p5.
54 Trost, Margaret: 'Inside Queensland's tropical rainforest' (photographs by John Trost and Douglass Baglin), *Geo*, Vol. 1 No. 2, 1979, pp133–138.
55 Roces, Alfredo: Editorial, *Geo*, Vol. 1 No. 4, 1979, pp6–7.

56 Benson, John (photography by Esther Beaton): 'Beautiful but vanishing flora', *Geo*, Vol. 13 No. 3, Sep–Nov 1991, pp52–61; p53.
57 Innocent, Christopher: 'The Boar War on Cape York Peninsula', *Geo*, Vol. 18 No. 1, Jan–Feb 1996, pp26–34.
58 Aw, Michael: 'Underwater tower power', *Geo*, Vol. 18 No. 1, Jan–Feb 1998, pp36–43.
59 Michaels, Calvin: 'Himalayan hemp', *Geo*, Vol. 18, No. 5, Sep–Oct 1996, pp28–38.
60 Stirling, Peter: 'Reversing decades of despoliation', Comment, *Geo*, Vol. 19, No. 5, Sep–Oct 1997, p17.
61 Jarrett, Ian and Emma Stirling: 'The return of the whale', *Geo*, Vol. 20 No. 4, August 1998, pp18–36.
62 Spanner, Geoff: 'Feeding off the dolphins', *Geo*, Vol. 18 No. 2, March–April 1996, pp54–60.
63 Smith, Dick: Editorial, *Australian Geographic* (*AG*) (Dick Smith's Journal of Discovery and Adventure). Vol. 1 No1, January–March 1986. Australian Geographic Pty Ltd, Terrey Hills, NSW 2084. p13.
64 Mann, Paul: 'Sewerage after the flush', *AG*, Vol. 1 No. 2, April–June 1986, pp94–108.
65 'Len Beadell', Tribute, *AG*, No. 8, Oct–Dec 1987, pp112–113.
66 Trainer, Ted: 'Is growth the solution—or the problem?', Opinion, *AG*, No. 9, Jan–Mar 1988, p27.
67 Spencer, Rodney: 'Why small is beautiful', *AG*, No. 22, April–June 1991, p31.
68 Campbell, Greg: 'Who should manage our heritage?', *AG*, No. 41, Jan–Mar 1996, p33.
69 Hopwood, Paul: 'Pet or perish?', *AG*, No. 44, Oct–Dec 1996, p33.
70 Smiedt, David: 'Battalions, boots and bulldust', *AG*, No. 42, p60.
71 Boyle, Brian: 'Mapping the universe', *AG*, No. 54, April–June 1999, p70.
72 Eastwood, Ken: 'The Grampians: welcome to wonderland', *AG*, No. 54, April–June 1999, p36.
73 *Bogong*. Journal of the Canberra and South-East Region Environment Centre. (Issue No. 1 published on April–May 1980, edited by Laurie Shane.) Canberra City, ACT.
74 (Anon): 'On the smoker as environmentalist', *Bogong*, Vol. 1 No. 2, May–June 1980, p5.
75 Allworth, David: 'The Brindabellas', *Bogong*, Vol. 1 No. 2, May–June 1980, pp6–7; Pfanner, David: 'Birding by bike in Canberra—a brief guide to the sport', pp8–9.
76 Fraser, Ian: 'Echoes from the thunder field', *Bogong*, Vol. 1 No. 3, July–August 1980, pp4–5.
77 Hopkins, Skyda: 'The blockade', *Bogong*, Vol. 4 No. 1, March–April 1983, pp10–11; Faunce, Tom: 'Australia and the "One World"—the High Court decision in the Franklin River case', Vol. 4 No. 3, Sep–Oct 1983, pp4–6.
78 'Green Shorts', *Bogong*, Vol. 12 No. 1, Autumn 1991, p23.
79 Falconner, Rodney: 'The silent invasion', *Bogong*, Vol. 12 No. 4, Summer 1991, pp4–6; Rawson, Anne: 'Going native', pp10–11; Falconner, 'Seriously invasive plants in the ACT', Vol. 13 No. 1, Autumn 1992, pp4–6.
80 Rowland, Moira: 'Nuclear power: The answer to greenhouse?', *Bogong*, Vol. 19 No. 1, Autumn 1998, pp13–18.
81 Hamilton, Clive: 'Australia's climate change "victory" a poison chalice', *Bogong*, Vol. 19 No. 1, Autumn 1998, pp3–6.
82 Burritt, Roger: 'A pretty penny', *Bogong*, Vol. 20 No. 1, Autumn 1999, pp17–19.
83 Goddard, Phyl: 'Populate and perish? Why Australia needs population policy', *Bogong*, Vol. 20 No. 1, Autumn 1999, pp3–6; Schreiner, Barbara: 'The art of travelling', pp15–16.
84 Osborne, Will and David Hunter: 'Frog declines in the Australian Alps', *Bogong*, Vol. 19 No. 3, Spring 1998.
85 Editorial, *Wildlife in Australia* (A magazine for the environment.), official quarterly publication of the Wildlife Preservation Society of Queensland (president: Judith Wright), Brisbane, Vol. 13 No. 4, December 1976, p109.
86 Editorial, *Wildlife in Australia*, Vol. 14 No. 1, March 1977, p3.
87 Editorial, *Wildlife in Australia*, Vol. 18 No. 1, March 1981, p1.
88 Comben, Pat: 'From the Director', Vol. 34 No. 1, Autumn 1997, p2; Hall, Leslie S, Kim Halpin and Patrina Birt: 'Fruitful discoveries', pp8–13.
89 Kendall, Irene: 'Wildlife Tales', *Wildlife Australia*, Vol. 35 No. 2, Winter 1998,

p10; Pitt, Sylvia: Untitled, Vol. 35 No. 1, Autumn 1998, pp33–35.
90 Hohenhaus, Ron: 'Building a safe haven for bats', *Wildlife Australia*, Vol. 35 No. 4, Summer 1998, pp15–18.
91 Caneris, Adrian and Jan Oliver: 'Backyard wildlife', *Wildlife Australia*, Vol. 36 No. 1, Autumn 1999, pp21–22.
92 Brown, Anthony: interview with Pollak, May 1997.
93 Pye, Geoff: 'Animal Clinic', *Wildlife Australia*, Vol. 34 No. 2, Winter 1997, p36.

Chapter 6 Films and Plays

1 Bennett, Bill: 'Politics is our fuel', *SMH*, 19 November 1997, p15.
2 Murray, Scott: interview with authors, 31 May 1999.
3 Murray interview.
4 Murray interview.
5 Walsh, Michael: telephone interview with Pollak, 9 September 1999.
6 Larkin, John: 'Nowra's hard look at the '80s', *Sunday Age* Agenda, 25 July 1993, p7.
7 Harris, Samela: 'Reluctant elitist writes with dedicated fervour', *Australian*, 20 May 1982, p8.
8 Sewell, Stephen: 'But what does it *mean*?' *24 Hours*, June 1994, pp68–71; 70.
9 Radic, Leonard: interview with authors in Melbourne, 26 May 1999.
10 Radic interview.
11 Radic interview.
12 McCallum, John: telephone interview with Pollak, 10 September 1999.
13 Brooksbank, Anne: *Avengers of the Reef.* Copyright 1973. Timon Productions. Script: Anne Brooksbank. Director: Chris McCullough. Producer: Noel Ferrier. Script at Sydney Office, National Film and Sound Archive. p205.
14 Leser, David: 'The two of us', *SMH Good Weekend*, 16 August 1997, p12.
15 Borg, Sonia: *Storm Boy*. Copyright 1976. South Australian Film Corporation. Director: Henri Safran. Producer: Matt Carroll. Script: Sonia Borg (from the novel by Colin Thiele). p20.
16 Borg, *Storm Boy*, pp57–58.
17 Borg, Sonia: interview with authors, 1 June 1999.
18 Borg interview.
19 McCausland, James: *Mad Max*. Copyright 1979. Kennedy Miller Entertainment Pty Ltd. Director: George Miller. Script: James McCausland, George Miller.
20 McCausland/Miller, *Mad Max*.
21 Martin, Adrian: 'Review of *Mad Max 2*', in Scott Murray (compiler and editor), *Australian Film 1978–1992 (A survey of theatrical features)*. Oxford University Press Australia. Melbourne, 1993. p78.
22 Hayes, Terry: *Mad Max 2 (The Road Warrior)*. Copyright 1981. Kennedy Miller Entertainment Pty Ltd. Director: George Miller. Producer: Byron Kennedy. Script: Terry Hayes, George Miller, Brian Hannant. p1.
23 Hayes/Miller/Hannant, *Mad Max 2*, p1c.
24 Hayes/Miller/Hannant, *Mad Max 2*, p15.
25 Hayes/Miller/Hannant, *Mad Max 2*, p19.
26 Hayes/Miller/Hannant, *Mad Max 2*, p31.
27 Martin, Adrian: 'Review of *Mad Max 2*', in Scott Murray (compiler and editor), *Australian Film 1978–1992 (A survey of theatrical features)*. Oxford University Press Australia. Melbourne, 1993. p78.
28 Hannant, Brian: interview with Pollak, 30 April 1999.
29 Hannant interview.
30 Hannant interview.
31 Hannant interview.
32 Hannant interview.
33 Hannant interview.
34 Hayes, Terry: *Mad Max III (Beyond Thunderdome)*. Copyright 1985. Kennedy Miller. Directors: George Miller and George Ogilvie. Producer: George Miller. Script: Terry Hayes and George Miller.
35 Sardi Jan: *Ground Zero*. Copyright 1987. Avenue Pictures/BDB Productions. Directors: Michael Pattinson, Bruce Myles. Producer: Michael Pattinson. Script: Jan Sardi, Mac Gudgeon. (Shooting script, developed with the assistance of Film Victoria, July 1986, by Mac Gudgeon, Michael Pattinson and Jan Sardi.) p36.
36 Sardi/Gudgeon, *Ground Zero*, p84.
37 Sardi/Gudgeon, *Ground Zero*, p84.
38 Sardi, Jan: telephone interview with Pollak, 9 September 1999.
39 Sardi interview.
40 Sardi interview.

41 Sardi interview.
42 Gudgeon, Mac: telephone interview with Pollak, 10 September 1999.
43 Bennett, Bill: *A Street to Die*. Copyright 1985. Mermaid Beach Productions and Multifilms. Director: Bill Bennett. Producer: Bill Bennett. Script: Bill Bennett. Final script copyright 1984. p34.
44 Bennett, *A Street to Die*, p36.
45 Bennett, *A Street to Die*, pp59–60.
46 Bennett, *A Street to Die*, pp66–67.
47 Bennett, Bill: 'Politics is our fuel', *SMH*, 19 November 1997, p15.
48 Loane, Sally: 'Riding the high road', *SMH*, 22 December 1993, p24.
49 Craig, Michael: *The Killing of Angel Street*. Forest Home Films Pty Ltd. Screenplay by Michael Craig and Evan Jones; additional material by Cecil Holmes; based on an original story by Michael Craig. In association with the Australian Film Commission. Director: Don Crombie. Producer: Anthony Buckley. Copyright 1980. p2.
50 Craig et al, *The Killing of Angel Street*, p8.
51 Craig et al, *The Killing of Angel Street*, pp12, 13.
52 Craig et al, *The Killing of Angel Street*, pp23–24.
53 Rosenberg, Marc: *Heatwave*. Copyright 1981. Preston Crothers in association with M & L Pty Ltd. Director: Phillip Noyce. Producer: Hilary Linstead. Script: Marc Rosenberg, Phillip Noyce (based on an original screenplay by Mark Stiles, Tim Gooding). p17.
54 Rosenberg et al, *Heatwave*, p20.
55 Hogan, Paul: *Crocodile Dundee*. 4th draft, 29th March, 1985. Script: Paul Hogan, Ken Shadie. p21. *Crocodile Dundee*. Copyright 1986. CBS Fox. Written by Paul Hogan, John Cornell and Ken Shadie. Producer: John Cornell. Director Peter Faiman.
56 Hogan/Shadie, *Crocodile Dundee*, p22.
57 Hogan/Shadie, *Crocodile Dundee*, p22.
58 Shadie, Ken: telephone interview with Pollak, 11 September 1999.
59 Shadie interview.
60 Hogan/Shadie, *Crocodile Dundee*, p32.
61 Duigan, John: *One Night Stand*. Film copyright 1984. Michael Edgley International and Hoyts present an Astra Film Production. Director: John Duigan. Script: John Duigan. Producer: Richard Mason. Operative draft copyright 1983, as provided by the Sydney Office of the National Film and Sound Archive. p70.
62 Duigan, *One Night Stand*, p34.
63 Duigan, John: telephone interview with Pollak, 31 August 1999.
64 Duigan interview.
65 Dermody, Susan Murphy: *Breathing Under Water*. Copyright 1991. Periscope Productions. Producer: Megan McMurchy. Director and screenplay: Susan Murphy Dermody. In association with Australian Film Commission and Channel Four Television. Post-production script. p17.
66 Murphy Dermody, *Breathing Under Water*, p24.
67 Murphy Dermody, *Breathing Under Water*, p43.
68 Ansara, Martha: *The Pursuit of Happiness*. Film copyright 1987. With assistance from the Australian Film Commission. Director: Martha Ansara. Producer: Martha Ansara, Jequerity Pty Ltd. Script: Martha Ansara, Alex Glasgow, Laura Black and cast. Post-production script, as provided by Ansara. pp3–5.
69 Ansara, *The Pursuit of Happiness*, p44.
70 Ansara, *The Pursuit of Happiness*, pp3–5.
71 Lewis, Mark: *Cane Toads (An unnatural history)*. Director and scriptwriter: Mark Lewis. Executive producer: Tristram Miall. Film Australia. 1987.
72 Lewis, Mark; email interview with Pollak, 12 September 1999.
73 Gibson, Ross: *Wild*. Copyright 1991. Huzzah Productions Pty Ltd (with assistance of Australian Film Commission). Director: Ross Gibson. Producer: John Cruthers. Script: Ross Gibson, in collaboration with John Cruthers and James Manche (shooting script, eighth draft, August 11, 1991. p6.
74 Gibson/Cruthers/Manche, *Wild*, p6.
75 Gibson/Cruthers/Manche, *Wild*, p6.
76 Gibson/Cruthers/Manche, *Wild*, p7.
77 Gibson/Cruthers/Manche, *Wild*, p21.
78 Gibson/Cruthers/Manche, *Wild*, p10.
79 Gibson/Cruthers/Manche, *Wild*, pp2, 37.
80 Kendell, Jeni: *Give Trees a Chance*. Copyright 1980. Terania Creek Protesters. Melbourne, Wilderness Society Education. National Library of Australia Film Collection. Producer and director: Jeni Kendell.
81 Kendell, Jeni: *Earth First*. Copyright 1987. Director: Jeni Kendell.

Producers: Jeni Kendell and John Seed. Script: Jeni Kendell, John Seed, Jack Thompson. Gaia Films. Nimbin, NSW. (Produced with the assistance of the Australian Conservation Foundation, Fund for Animals, Rainforest Information Centre and the Australian Film Commission.)
82 Kendell et al, *Earth First.*
83 Fiske, Pat: *Rocking the Foundations (A History of the NSW Builders Labourers Federation 1940–1974).* Produced, directed, narrated and script by Pat Fiske. Bower Bird Films. Sydney 1985.
84 Fiske, *Rocking the Foundations.*
85 Smith, Margaret: 'A topical twist to a new look at the BLF', *The Australian Financial Review*, 20 November 1985.
86 Radic, Leonard: *The State of Play (The revolution in the Australian theatre since the 1960s).* Penguin Books. Ringwood, Vic, 1991. p203.
87 Romeril, John: *The Radioactive Horror Show.* 1977. Unpublished; various versions of the play in Romeril Papers held at the Australian Defence Force Academy Library in Canberra; BRN265908 G54, mostly in Box 2, Wallet 5, Part 4, Cons 1. (First major production at Pram Factory, Carlton, Melbourne, on 7 July 1977; directed by Carol Porter.)
88 Romeril, *The Radioactive Horror Show.*
89 Romeril, *The Radioactive Horror Show.*
90 Romeril, John: *Bring Down the House.* 1992. Completed script in Romeril Papers at ADFA, Box 1, Wallet 2. (Production at Arena Theatre Company, 1992.) pp17, 26–27, 32.
91 Romeril, John: *Lost Weekend.* 1988. Romeril Papers (Box 3, Wallet 6).
92 'Interview with Katharine Brisbane', in Jennifer Palmer (editor): *Contemporary Australian Playwrights.* Adelaide University Union Press. Adelaide, 1979. p26.
93 'Brisbane', Palmer, *Contemporary Australian Playwrights*, pp30–31.
94 Davidson, Jim: *Sideways from the Page (The* Meanjin *Interviews).* Fontana/Collins. Melbourne, 1983. pp122, 123–124. (Original interview in *Meanjin*, 11 August 1978.)
95 Lyssa, Alison: *The Boiling Frog.* Currency Press (in association with the Nimrod Theatre Company, Sydney). Paddington, NSW, 1984. (First performed at the Nimrod Theatre, Sydney, on 11 July 1984; directed by John Bell.) p20.
96 Lyssa, *The Boiling Frog*, p33.
97 Curb, Rosemary: 'Alison Lyssa Speaks', *Belles-Lettres* (US), Vol 2 No 1, September 1986, p6.
98 Kippax, H.G.: 'Nimrod gives us the most off-putting title of the year'. *SMH*, 13 July 1984.
99 Moses, John: 'Too much croaking, not enough passion'. *Australian.* 13 July 1984.
100 Nowra, Louis: *Sunrise.* Currency Press. Paddington, NSW, 1983. (First performed by Lighthouse, the State Theatre Company of SA, on 19 November 1983; directed by Jim Sharman.) p16.
101 Nowra, *Sunrise*, pp39–40.
102 Nowra, *Sunrise*, p40.
103 Nowra, *Sunrise*, p40.
104 White, Patrick: *Big Toys.* Currency Press. Sydney, 1978. First performed at the Parade Theatre, Sydney, by the Old Tote Theatre Company on 27 July 1977, directed by Jim Sharman. (Preface by Katharine Brisbane, pxiii.)
105 White, *Big Toys*, p56.
106 'Patrick White' in Dorothy Green and David Headon (editors), *Imagining the Real: Australian Writing in the Nuclear Age.* ABC Enterprises. Sydney, 1987. pp15–18; pp17–18.
107 White, Patrick: 'Bitter old man prays for the failed, the deranged, the poor', *SMH*, 22 July 1988.
108 Marr, David: *Patrick White: A Life.* Random House Australia. Milsons Point, NSW, 1991. p294.
109 Marr, *Patrick White*, p508.
110 White, Patrick: *Patrick White Speaks.* (Edited by Paul Brennan and Christine Flynn.) Primavera Press. Leichhardt, NSW, 1989. p27.
111 Marr, *Patrick White*, p602.
112 Hewett, Dorothy: *The Man from Mukinupin.* The Currency Press. Paddington, NSW, 1985 (first published 1979). (First performed by the National Theatre Company at the Playhouse, Perth, on 31 August 1979; directed by Stephen Barry; music by Jim Cotter.) p7.
113 Hewett, *The Man from Mukinupin*, p62.
114 Hewett, *The Man from Mukinupin*, pp111–112.
115 Hewett, Dorothy: *Golden Valley/Song of the Seals.* Currency Press. Paddington, NSW, 1985. p25.

116 Hewett, *Golden Valley/Song of the Seals*, p50.
117 Hewett, Dorothy: *The Fields of Heaven*. (Unpublished—manuscript from Hewett.) Copyright 1982. (As performed by Sydney Theatre Company.) p12.
118 Kavanagh, Paul: 'An interview with Dorothy Hewett', *Southerly*, Vol 44 No 2, June 1984, pp123–142; p125.
119 Hewett, *The Fields of Heaven*, p16.
120 Hewett, *The Fields of Heaven*, p16.
121 Hewett, *The Fields of Heaven*, p42.
122 Webby, Elizabeth: 'Dorothy Hewett: a profile', *Dialogue* (publication of the National Playwrights' Centre), Issue 16, July 1993, p8.
123 Hewett, Dorothy: *Pandora's Cross*. First performed at the Paris Theatre, Sydney, on 29 June 1978, directed by Jim Sharman. Act 1 published in *Theatre Australia*, Vol 3 No 2, September 1978, pp34–40; Act II in *Theatre Australia*, Vol 3 No 3, October 1978, pp33–39.
124 Hewett, *Pandora's Cross*.
125 Hewett, *Pandora's Cross*.
126 Webby, 'Dorothy Hewett: a profile', *Dialogue*, Issue 16, July 1993, p8.
127 Sewell, Stephen: *The Blind Giant is Dancing*. Currency Press. Sydney, 1995 (revised edition; originally published in 1983). (First performed by Lighthouse, the State Theatre Company of SA, at the Playhouse, Adelaide Festival Centre, on 15 October 1983; directed by Neil Armfield.) p59.
128 Sewell, *The Blind Giant is Dancing*, p59.
129 Sewell, *The Blind Giant is Dancing*, p78.
130 Sewell, *The Blind Giant is Dancing*, p92.
131 Sewell, *The Blind Giant is Dancing*, 'Author's preface', pp xiii–xiv.
132 Hoad, Brian: 'Let this be a warning yet again', *Bulletin*, 29 August 1995, p87.
133 Sewell, Stephen: *Dreams in an Empty City*. Currency Press Pty Ltd. Paddington, NSW, 1986. (First performed by the State Theatre Company of SA at the Playhouse of the Adelaide Festival Centre on 1 March 1986; directed by Neil Armfield.) p23.
134 Sewell, *Dreams in an Empty City*, p31.
135 Sewell, *Dreams in an Empty City*, p80.
136 Morley, Sheridan: Untitled, *Punch*, 30 September 1988, p53.
137 Hiley, Jim: 'Officer material—two types of Australian amateur dramatics', *The Listener*, 22 September 1988, p33.
138 Eccles, Jeremy: 'Diet of hate', *Bulletin*, 6 December 1988, p149.
139 Hibberd, Jack: *A Stretch of the Imagination*. The Currency Press. Sydney, 1973. (First performed at the Pram Factory, Melbourne, for the Australian Performing Group, on 8 March 1972, with Peter Cummins as Monk O'Neill; directed by Jack Hibberd.) pp4,9.
140 Hibberd, *A Stretch of the Imagination*, p21.
141 Hibberd, *A Stretch of the Imagination*, p40.
142 Hibberd, *A Stretch of the Imagination*, pp40–41.
143 Hibberd, Jack: 'Hibberd', *Age* Opinion, 13 March 1996, p13.
144 Kippax, H.G.: 'Satire, harsh but enriching', *SMH*, 4 July 1977, p8.
145 Hewett, Dorothy, 'Shirts, Prams and Tomato Sauce: The All-Australian Theatre', in Peter Holloway (editor): *Contemporary Australian Drama*. Currency Press. Paddington, NSW, 1987 (wholly revised; originally published 1981). pp108–119; p112. (Originally published in *Meanjin*, No 3, 1976.); Fitzpatrick, Peter, 'After the Wave: Australian Drama Since 1975', in Holloway, *Contemporary Australian Drama*, pp161–180; p170. (Originally published in *New Theatre Quarterly*, Vol 2, 5 February 1986.)
146 Nugent, Ann: 'Gillies as Monk: recommended', *CT*, 9 May 1991, p2.
147 Barron, Paul: 'A challenging piece of drama', *CT*, 18 November 1974, p8.
148 McGillick, Paul: 'A chance not to be missed', *AFR Weekend Review*, 5 October 1990, p9.
149 De Groen, Alma: *Going Home/Vocations*. Currency Press. Paddington, NSW, 1983. (*Vocations* first performed by the Melbourne Theatre Company at its Russell Street Theatre on 24 August 1982, directed by Graeme Blundell; later revised and first presented in its published form at La Boite, Brisbane, on 23 February 1983, directed by Andrew Ross.) p133.
150 De Groen, *Vocations*, p134.

151 Perkins, Elizabeth: *The Plays of Alma De Groen*. Rodopi. Amsterdam/ Atlanta (Georgia), 1994. (Monograph No 6 in the series entitled 'Australian Playwrights', series editor Ortrun Zuber-Skerritt.) p2.
152 De Groen, Alma: email interview with Pollak, 27 April 1999.
153 Robinson, Peter: 'Poetry in a bleak future', *CT* Panorama, 18 July 1998, p15.
154 De Groen, Alma: *The Woman in the Window*. Final draft, 23 September 1998. Provided by De Groen. p45.
155 De Groen, *The Woman in the Window*, p50.
156 De Groen, *The Woman in the Window*, pp64–65.
157 De Groen, Alma and Ian Mackenzie: *Stories in the Dark*. 1996. Australian entry into the 1996 Prix Italia, Drama Section. First broadcast on ABC Radio National on 30 July 1995; final script February 1995, p9.
158 De Groen/Mackenzie, *Stories in the Dark*, p16.
159 De Groen/Mackenzie, *Stories in the Dark*, p17.
160 De Groen/Mackenzie, *Stories in the Dark*, p19.
161 Mackenzie, Ian: email interview with Pollak, 30 April 1999.
162 Fitzpatrick, Peter: *After 'The Doll'. (Australian drama since 1955)*. Edward Arnold (Australia) Pty Ltd. Melbourne, 1979. ('Studies in Australian Culture' series; general editor John Colmer.) p96.
163 Buzo, Alexander: *Batman's Beach-head*. (Based on *An Enemy of the People*, by Henrik Ibsen.) Unpublished (manuscript provided by Buzo). First performed in 1973, to celebrate the 20th anniversary of the Melbourne Theatre Company, at the Comedy Theatre, Melbourne. Directed by John Sumner. p15.
164 Buzo, *Batman's Beach-head*, p3.
165 Buzo, *Batman's Beach-head*, p23.
166 Buzo, Alex: *Tom*. Angus & Robertson Publishers. Sydney, 1975. (First performed by the Melbourne Theatre Company at the Russell Street Theatre on 21 November 1972; directed by Malcolm Robertson.) p13.
167 Buzo, *Tom*, p25.
168 Lines, William J.: *Taming the Great South Land (A history of the conquest of nature in Australia)*. Allen & Unwin Pty Ltd. North Sydney, NSW, 1991. p230.
169 Buzo, *Tom*, p13.
170 McCallum, John: *Buzo*. Methuen Australia Pty Ltd. North Ryde, NSW, 1987 (monograph, part of 'Methuen Australian Drama' series). p63.
171 Buzo, Alex: 'Armidale: No Palo Alto' in Geoffrey Dutton (editor): *Country Childhoods*. University of Queensland Press. St Lucia, Qld, 1992. pp40–49; p47.
172 Buzo, 'Armidale: No Palo Alto', Dutton, *Country Childhoods*. p42.
173 Williams, Daniel: 'Alex stays as the suburbs go', *SMH* Metro, 1 September 1989, p2.
174 Williamson, David: *Money and Friends*. Currency Press. Sydney, 1992. (First performed at Brisbane's Suncorp Theatre by the Royal Queensland Theatre Company on 28 November 1991, in celebration of the State company's 21st year; directed by Aubrey Mellor.) p4.
175 Williamson, *Money and Friends*, p4.
176 Williamson, *Money and Friends*, p14.
177 Kiernan, Brian: *David Williamson: A Writer's Career*. Currency Press Ltd. Paddington, NSW, 1996. p42.
178 Kiernan, *David Williamson*, p11.
179 Radic, *The State of Play*, p2.
180 Billington, Michael: 'The power and the tawdry', *The Guardian* (London), 10 September 1988, p35.
181 Kellaway, Kate: 'The art of losing', *The Observer* (London), 11 September 1988, p38.
182 Milliken, Robert: 'Nowra carves into the upper crust', *The National Times*, 18–24 November 1983, p34.
183 Williamson, David: 'Letter to Peter Carey', 10 June 1993; in Brenda Niall and John Thompson (editors), *The Oxford Book of Australian Letters*, Oxford University Press, Melbourne, 1998, p303.

Chapter 7 Songs and Poems

1 O'Connor, Mark: 'Greening the psyche of white Australia', *Weekend Australian*, 10–11 February 1990, p wkd 8.
2 Totaro, Paolo: 'Saving the land with poetry', *Bulletin*, 30 January/ 6 February 1990, pp188–192; p192.
3 Deere, Christopher: 'A poet who turns Australians on to their environmental heritage', *CT*, 4 January 1997, p C8.

4 Beck, Chris: 'On the couch: an interview with Vince Jones', *Age* Saturday Extra, 2 November 1996, p4.
5 Jones, Vince: 'On the Brink of It', *On the Brink of It.* A Suitable Recording, distributed by EMI Music (Aust) Ltd, Lyrics copyright Chappell and Intersong Music Group (Australia) Ltd, 1985.
6 Jones, Vince: 'Sensual Item', *Tell Me a Secret.* Suitable Management, Lyrics copyright Boosey and Hawkes Australia Pty Ltd, Matthews Music, Chappell and Intersong Music Group, 1986.
7 Jones, 'Too Much Too Soon', *Tell Me a Secret.*
8 Birnbauer, Bill: 'The misty mountains and all that jazz' *Age*, 24 May 1994, p15.
9 Cosic, Miriam, 'Vince's vision', *SMH*, 30 August 1995, p24.
10 Birnbauer, Bill: 'The misty mountains and all that jazz', *Age*, 24 May 1994, p15.
11 Buchanan, Dominic: 'Twilight Jones', *SMH* Metro, 27 August 1993, p20.
12 Daly, Mike, 'Mellowing with life on the farm', *Age*, 5 November 1993, p4.
13 Shand, John: 'Keeping up with Mr Jones', *SMH* Metro, 19 June 1998, p23.
14 McQueen, Humphrey: Introduction to *The Redgum Song Book.* Tombola Publishing. Bondi Junction, NSW, 1981.
15 Schumann, John: telephone conversation with MacNabb, January 1999.
16 Schumann conversation.
17 McDonald, Hugh: 'Spirit of the Land', lyrics reprinted in *The Very Best of Redgum.* Rondor Music, Woolloomooloo, NSW; Chappell & Intersong Music, 1984.
18 Hastwell, Geoff: John Schumann talk to SCALA (The Songwriters, Composers and Lyricists Association Inc) annual general meeting, 1990. SCALA webpage, 1998.
19 SCALA webpage.
20 Schumann, John, preamble, to 'I was only 19', *Caught in the Act*, Epic (CBS Records Australia Ltd). Sydney, 1983.
21 Schumann, 'The Last Frontier', *Caught in the Act.*
22 Redgum, 'Nuclear Cop', *Caught in the Act.*
23 Schumann, 'Where Ya Gonna Run To', *Caught in the Act.*
24 Schwartz, Larry: 'The voice in the wilderness', *Sunday Age* Agenda, 15 December 1996, p1.
25 Hannan, Michael: telephone interview with MacNabb, 5 August 1999.
26 Holmes, Peter: 'The midnight hour', *Sun-Herald* Time Out, 2 November 1997, p4.
27 Holmes, 'The midnight hour', *Sun-Herald* Time Out, 2 November 1997, p4.
28 Verrender, Ian, 'Midnight Oil', *SMH* Spectrum, 14 September 1996, p5.
29 Schwartz, Larry: 'The voice in the wilderness', *Sunday Age* Agenda, 15 December 1996, p1.
30 David, Andrew: Untitled, *Rolling Stone*, January 1999, p63.
31 David, Untitled, *Rolling Stone*, January 1999, p63.
32 Moginie, Jim: 'Outside World', *10,9,8,7,6,5,4,3,2,1*, Columbia. Sydney, 1982.
33 Hirst/Moginie, 'Only the Strong', *10,9,8,7,6,5,4,3,2,1.*
34 Hirst/Moginie/Garrett, 'Read About It', *10,9,8,7,6,5,4,3,2,1.*
35 Marx, Jack: 'Kings of the mountain' *SMH* Metro, 12 March 1993, p1.
36 Thomas, Brett: 'The Oils swing to the past', *Sun-Herald*, 4 April 1993, p107.
37 Moginie, Jim: 'Earth and Sun and Moon', *Earth and Sun and Moon.* Columbia. Sydney, 1993.
38 Garrett/Moginie, 'Bushfire', *Earth and Sun and Moon.*
39 Hirst/Moginie, 'Truganini', *Earth and Sun and Moon.*
40 Creswell, Toby: 'Green dreams', *Rolling Stone*, December 1989, p17.
41 Creswell, 'Green dreams', *Rolling Stone*, December 1989, p17.
42 Creswell, Toby: 'Bullet the blue sky', *Rolling Stone*, March 1990, p44.
43 Midnight Oil: 'Blue Sky Mine', *Blue Sky Mining.* CBS. Sydney, 1990.
44 Moginie/Garrett, 'The Stars of Warburton', *Blue Sky Mining.*
45 Schwartz, Larry: 'The voice in the wilderness', *Sunday Age* Agenda, 15 December 1996, p1.
46 Safe, Mike: 'Rock of middle ages', *Australian Magazine*, 27 June 1998, p32.
47 Leser, David: 'Faith, hype and charity', *HQ*, November 1990, pp70–77.
48 Leser, 'Faith, hype and charity', *HQ*, November 1990, pp70–77.
49 McUtchen, Andrew: 'It's still oil or nothing', *Herald Sun* Business, 6 August 1998, p40.

50 Scatena, Dino: 'Anger is an energy', *Daily Telegraph*, 16 July 1997, p49.
51 Schwartz, Larry: 'The voice in the wilderness', *Sunday Age* Agenda, 15 December 1996, p1.
52 Safe, Mike: 'Rock of middle ages', *Australian Magazine*, 27 June 1998, p32.
53 Midnight Oil: 'Concrete', *Redneck Wonderland*. Sony Music. Sydney, 1998.
54 O'Donnell, John: 'Midnight Oil', *Rolling Stone*, July 1991, p54.
55 Garrett, Peter: speaking at Gleebooks, Sydney, for the launch of *An Environmental History of Australia*, by Drew Hutton and Libby Connors, 4 March 1999.
56 Bogle, Eric: *The Official Bogleography*. Larrikin Entertainment. Newtown, NSW, 1993.
57 Childs, Robert: 'An interview with Eric Bogle', *SCALA News*, 1990, SCALA website.
58 Coupe, Stuart: 'Eric Bogle: folk hero', *Sun-Herald*, 28 December 1986.
59 Bogle, Eric: 'Hard, Hard Times', *When the Wind Blows*. Larrikin Records. Sydney, 1984.
60 Bogle, Eric: interview with MacNabb, 23 April 1999.
61 Bogle, 'Bushfire', *When the Wind Blows*.
62 Bogle, 'When the Wind Blows', *When the Wind Blows*.
63 Bogle interview.
64 MacMillan, Sally: 'Soul Mates', *Australian*, 23 June 1990.
65 Jarvis, Susan: 'Eric Bogle—no longer a voice in the wilderness', *Sun-Herald* Real Australia, 12 August 1990, p97.
66 Bogle interview.
67 Bogle, Eric: 'Wilderness', *Voices in the Wilderness*. Larrikin Records. Sydney, 1990.
68 Bogle interview.
69 Mordue, Mark: 'Kev Carmody', *Rolling Stone*, April 1993, p62.
70 Carmody, Kev: 'Living South of the Freeway', *Bloodlines*. Festival Records. Sydney, 1993.
71 Carmody, Kev: 'Comrade Jesus Christ', *Pillars of Society*. Larrikin Records. Sydney, 1988.
72 O'Donnell, John: 'The black and the blues', *Rolling Stone*, January 1989.
73 Adams, David: 'Carmody's message has sting of reality', *Age* Entertainment Guide, 20 May 1994, p11.
74 Casimir, Jon and Tony Hewett: 'Rock and rights', *SMH* Metro, 7 July 1989, p1.
75 Jarvis, Susan: 'Social issues are the pillars of Kev's music', *Sun-Herald* Real Australia, 13 May 1990, p95.
76 Carmody, Kev: 'Thou Shalt Not Steal', *Shaking the Pillars*. Larrikin Records. Sydney, 1988.
77 Barber, Lynden: 'Angry Carmody shakes the pillars', *SMH*, 21 December 1988, p16.
78 Casimir, Jon and Tony Hewett: 'Rock and rights', *SMH* Metro, 7 July 1989, p1.
79 Drummond, Pat: telephone interview with MacNabb, 12 April 1999.
80 Drummond interview.
81 Drummond, Pat: 'The Spirit of the Southern Shore', *Through the Cracks—Live at the Clarendon*. Shoestring Productions Pty Ltd. 1998. Lyrics on Pat Drummond website.
82 Drummond, 'Acceptable Losses', *Through the Cracks*.
83 Drummond, Geoff: 'Shoalhaven Man', *The Local Rag*. 1990. Lyrics on Pat Drummond website.
84 Caswell, Allan: telephone interview with MacNabb, 9 June 1999.
85 Caswell interview.
86 Caswell, Allan: 'Moruroa'. Roxy Records. 1984.
87 Caswell interview.
88 Caswell interview.
89 Caswell, Allan and Don Spencer: 'Trees', *Cilla's World*. Virgin. London, 1990.
90 Caswell interview.
91 Liddiard, Brenda: telephone interview with MacNabb, 2 February 1999.
92 Liddiard, Brenda: 'Save the Daintree', *Songs of Protest and Survival*. Subliminal Studios. Auckland, 1991.
93 Jarvis, Susan: telephone interview with MacNabb, March 1999.
94 Hannan, Michael: telephone interview with MacNabb, 5 August 1999.
95 Parks, Andy: *Lock On! Songs to Save Australia's Rainforests*. Raw Power Productions. Dunoon, NSW, 1999.
96 Yeates, Lisa: telephone interview with MacNabb, 21 January 1999.
97 Yeates interview.
98 Moony, Saul: 'Lurk Loiter, Hide, Secrete'. 1983.
99 Gately, Mark: 'Mallee Boy conjures up taste of the Aussie bush', *SMH* Northern Herald, 27 November 1986, p14.
100 Elder, Bruce: 'John Williamson, a

man of contradictions', *Nature and Health* magazine, Vol II No 2, Winter 1990, pp53–56.
101 Jarvis, Susan: 'Hey, true blue', *Sun-Herald* Real Australia, 3 December 1995, p95.
102 Williamson, John: *True Blue, Stories and Songs about Australia.* Angus & Robertson. North Ryde, NSW, 1995.
103 Greason, David: 'John Williamson', *Australian Playboy*, August 1990, pp22–28.
104 Williamson, John: 'Rip Rip Woodchip', *Warragul.* EMI. Sydney, 1989.
105 Sampson, John: 'Axe greenie Grand Final song, say loggers', *SMH*, 21 September 1989, p1.
106 Elder, Bruce: 'John Williamson, a man of contradictions', *Nature and Health* magazine, Vol II No 2, Winter 1990, pp53–56.
107 Greason, David: 'John Williamson', *Australian Playboy*, August 1990, pp22–28.
108 Apelt, Ruth: '1969', *Told You Twice.* Grevillea Studios. Brisbane, 1994.
109 Apelt, 'Tourist-Led Recovery', *Told You Twice.*
110 Apelt, 'Mother Nature Blues', *Told You Twice.*
111 Apelt, Ruth: telephone interview with MacNabb, 12 April 1999.
112 Stewart, Paul: 'Memories of the Streets', *Sunday Herald Sun*, 3 March 1996, p119.
113 Swales, Penelope: 'Lost in London', *Justify Your Longings to the Doctor.* Black Market Music. Melbourne, 1998.
114 Swales, 'Bougainville', *Justify Your Longings to the Doctor.*
115 Swales, 'Bougainville', *Justify Your Longings to the Doctor.*
116 Swales, Penelope: telephone interview with MacNabb, 8 June 1999.
117 Swales interview.
118 Swales interview.
119 Davidson, Jim: *Sideways from the Page (The* Meanjin *interviews).* Fontana/Collins. Melbourne, 1983. p59. (Original interview in *Meanjin* on 8 July 1977.)
120 'Kath Walker' in Dorothy Green and David Headon (editors): *Imagining the Real: Australian Writing in the Nuclear Age.* ABC Enterprises. Sydney, 1987. pp53–54.
121 Wright, Judith: 'Interface III', *Fourth Quarter.* Angus & Robertson Publishers. Sydncy/Mclbourne, 1976. pp7–8.
122 Wright, Judith: 'Rockpool', *Phantom Dwelling.* Angus & Robertson Publishers. North Ryde, NSW, 1985. p41.
123 Wright, 'Patterns', *Phantom Dwelling*, p52.
124 Frith, Marion: 'Sounds of silence', *CT*, 19 March 1994, p C3.
125 Harrex, Syd: telephone interview with Pollak, 28 July 1999.
126 Caesar, Adrian: telephone interview with Pollak, 3 August 1999.
127 Moffitt, Ian: 'The poet of the republic', *Bulletin*, 15 January 1977, pp36–37; p36.
128 Gould, Alan: ''With the distinct timbre of an Australian voice'—the poetry of Les A. Murray', *Antipodes*, Vol 6 No 2, December 1992, pp121–129; 'Wit and passion with a paradox', *The Age* Saturday Extra, 3 December 1983, p15.
129 Tulip, James: 'Australian poetry's Uluru and Iturkawara', *Southerly*, Vol. 57 No 4, Summer 1997–98, pp191–198; p191.
130 Bourke, Lawrence: *A Vivid Steady State (Les Murray and Australian poetry).* NSW University Press and New Endeavour Press. Kensington/Strawberry Hills, NSW, 1992. p82.
131 Bourke, *A Vivid Steady State*, p90.
132 Kavanagh, Paul and Peter Kuch: *Conversations (interviews with Australian writers).* Angus & Robertson. North Ryde, NSW, 1991. p202.
133 Murray, Les A.: *The Boys Who Stole the Funeral.* Angus & Robertson. Sydney/Melbourne, 1980. p25.
134 Murray, *The Boys Who Stole the Funeral*, p29.
135 Murray, *The Boys Who Stole the Funeral*, p41.
136 Murray, Les A.: *The People's Otherworld.* Angus & Robertson. Sydney/Melbourne, 1983. p10.
137 Murray, *The People's Otherworld*, pp21, 26.
138 Murray, Les A.: *Dog Fox Field.* Collins/Angus & Robertson. North Ryde, NSW, 1990. pp9–10.
139 Walcott, Derek: 'Crocodile Dandy', *The New Republic*, 6 February 1989, pp25–28; p25.
140 Murray, Les: *Subhuman Redneck Poems.* Duffy & Smellgrove. Potts Point, NSW, 1996. pp41–42.
141 Murray, 'Inside Ayers Rock', *Subhuman Redneck Poems*, p33.
142 Tulip, James: 'Australian poetry's

Uluru and Iturkawara', *Southerly*, Vol 57 No 4, Summer 1997–98, pp191–198; p196.
143 Taylor, Andrew: *Reading Australian Poetry*. University of Queensland Press. St Lucia, Qld, 1987. pp144–145.
144 Goldsworthy, Peter: 'Our one big river, a force of nature', *The Adelaide Review*, October 1996, pp27–28; p28.
145 Goldsworthy, *The Adelaide Review*, p27.
146 Sorensen, Rosemary: 'Murray and poetic justice', *CML* Weekend, 21 March 1998, p7.
147 Logan, William: Untitled. *Poetry* magazine, June 1992, pp162–166; *New Criterion*, June 1997.
148 Kuch, Peter: *Bruce Dawe*. (Oxford Australian Writers Series). Oxford University Press. Melbourne, 1995. pp23–24, 26.
149 Dawe Archive, ADFA Library, Box 1, Folder 7, as reprinted in Haskell, Dennis: 'Bruce Dawe, the ordinary and extraordinary bloke', *Southerly*, Vol 57 No 3, Spring 1997, pp171–181; p176.
150 Dawe, Bruce: 'In a Time of Drought', *Sometimes Gladness* (Collected Poems 1954 to 1997—5th edition). Addison Wesley Longman Australia Pty Ltd. South Melbourne, Vic, 1997. p287.
151 Dawe, 'When First the Land Was Ours', *Sometimes Gladness*, pp260–261.
152 Dawe, 'When First the Land Was Ours', *Sometimes Gladness*, pp260–261.
153 Dawe, 'Search and Destroy', *Sometimes Gladness*, p149.
154 Dawe, 'You and Sarajevo', *Sometimes Gladness*, p274.
155 Kavanagh & Kuch, *Conversations*, pp44–45.
156 Goodwin, Ken: *Adjacent Worlds (a literary life of Bruce Dawe)*. Longman Cheshire. Melbourne, 1988. pp164, 133.
157 Kavanagh & Kuch, *Conversations*, pp34–35.
158 Hewett, Dorothy: *Collected Poems 1940–1995*. (Edited by William Grono.) Fremantle Arts Centre Press. South Fremantle, WA, 1995. *Alice in Wormland*, p250, pp307–396.
159 Hewett, *Peninsula*, pp307–396, in *Collected Poems*; 'The Witnesses', pp333–334.
160 Hewett, *Peninsula, Collected Poems*; 'Pylons', pp334–335; 'The Frogs', pp335–336.
161 Hewett, *Peninsula, Collected Poems*; 'South–West', p338.
162 Lawrence, Anthony: 'Mountview', *The Darkwood Aquarium*. Penguin Books Australia Ltd. Ringwood, Vic, 1993. pp6–8; p6.
163 Lawrence, 'The Deadline', *The Darkwood Aquarium*, p43.
164 Lawrence, 'The Fan', *The Darkwood Aquarium*, pp62–63.
165 O'Connor, Mark: 'The Rainbow Serpent—ix: Man', *The Eating Tree*. Angus & Robertson. Sydney, 1980. p52.
166 O'Connor, Mark: 'In the Gardiner Valley', *Fire-Stick Farming (selected poems 1972–90)*. Hale & Iremonger Pty Ltd. Sydney, 1990. p160.
167 Deere, Christopher: 'A poet who turns Australians on to their environmental heritage', *CT*, 4 January 1997, p C8.
168 Aubin, Tracey: ''Reef poet' beat the word barrier', *SMH*, 16 August 1989.
169 O'Connor, Mark: 'The poetry of the north: finding the words', *Westerly*, Vol 31 No 1, March 1986, pp18–27; pp21, 26.
170 Metcalfe, Jon: 'Poetry for our land', *The Cairns Post*, 2 July 1992, p9.
171 Brissenden, R.F. (Robert Francis): 'By the Shore', *The Whale in Darkness*. Australian National University Press. Canberra, 1980. p4.
172 Brissenden, 'Leaving Jakarta', *The Whale in Darkness*, p29.
173 Brooks, David: Introduction to R.F. Brissenden, *Suddenly Evening, (the selected poems of R.F. Brissenden)*. McPhee Gribble/Penguin. Ringwood, Vic, 1993. page x.
174 Gilbert, Kevin: 'The Black Drunkard', *People Are Legends*. University of Queensland Press. St Lucia, Qld, 1978. p19.
175 'Kevin Gilbert', in Dorothy Green and David Headon (editors): *Imagining the Real: Australian Writing in the Nuclear Age*. ABC Enterprises. Sydney, 1987. pp76–84; pp82–83.
176 Russell, W. Les: 'The Developers' in Kevin Gilbert (editor), *Inside Black Australia (an anthology of Aboriginal poetry)*. Penguin Books. Ringwood, Vic, 1988. p3.
177 Russell, 'The Nuclear Winter', *Inside Black Australia*, pp6–7.
178 Gilbert, 'The New True Anthem', *Inside Black Australia*, pp197–198.
179 Gilbert, 'Celebrators '88', *Inside Black Australia*, p198.

180 Hefner, Robert: 'Bob's spirit lives on in his poetry', *CT*, 14 April 1991, p23.

Chapter 8 Children

1 Aitken, Margaret: telephone interviews with Pollak, October 1997.
2 Lees, Stella and Pam Macintyre (editors): *The Oxford Companion to Australian Children's Literature*. Oxford University Press (in association with ALIA Press). Melbourne, 1993. p292.
3 McVitty, Walter: *Innocence and Experience (Essays on contemporary Australian children's writers)*. Thomas Nelson Australia. Melbourne, 1981. p3.
4 McVitty, *Innocence and Experience*, pp3–4.
5 McVitty, *Innocence and Experience*, p4.
6 Nieuwenhuizen, Agnes: *No Kidding (Top writers for young people talk about their work)*. Pan Macmillan Publishers Australia. Chippendale, NSW, 1991. pxix.
7 Niall, Brenda: 'Children's Literature', *Australian Literary Studies*, Vol. 13 No. 4, October 1988, pp547–559; p553.
8 Boetje, Marion: 'Born to read—the choice of books for young readers', *GR*, No. 79, June 1990.
9 Walker, Kath (writer and illustrator): *Father Sky and Mother Earth*. The Jacaranda Press. Gladesville, NSW, 1981. p26.
10 Thiele, Colin: *With Dew on My Boots (A childhood revisited)*. Walter McVitty Books. Montville, Qld, 1997. p128.
11 Thiele, *With Dew on My Boots*, p128.
12 Thiele, Colin: *Pinquo*. (Illustrated by Mary Milton.) Rigby Publishers. Adelaide, 1983. p15.
13 Thiele, *Pinquo*, p48.
14 Thiele, *Pinquo*, p50.
15 Thiele, *Pinquo*, p58.
16 Thiele, Colin: *Brahminy (The story of a boy and a sea eagle)*. Walter McVitty Books. Montville, Qld, 1995. pp25–26.
17 Thiele, *Brahminy*, pp45–46.
18 Thiele, *Brahminy*, p71.
19 Thiele, Colin: *High Valley*. Walter McVitty Books. Montville, Qld, 1996. p58.
20 Thiele, *High Valley*, pp58–59.
21 Thiele, *High Valley*, p59.
22 Thiele, *High Valley*, p71.
23 Thiele, *High Valley*, pp77–78.
24 Thiele, Colin: *Little Tom Little*. (Photography by David Simpson.) Rigby Publishers Ltd. Adelaide, 1981. p11.
25 Thiele, *Little Tom Little*, p20.
26 Thiele, Colin: *Fight Against Albatross Two*. (Cover illustration by Donna Gynell; text illustration by Deborah Crooks.) Rigby Publishers. Dee Why West, NSW, 1986 (1974). (First published as *Albatross Two*, 1974.) p9.
27 Thiele, *Fight Against Albatross Two*, p10.
28 Thiele, *Fight Against Albatross Two*, p17.
29 Thiele, *Fight Against Albatross Two*, p61.
30 Thiele, *Fight Against Albatross Two*, p124.
31 Wighton, Rosemary: 'Colin Thiele: from sun to shadow', *Reading Time*, No. 66, January 1978; p7.
32 Wighton, 'Colin Thiele', *Reading Time*, No. 66, January 1978; p7.
33 Wrightson, Patricia: *The Nargun and the Stars*. Hutchinson Australia. Hawthorn, Vic, 1973 (1988). p22.
34 Wrightson, *The Nargun and the Stars*, p27.
35 Hillel, Margot: 'Patricia Wrightson Talks', *Australian Book Review*, No. 167, December–January 1994–95, pp58–59.
36 Niall, Brenda: 'Children's Literature', *Australian Literary Studies*, Vol. 13 No. 4, October 1988, pp547–559; pp553–554.
37 Matthews, Stephen: 'A view from the pinnacle of children's literature', *CT*, 22 June 1996. p20.
38 McVitty, *Innocence and Experience*, p100.
39 Wrightson, Patricia: *Moon-Dark*. (Illustrated by Noela Young.) Margaret K. McElderry Books (Macmillan Publishing Company). New York, 1987 (originally published Century Hutchinson Ltd, London). pp89, 47–48.
40 Matthews, Stephen, 'A view from the pinnacle of children's literature', *CT*, 22 June 1996. p20.
41 Brinsmead, H.F. (Hesba): *Echo in the Wilderness*. Oxford University Press. London, 1972. p12.
42 Brinsmead, *Echo in the Wilderness*, p12.
43 Brinsmead, *Echo in the Wilderness*, p22.
44 Brinsmead, *Echo in the Wilderness*, p52.
45 Brinsmead, H.F. (Hesba): *The Sand Forest*. Angus & Robertson. North Ryde, NSW, 1985. p27.
46 Brinsmead, *The Sand Forest*, p9.
47 Brinsmead, *The Sand Forest*, p54.
48 Brinsmead, H.F. (Hesba): *Someplace Beautiful*. (Illustrated by Betina Ogden.)

Hodder & Stoughton (Australia) Pty Ltd. Lane Cove, NSW, 1986. p36.
49 Brinsmead, *Someplace Beautiful*, p37.
50 McVitty, *Innocence and Experience*, pp135, 141.
51 'Notes by Hesba Brinsmead', McVitty, *Innocence and Experience*, p160.
52 Park, Ruth: *My Sister Sif*. Viking Kestrel (Penguin Books Australia Ltd). Ringwood, Vic, 1986. p136.
53 Park, *My Sister Sif*, p28.
54 Park, *My Sister Sif*, p4.
55 Park, *My Sister Sif*, pp20–21.
56 Park, *My Sister Sif*, p33.
57 Park, *My Sister Sif*, p56.
58 Park, *My Sister Sif*, pp116–117, 91.
59 McVitty, Walter: 'Fascinating dip into the future', *Weekend Australian Magazine*, 20–21 December 1986, p13; Stodart, Eleanor: 'Adventures with the sea people', *CT*, 20 December 1986, pB2.
60 Mattingley, Christobel: *The Battle of the Galah Trees*. (Illustrated by Gareth Floyd.) Brockhampton Press. Leicester, UK, 1973. pp84–85.
61 Mattingley, *The Battle of the Galah Trees*, p88.
62 Mattingley, *The Battle of the Galah Trees*, p89.
63 Downes, Stephen: 'An ageless sense of childhood', *Age*, 11 October 1982.
64 Wheatley, Nadia: *The Blooding*. Viking Kestrel. Ringwood, Vic, 1987.
65 Aitken interview.
66 Nieuwenhuizen, *No Kidding*, p303; Gascoigne, Toss: 'Quality from an island retreat', *CT*, 24 August 1988. p30.
67 Gascoigne, 'Quality from an island retreat', *CT*, 24 August 1988. p30.
68 Gascoigne, 'Quality from an island retreat', *CT*, 24 August 1988. p30.
69 Barnett, Gillian: *The Inside Hedge Story*. (Illustrations by Geoff Todd.) Oxford University Press. Melbourne, 1981. p40.
70 Barnett, *The Inside Hedge Story*, p112.
71 Barnett, Gillian: telephone interview with Pollak, 5 September 1999.
72 Barnett interview.
73 Barnett interview.
74 Barnett interview.
75 Covernton, Helen and Jane: 'Writer of mystery and mischief', *Australian*, 15 August 1997, p14.
76 Macdonald, Caroline: *The Lake at the End of the World*. Viking Kestrel (Penguin Books Australia Ltd). Ringwood, Vic, 1988. pp95–96.
77 McInerney, Sally: 'Logical selection and one glaring omission', *SMH Spectrum*, 3 June 1989, p85.
78 Harding, Lee: *Waiting for the End of the World*. Hyland House Publishing Pty Ltd. Melbourne, 1983. pp27–28.
79 Harding, *Waiting for the End of the World*, p137.
80 Nieuwenhuizen, *No Kidding*, pp94–95.
81 Lees/Macintyre, *The Oxford Companion to Australian Children's Literature*, p383.
82 Morimoto, Junko: *My Hiroshima*. Viking (Penguin Books Australia). Ringwood, Vic, 1987.
83 Morimoto, *My Hiroshima*.
84 'Junko Morimoto', in Dunkle, Margaret: *The Story Makers*. Oxford University Press. Melbourne, 1987. p52.
85 Thiele, Colin: *The Sknuks*. (Illustrated by Mary Milton). Rigby Ltd. Adelaide, 1977.
86 Thiele, *The Sknuks*.
87 Thiele, *The Sknuks*.
88 Thiele, *The Sknuks*.
89 Oktober, Tricia (writer and illustrator): *Oil Spill*. A Hodder Children's Book. Rydalmere, NSW, 1996.
90 Oktober, *Oil Spill*.
91 Oktober, Tricia: telephone interview with Pollak, 3 September 1999.
92 Oktober interview.
93 Novak, Jiri Tibor: telephone interview with Pollak, 31 August 1999.
94 Novak, Jiri Tibor: *Birdman*. A Mark Macleod Book. (Random House Australia.) Milsons Point, NSW, 1994.
95 Greenwood, Ted (writer and illustrator): *Joseph and Lulu and the Prindiville House Pigeons*. Angus & Robertson. Sydney, 1972.
96 Jennings, Paul: 'Obituary for Edward (Ted) Greenwood', *Age*, 25 May 1999, p18.
97 Cato, Nancy: *Nin and the Scribblies*. (Photographs by Stanley Breeden.) The Jacaranda Press. Milton, Qld, 1976. p24.
98 Wagner, Jenny: *Goanna*. (Illustrated by Noela Hills.) Puffin Books (Penguin Books Australia Ltd). Ringwood, Vic, 1989.
99 Oliver, Narelle (writer and illustrator): *High Above the Sea*. Jam Roll Press Pty Ltd. Nundah, Qld, 1991.
100 Oliver, Narelle (writer and illustrator): *Leaf Tail*. McCulloch Publishing Pty Ltd. Carlton, Vic, 1989.

101 Oliver, Narelle (writer and illustrator): *The Best Beak in Boonaroo Bay*. A Lothian Book. Melbourne, 1993.
102 Dunkle, Margaret and Robert Ingpen: *Conservation (A thoughtful way of explaining conservation to children)*. Hill of Content. Melbourne, 1987.
103 Dunkle/Ingpen, *Conservation*.
104 Dunkle/Ingpen, *Conservation*.
105 Dunkle/Ingpen, *Conservation*.
106 Dunkle, Margaret: telephone interview with Pollak, 4 September 1999.
107 Aitken interview.
108 Baker, Jeannie: *Where the Forest Meets the Sea*. Walker Books Ltd (Julia MacRae Books). Sydney, 1989.
109 Clark, Lucy: 'Tour of beauty', *Daily Telegraph*, 31 January 1998, p99.
110 Clark, 'Tour of beauty', *Daily Telegraph*, 31 January 1998, p99.
111 Baker, Jeannie: *The Story of Rosy Dock*. Random House Australia Pty Ltd (A Mark Macleod Book). Milsons Point, NSW, 1995.
112 Sorensen, Meg and Sally McInerney: 'Panic station', *SMH Spectrum*, 16 December 1995, p13.
113 Baker, Jeannie: *Window*. Random Century Australia (Pty) Ltd (Julia MacRae Books). Milsons Point, NSW, 1991.
114 Baker, *Window*.
115 'Junko Morimoto', in Dunkle, Margaret: *The Story Makers*. Oxford University Press. Melbourne, 1987. p52.
116 Oktober interview.
117 McVitty, Walter: 'Fascinating dip into the future', *Weekend Australian Magazine*, 20–21 December 1986, p13.

BIBLIOGRAPHY
Chapter 2: Novels

Astley, Thea: *An Item from the Late News*. University of Queensland Press. St Lucia, Qld, 1982.

Astley, Thea: *It's Raining in Mango (Pictures for the family album)*. Viking (Penguin Books Australia Ltd). Ringwood, Vic, 1987.

Astley, Thea: *Vanishing Points*. William Heinemann Australia. Port Melbourne, Vic, 1992.

Bedford, Jean: *To Make a Killing*. Angus & Robertson. Pymble, NSW, 1992.

Brissenden, R.F. (Robert Francis): *Poor Boy*. Allen & Unwin Australia Pty Ltd. North Sydney, NSW, 1987.

Brissenden, R.F.: *Wildcat*. Allen & Unwin Australia Pty Ltd. North Sydney, NSW, 1991.

Carey, Peter: *The Fat Man in History*. University of Queensland Press. St Lucia, Qld, 1974.

Carey, Peter: *War Crimes*. University of Queensland Press. St Lucia, Qld, 1981 (first published 1979).

Carey, Peter: *Bliss*. University of Queensland Press. St Lucia, Qld, 1982 (first published 1981).

Carey, Peter: *The Tax Inspector*. University of Queensland Press. St Lucia, Qld, 1991.

Corris, Peter: *Wet Graves*. A Bantam Book. Moorebank, NSW, 1991.

Corris, Peter: *The Japanese Job*. Angus & Robertson. Pymble, NSW, 1992.

Corris, Peter: *The Washington Club*. A Bantam Book (Transworld Publishers). Moorebank, NSW, 1997.

Corris, Peter: *The Reward*. A Bantam Book. Moorebank, NSW, 1997.

D'Alpuget, Blanche: *Turtle Beach*. Viking (Penguin Books Australia Ltd). Ringwood, Vic, 1981.

D'Alpuget, Blanche: *White Eye*. Viking (Penguin Books Australia Ltd). Ringwood, Vic, 1993.

Dowling, Terry: *Rynosseros*. Aphelion Publications. North Adelaide, 1990.

Dowling, Terry: *Wormwood*. Aphelion Publications. North Adelaide, 1991.

Dowling, Terry: *Twilight Beach*. Aphelion Publications. North Adelaide, 1993.

Dowling, Terry and Van Ikin (editors): *Mortal Fire (best Australian sf)*. Coronet Books (Hodder & Stoughton Australia). Rydalmere, NSW, 1993.

Herbert, Xavier: *Poor Fellow My Country*. Fontana/Collins. Sydney, 1983. (First published William Collins, 1975; first published Fontana 1976.)

Houbein, Lolo: *Walk a Barefoot Road*. ABC Enterprises. Crows Nest, NSW, 1988.

Ikin, Van: see Terry Dowling.

Johnston, Dorothy: *Maralinga My Love*. McPhee Gribble/Penguin Books. Fitzroy, Vic, 1988.

Johnston, Dorothy: *One for the Master*. Wakefield Press. Kent Town, SA, 1997.

Lord, Gabrielle: *Jumbo*. The Bodley Head Ltd. London, 1986.

Lord, Gabrielle: *Salt*. McPhee Gribble. Ringwood, Vic, 1990.

McMullen, Sean: *Call to the Edge*. Aphelion Publications. North Adelaide, 1992.

McNamara, Peter and Margaret Winch (editors): *Alien Shores (An anthology of Australian science fiction)*. Aphelion Publications. North Adelaide, SA, 1994.

McQueen, James: *Hook's Mountain*. The Macmillan Company of Australia Ltd. Crows Nest, NSW, 1982.

Moffitt, Ian: *Gilt Edge*. Pan Australia. Sydney, 1991.

Nowra, Louis: *Red Nights*. Picador/Pan Macmillan. Sydney, 1997.

Scott, Rosie: *Nights with Grace*. Minerva Australia. Port Melbourne, Vic, 1991. (First published William Heinemann Australia. Melbourne, 1990.)

Scott, Rosie: *Feral City*. Minerva Australia. Port Melbourne, Vic, 1993. (First published William Heinemann Australia. Melbourne, 1992).

Simons, Margaret: *The Ruthless Garden*. Minerva Australia. Port Melbourne, Vic, 1993. (First published Angus & Robertson Bookworld, 1993.)

Turner, George: *The Sea and Summer*. Faber & Faber Ltd. London, 1987.

Turner, George: *Brain Child*. William Morrow & Company Inc. New York, 1991.

White, Patrick: *The Eye of the Storm*. Vintage (Random House). London, 1973.

White, Patrick: *A Fringe of Leaves*. Vintage (Random House). London, 1977. (First published by Jonathan Cape Ltd, London, 1976.)

Winch, Margaret: see Peter McNamara, *Alien Shores*.

Winton, Tim: *Shallows*. George Allen & Unwin. North Sydney, NSW, 1984.

Wongar, B.: *Walg (A novel of Australia)*. Macmillan. South Melbourne, Vic, 1986. (First published 1983.)

Wongar, B.: *Karan (A novel of the Australian hinterland)*. Dodd, Mead & Co Inc. New York, 1985.

Wongar, B.: *Gabo Djara*. Macmillan Company of Australia Pty Ltd. South Melbourne, Vic, 1988. (First published by Dodd Mead and Co, New York, 1987).

Wright, Steve: *A Drop in the Ocean*. Pan Macmillan Publishers Australia. Chippendale, NSW, 1991.

Chapter 3: Non-Fiction

Beder, Sharon: *Toxic Fish and Sewer Surfing (How deceit and collusion are destroying our great beaches)*. Allen & Unwin. North Sydney, NSW, 1989.

Beder, Sharon: *Global Spin (The corporate assault on environmentalism)*. Scribe Publications Pty Ltd. Carlton North, Vic, 1997.

Birch, Charles: *On Purpose*. University of NSW Press. Kensington, NSW, 1990 (1991).

Birch, Charles: *Regaining Compassion (for humanity and nature)*. University of NSW Press. Kensington, NSW, 1993.

Blakeway, Denys: see Sue Lloyd-Roberts, *Fields of Thunder*.

Bolton, Geoffrey: *Spoils and Spoilers (A history of Australians shaping their environment)*. 2nd edition. Allen & Unwin. North Sydney, NSW, 1992. (First edition 1981.)

Brinsmead-Hungerford, Hesba: *I Will Not Say the Day Is Done*. Alternative Publishing Co-operative Ltd. Chippendale, NSW, 1983.

Cawcutt, Linda and Catherine Watson: *Pesticides: The New Plague*. Friends of the Earth. Collingwood, Vic, 1984.

Collins, Paul: *God's Earth (Religion as if matter really mattered)*. A Dove publication (an imprint of Harper/Collins Religious). North Blackburn, Vic, 1995.

Flannery, Timothy Fridtjof: *The Future Eaters (An ecological history of the Australasian lands and people)*. Reed Books. Chatswood, NSW, 1994.

Lines, William J.: *Taming the Great South Land (A history of the conquest of nature in Australia)*. Allen & Unwin Pty Ltd. North Sydney, NSW, 1991.

Lines, William J.: *An All-consuming Passion (Origins, modernity, and the Australian life of Georgiana Molloy)*. Allen & Unwin. St Leonards, NSW, 1994.

Lines, William J.: *False Economy (Australia in the 20th century)*. Fremantle Arts Centre Press. South Fremantle, WA, 1998.

Lines, William J.: *A Long Walk in the Australian Bush*. University of NSW Press Ltd. Sydney, 1998.

Lloyd-Roberts, Sue and Denys Blakeway: *Fields of Thunder (Testing Britain's bomb)*. George Allen and Unwin Australia Pty Ltd. North Sydney, NSW, 1985.

Milliken, Robert: *No Conceivable Injury (The story of British and Australia's atomic cover-up)*. Penguin Books Australia Ltd. Ringwood, Vic, 1986.

Murray-Smith, Stephen: *Sitting on Penguins (Australia and the Antarctic)*. Mandarin Australia. Port Melbourne, Vic, 1991. (First published by Hutchinson Australia, 1988.)

Noonuccal, Oodgeroo (previously Kath Walker): *Stradbroke Dreamtime*. Angus & Robertson. Pymble, NSW, 1992. (First published 1972.)

Robotham, F.P.J: see Adrian Tame, *Maralinga*.

Rolls, Eric: *A Million Wild Acres (Two hundred years of man and an Australian forest)*. Thomas Nelson Australia. Sydney, 1981.

Rolls, Eric: *They All Ran Wild (The animals and plants that plague Australia)*.

Angus & Robertson. Sydney, 1984. (First published A&R 1968.)

Smith, David: *Continent in Crisis (A natural history of Australia)*. Penguin Books. Ringwood, Vic, 1990.

Smith, David: *Saving a Continent*. (Expanded and updated colour edition of *Continent in Crisis*.) University of NSW Press. Kensington, NSW, 1994.

Tame, Adrian and Robotham, F.P.J: *Maralinga (British A-bomb Australian legacy)*. Fontana/Collins. Melbourne, 1982.

Thompson, Peter: *Bob Brown of the Franklin River*. George Allen & Unwin Pty Ltd. North Sydney, NSW, 1984.

Trezise, Percy: *Last Days of a Wilderness*. William Collins (Aust) Ltd. Sydney, 1973.

Walker, Kath: see Oodgeroo Noonuccal.

Watson, Catherine: see Linda Cawcutt, *Pesticides: The New Plague*.

Watson, Ian: *Fighting Over the Forests*. Allen & Unwin. North Sydney, 1990.

Chapter 4: Illustrated Books

Bell, Chris: *Beyond the Reach (Cradle Mountain–Lake St Clair National Park)*. Laurel Press. Sandy Bay, Tas, 1990.

Brown, Bob: *Wild Rivers, (Franklin/Denison/Gordon)*. Photographs by Peter Dombrovskis; drawings by Peter Jackson. Peter Dombrovskis Pty Ltd. Sandy Bay, Tas, 1983.

Buivids, Eddie: see Jeni Kendell, *Earth First*.

Cato, Nancy: *The Noosa Story (A study in unplanned development)*. The Jacaranda Press. Milton, Qld, 1979.

Chester, Quentin and Alasdair McGregor: *The Kimberley (Horizons of stone)*. Photographs by Rob Jung, Alasdair McGregor and Quentin Chester. Hodder & Stoughton (Australia) Pty Ltd. Rydalmere, NSW, 1992.

Chester, Quentin and Alasdair McGregor: *Australia's Wild Islands*. Hodder & Stoughton. Rydalmere, NSW, 1997.

Coleman, Neville: see Mark O'Connor, *Poetry in Pictures*.

Fenton, Janet: see Helen Gee, *The South-West Book*.

Frith, Dawn and Clifford: *A Walk in the Rainforest (An introduction to the environment)*. Illustrated by David Stacey. Tropical Australia Graphics. Townsville, Qld, 1985.

Frith, Dawn and Clifford: *Cape York Peninsula—a natural history*. Wilderness photography by Kerry Trapnell. Reed Books. Chatswood, NSW, 1995.

Gee, Helen and Janet Fenton (editors): *The South-West Book (A Tasmanian wilderness)*. Collins/Australian Conservation Foundation. Sydney/Melbourne, 1978 (3rd impression, 1983).

Hawes, Martin (text and photographs): *Above Me Only Sky (A portrait of the Tasmanian wilderness)*. Drinkwater Publishing. Hobart, 1981.

Jarver, Peter (author and photographer): *Kakadu Country*. Thunderhead Publishing. Darwin, 1988.

Kendell, Jeni and Eddie Buivids: *Earth First (The struggle to save Australia's rainforest)*. ABC Enterprises. Sydney, 1987.

Kennedy, Michael (editor): *Australia's Endangered Species (The extinction dilemma)*. Simon & Schuster Australia. Brookvale, NSW, 1990.

McGregor, Alasdair: see Quentin Chester, *The Kimberley, Australia's Wild Islands*.

Milliken, Robert: *On the Edge (The changing world of Australia's farmers)*. Photographs by Lorrie Graham. Simon & Schuster Australia. East Roseville, NSW, 1992.

Nankin, Harry: *Victoria's Alps (An Australian endangered heritage)*. Australian Conservation Foundation/William Collins Pty Ltd. Sydney/Melbourne, 1983.

Neilson, David: *South West Tasmania (A land of the wild)*. Rigby Ltd. Adelaide, 1975.

O'Connor, Mark and Neville Coleman: *Poetry in Pictures (The Great Barrier Reef)*. Hale & Iremonger. Sydney, 1985.

Russell, Rupert: *Daintree (Where the rainforest meets the reef)*. Photography by Leo Meier and by Cliff and Dawn Frith. Published jointly by Kevin Weldon and Associates Pty Ltd and the Australian Conservation Foundation. McMahons Point, NSW, 1985.

Sanders, Norm: *A Time to Care (Tasmania's endangered wilderness)*. Photographs: Chris Bell; foreword: Bob Brown. Chris Bell. Blackmans Bay, Tas, 1980.

Vandenbeld, John: *Nature of Australia (A portrait of the island continent)*. William Collins Pty Ltd and ABC Enterprises. Sydney, 1988.

Chapter 5: Magazines

Australian Geographic (Dick Smith's journal of discovery and adventure). Vol 1 No 1, January-March 1986. Australian Geographic Pty Ltd, Terrey Hills, NSW.

Australian Wild: see *Wild*.

Bogong. (Journal of the Canberra and South-East Region Environment Centre). Issue No 1 published April–May 1980. Edited by Laurie Shane. Canberra. From Vol 19 No 3, Spring 1998, edited by Stephanie Haygarth.

Earth Garden. No 1, 1972. Edited and published by Keith Vincent Smith and Irene Smith. Balmain, NSW. In 1999, edited by Alan T. Gray, Balmain, NSW.

Geo (Australia's National Geographic). Vol 1 No 1, 1979. Paul Hamlyn Pty Ltd Magazine Division. Dee Why West, NSW. Since 1993, published by GEO Productions, Chatswood, NSW.

Grass Roots. Issue No 1, April–June 1973. Edited by David and Meg Miller. Shepparton, Vic. In 1999 published by Grass Roots Publishing Pty Ltd, Seymour, Vic, edited by Megg Miller and Mary Horsfall.

Simply Living. Issue Vol 1 No 1, 1975. Founding editors David Stewart and John Stewart. Shree Media Productions Pty Ltd, Mona Vale, NSW. In the late 1990s, published by Otter Publications with Phillip Keir as publisher and Matthew W. McPherson as editor.

Wildlife Australia. See *Wildlife in Australia.*

Wildlife in Australia. (A magazine for the environment.) Originally official quarterly publication of the Wildlife Preservation Society of Queensland (president: Judith Wright). In late 1990s published by the Queensland Museum, Brisbane.

Wild (originally *Australian Wild.*) Vol 1 No 1, July/Aug/Sep (Winter) 1981. Prahran, Vic. Wild Publications Pty Ltd. Editor and publisher: Chris Baxter. Became *Wild* in Vol 7 No 1, Summer 1987.

Chapter 6: Films

Ansara, Martha: *The Pursuit of Happiness*. Film copyright 1987. With assistance from the Australian Film Commission. Director: Martha Ansara. Producer: Martha Ansara, Jequerity Pty Ltd. Script: Martha Ansara, Alex Glasgow, Laura Black and cast. Post-production script, as provided by Ansara.

Bennett, Bill: *A Street to Die*. Copyright 1985. Mermaid Beach Productions and Multifilms. Director: Bill Bennett. Producer: Bill Bennett. Script: Bill Bennett. Final script copyright 1984.

Borg, Sonia: *Storm Boy*. Copyright 1976. South Australian Film Corporation. Director: Henri Safran. Producer: Matt Carroll. Script: Sonia Borg (from the novel by Colin Thiele).

Brooksbank, Anne: *Avengers of the Reef*. Copyright 1973. Timon Productions. Script: Anne Brooksbank. Director: Chris McCullough. Producer: Noel Ferrier. Script at Sydney Office, National Film and Sound Archive.

Craig, Michael: *The Killing of Angel Street*. Copyright 1980. Forest Home Films Pty Ltd. Script: Michael Craig and Evan Jones; additional material by Cecil Holmes; based on an original story by Michael Craig. In association with the Australian Film Commission. Director: Don Crombie. Producer: Anthony Buckley.

Dermody, Susan Murphy: *Breathing Under Water*. Copyright 1991. Periscope Productions. Producer: Megan McMurchy. Director and screenplay: Susan Murphy Dermody. In association with Australian Film Commission and Channel Four Television. Post-production script.

Duigan, John: *One Night Stand*. Film copyright 1984. Michael Edgley International and Hoyts present an Astra Film Production. Director: John Duigan. Script: John Duigan. Producer: Richard Mason. Operative draft copyright 1983, as provided by the Sydney Office of the National Film and Sound Archive.

Fiske, Pat: *Rocking the Foundations (A History of the NSW Builders Labourers Federation 1940–1974)*. Copyright 1985. Produced, directed, narrated and script by Pat Fiske. Bower Bird Films. Sydney.

Gibson, Ross: *Wild*. Copyright 1991. Huzzah Productions Pty Ltd (with assistance of Australian Film Commission). Director: Ross Gibson.

Producer: John Cruthers. Script: Ross Gibson, in collaboration with John Cruthers and James Manche. Shooting script, eighth draft, 11 August 1991.

Gudgeon, Mac: see Jan Sardi, *Ground Zero*.

Hannant, Brian: see Terry Hayes and George Miller, *Mad Max 2 (The Road Warrior)*.

Hayes, Terry: *Mad Max 2 (The Road Warrior)*. Copyright 1981. Kennedy Miller Entertainment Pty Ltd. Director: George Miller. Producer: Byron Kennedy. Script: Terry Hayes, George Miller, Brian Hannant.

Hayes, Terry: *Mad Max 3 (Beyond Thunderdome)*. Copyright 1985. Kennedy Miller. Directors: George Miller and George Ogilvie. Producer: George Miller. Script: Terry Hayes and George Miller.

Hogan, Paul: *Crocodile Dundee*. Copyright 1986. 4th draft, 29th March, 1985. Script: Paul Hogan, Ken Shadie (with additional material by John Cornell). CBS Fox. Producer: John Cornell. Director Peter Faiman.

Jones, Evan: see Michael Craig, *The Killing of Angel Street*.

Kendell, Jeni: *Give Trees a Chance*. Copyright 1980. Terania Creek Protesters. Melbourne, Wilderness Society Education. National Library of Australia Film Collection. Producer and director: Jeni Kendell.

Kendell, Jeni: *Earth First*. Copyright 1987. Director: Jeni Kendell. Producers: Jeni Kendell and John Seed. Script: Jeni Kendell, John Seed, Jack Thompson. Gaia Films. Nimbin, NSW. (Produced with the assistance of the Australian Conservation Foundation, Fund for Animals, Rainforest Information Centre and the Australian Film Commission.)

Lewis, Mark: *Cane Toads (An unnatural history)*. Copyright 1987. Film Australia. Director and script: Mark Lewis. Executive producer: Tristram Miall.

McCausland, James: *Mad Max*. Copyright 1979. Kennedy Miller Entertainment Pty Ltd. Director: George Miller. Script: James McCausland, George Miller.

Miller, George: see James McCausland, *Mad Max*; Terry Hayes and Brian Hannant, *Mad Max 2 (The Road Warrior)*; and Terry Hayes, *Mad Max 3 (Beyond Thunderdome)*.

Noyce, Phillip: see Marc Rosenberg, *Heatwave*.

Rosenberg, Marc: *Heatwave*. Copyright 1981. Preston Crothers in association with M & L Pty Ltd. Director: Phillip Noyce. Producer: Hilary Linstead. Script: Marc Rosenberg, Phillip Noyce (based on original screenplay by Mark Stiles, Tim Gooding).

Sardi, Jan: *Ground Zero*. Copyright 1987. Avenue Pictures/BDB Production. Directors: Michael Pattinson, Bruce Myles. Producer: Michael Pattinson. Script: Mac Gudgeon, Jan Sardi. Shooting script, developed with the assistance of Film Victoria, July 1986 by Mac Gudgeon, Michael Pattinson and Jan Sardi.

Chapter 6: Plays

Buzo, Alex: *Tom.* Angus & Robertson Publishers. Sydney, 1975. First performed by the Melbourne Theatre Company at the Russell Street Theatre on 21 November 1972; directed by Malcolm Robertson.

Buzo, Alex: *Batman's Beach-head.* (Based on *An Enemy of the People*, by Henrik Ibsen.) Unpublished (manuscript provided by Buzo). First performed in 1973, to celebrate the 20th anniversary of the Melbourne Theatre Company, at the Comedy Theatre, Melbourne; directed by John Sumner.

De Groen, Alma: *Going Home/Vocations.* Currency Press. Paddington, NSW, 1983. *Vocations* first performed by the Melbourne Theatre Company at its Russell Street Theatre on 24 August 1982, directed by Graeme Blundell; later revised and first presented in its published form at La Boite, Brisbane, on 23 February 1983, directed by Andrew Ross.

De Groen, Alma: *The Woman in the Window.* Final draft, 23 September 1998. Manuscript provided by De Groen.

De Groen, Alma and Ian Mackenzie: *Stories in the Dark.* 1996. Australian entry into the 1996 Prix Italia, Drama Section. First broadcast on ABC Classic FM, 2 July 1998. Manuscript provided by De Groen and Mackenzie.

Hewett, Dorothy: *The Man from Mukinupin.* The Currency Press. Paddington, NSW, 1985 (first published 1979). First performed by the National Theatre Company at the Playhouse, Perth, on 31 August 1979, directed by Stephen Barry. Music by Jim Cotter.

Hewett, Dorothy: *Golden Valley/Song of the Seals.* The Currency Press. Paddington, NSW, 1985. *Golden Valley* first performed by the Magpie Company at the Playhouse Festival Centre, Adelaide, on 8 May 1981, directed by Malcolm Moore; *Song of the Seals* first performed by the Magpie Company of the State Theatre Company of SA at the Playhouse, Festival Centre, Adelaide, on 6 May 1983, directed by Malcolm Moore. Music by Jim Cotter.

Hewett, Dorothy: *The Fields of Heaven.* 1982 Unpublished manuscript provided by Hewett, as performed by Sydney Theatre Company.

Hewett, Dorothy: *Pandora's Cross.* First performed at the Paris Theatre, Sydney, on 29 June 1978; directed by Jim Sharman. Act 1 published in *Theatre Australia*, Vol 3 No 2, September 1978, pp34–40; Act II in *Theatre Australia*, Vol 3 No 3, October 1978, pp33–39.

Hibberd, Jack: *A Stretch of the Imagination.* The Currency Press. Sydney, 1973. First performed at the Pram Factory, Melbourne, for the Australian Performing Group, on 8 March 1972; directed by Jack Hibberd.

Lyssa, Alison: *The Boiling Frog.* Currency Press (in association with the Nimrod Theatre Company, Sydney). Paddington, NSW, 1984. First performed at the Nimrod Theatre, Sydney, on 11 July 1984; directed by John Bell.

Mackenzie, Ian: see Alma de Groen, *Stories in the Dark.*

Nowra, Louis: *Sunrise*. Currency Press. Paddington, NSW, 1983. First performed by Lighthouse, the State Theatre Company of SA, on 19 November 1983; directed by Jim Sharman.

Romeril, John: *The Radioactive Horror Show*. 1977. Unpublished; various versions of the play in Romeril Papers held at the Australian Defence Force Academy library in Canberra; BRN265908 G54, mostly in Box 2, Wallet 5, Part 4, Cons 1. First major production at Pram Factory, Carlton, Melbourne, on 7 July 1977; directed by Carol Porter.

Romeril, John: *Lost Weekend*. 1988. Romeril Papers (Box 3, Wal 6).

Romeril, John: *Bring Down the House*. 1992. Completed script in Romeril Papers, Box 1, Wallet 2. (First production at Arena Theatre Company, 1992.)

Romeril, John: *Red Sun—Red Earth*. 1996. First performed in Nago City, Okinawa, Japan, in February 1996; then in March 1996 at the Bank of South Australia Amphitheatre, Elder Park. Script provided by Romeril.

Romeril, John: *XPO—the Human Factor*. Queensland Theatre Company. Performance version, as at 2 December 1998.

Sewell, Stephen: *The Blind Giant is Dancing*. Currency Press. Sydney, 1995 (revised edition; originally published in 1983). First performed by Lighthouse, the State Theatre Company of SA, at the Playhouse, Adelaide Festival Centre, on 15 October 1983; directed by Neil Armfield.

Sewell, Stephen: *Dreams in an Empty City*. Currency Press Pty Ltd. Paddington, NSW, 1986. First performed by the State Theatre Company of SA at the Playhouse of the Adelaide Festival Centre on 1 March 1986; directed by Neil Armfield.

White, Patrick: *Big Toys*. The Currency Press. Sydney, 1978. First performed at the Parade Theatre, Sydney, by the Old Tote Theatre Company on 27 July 1977; directed by Jim Sharman.

Williamson, David: *Money and Friends*. The Currency Press. Sydney, 1992. First performed at Brisbane's Suncorp Theatre by the Royal Queensland Theatre Company on 28 November 1991, in celebration of the State company's 21st year; directed by Aubrey Mellor.

Chapter 7: Songs

Apelt, Ruth: *Told You Twice*. Grevillea Studios. Brisbane, 1994.

Bogle, Eric: *Scraps of Paper*. Larrikin Records. Sydney, 1982.

Bogle, Eric: *When the Wind Blows*. Larrikin Records. Sydney, 1984.

Bogle, Eric: *Voices in the Wilderness*. Larrikin Records. Sydney, 1990.

Bogle, Eric: *The Official Bogleography*. Larrikin Entertainment, Newtown, NSW, 1993.

Carmody, Kev: *Pillars of Society*. Larrikin Records. Sydney, 1988.

Carmody, Kev: *Bloodlines*. Festival Records. Sydney, 1993.

Caswell, Allan: *Moruroa*. Roxy Records. 1984.

Caswell, Allan: *Cilla's World*. Virgin Records. London, 1990.

Drummond, Pat: *Through the Cracks—Live at the Clarendon*. Shoestring Productions. 1998. Lyrics on Pat Drummond website.

Drummond, Pat: *The Local Rag*. 1990. Lyrics on Pat Drummond website.

Jones, Vince: *On the Brink of It*. A Suitable Recording, distributed by EMI (Aust) Ltd, lyrics Chappell and Intersong Music Group (Australia) Ltd, 1985.

Jones, Vince: *Tell Me a Secret*. A Suitable Recording, distributed by EMI (Aust) Ltd, lyrics Chappell and Intersong Music Group (Australia) Ltd, Boosey and Hawkes Australia Pty Ltd, 1986.

Liddiard, Brenda: *Songs of Protest and Survival*. Subliminal Studios. Auckland, 1991.

Midnight Oil: *10,9,8,7,6,5,4,3,2,1*. Columbia. Sydney, 1982.

Midnight Oil: *Blue Sky Mining*. CBS. Sydney, 1990.

Midnight Oil: *Earth and Sun and Moon*. Columbia. Sydney, 1993.

Midnight Oil: *Redneck Wonderland*. Sony Music. Sydney, 1998.

Parks, Andy: *Lock On! Songs to Save Australia's Rainforests*. Raw Power Productions. Dunoon, NSW, 1999.

Redgum: *Caught in the Act*. Epic (CBS Records Australia Ltd). Sydney, 1983.

Redgum: *The Very Best of Redgum*. Rondor Music. Woolloomooloo, NSW; Chappell & Intersong Music, 1984.

Redgum: *The Redgum Song Book*. Tombola Publishing. Bondi Junction, NSW, 1981.

Swales, Penelope: *Justify Your Longings to the Doctor*. Black Market Music. Melbourne, 1998.

Williamson, John: *Warragul*. EMI. Sydney, 1989.

Williamson, John: *Mulga to Mangoes*. EMI. Sydney, 1994.

Williamson, John: *True Blue: Stories and Songs about Australia*. Angus & Robertson. North Ryde, NSW, 1995.

Chapter 7: Poems

Brissenden, R.F. (Robert Francis): *The Whale in Darkness*. Australian National University Press. Canberra, 1980.

Brissenden, R.F.: *Suddenly Evening, (The selected poems of R.F. Brissenden)*. (Edited and introduced by David Brooks.) McPhee Gribble/Penguin. Ringwood, Vic, 1993.

Dawe, Bruce: *Sometimes Gladness (Collected Poems 1954 to 1997—5th edition)*. Addison Wesley Longman Australia Pty Ltd. South Melbourne, Vic, 1997.

Gilbert, Kevin: *People are Legends*. University of Queensland Press. St Lucia, Qld, 1978.

Gilbert, Kevin (editor): *Inside Black Australia (An anthology of Aboriginal poetry)*. Penguin Books. Ringwood, Vic, 1988.

Hewett, Dorothy: *Collected Poems 1940–1995*. (Edited by William Grono.) Fremantle Arts Centre Press. South Fremantle, WA, 1995.

Lawrence, Anthony: *The Darkwood Aquarium*. Penguin Books Australia Ltd. Ringwood, Vic, 1993.

Murray, Les A.: *The Boys Who Stole the Funeral*. Angus & Robertson Publishers. Sydney/Melbourne, 1980.

Murray, Les A.: *The People's Otherworld*. Angus & Robertson Publishers. Sydney/Melbourne, 1983.

Murray, Les A.: *Dog Fox Field*. Collins/Angus & Robertson Publishers Australia. North Ryde, NSW, 1990.

Murray, Les: *Subhuman Redneck Poems*. Duffy & Smellgrove. Potts Point, NSW, 1996.

O'Connor, Mark: *The Eating Tree*. Angus & Robertson Publishers. Sydney, 1980.

O'Connor, Mark: *Fire-Stick Farming (Selected poems 1972–90)*. Hale & Iremonger Pty Ltd. Sydney, 1990.

Wright, Judith: *Fourth Quarter*. Angus & Robertson Publishers. Sydney/Melbourne, 1976.

Wright, Judith: *Phantom Dwelling*. Angus & Robertson Publishers. North Ryde, NSW, 1985.

Chapter 8: Children

Baker, Jeannie: *Window*. Random Century Australia (Pty) Ltd (Julia MacRae Books). Milsons Point, NSW, 1991.

Baker, Jeannie: *Where the Forest Meets the Sea*. Walker Books Ltd (Julia MacRae Books). Sydney, 1989.

Baker, Jeannie: *The Story of Rosy Dock*. Random House Australia Pty Ltd (a Mark Macleod Book). Milsons Point, NSW, 1995.

Barnett, Gillian: *The Inside Hedge Story*. (Illustrations by Geoff Todd.) Oxford University Press. Melbourne, 1981.

Brinsmead, H.F. (Hesba): *Echo in the Wilderness*. Oxford University Press. London, 1972.

Brinsmead, H.F. (Hesba): *The Sand Forest*. Angus & Robertson Publishers. North Ryde, NSW, 1985.

Brinsmead, H.F. (Hesba): *Someplace Beautiful*. (Illustrated by Betina Ogden.) Hodder & Stoughton (Australia) Pty Ltd. Lane Cove, NSW, 1986.

Cato, Nancy: *Nin and the Scribblies*. Photographs by Stanley Breeden. The Jacaranda Press. Milton, Qld, 1976.

Dunkle, Margaret: see Robert Ingpen, *Conservation*.

Greenwood, Ted (writer and illustrator): *Joseph and Lulu and the Prindiville House Pigeons*. Angus & Robertson. Sydney, 1972.

Harding, Lee: *Waiting for the End of the World*. Hyland House Publishing Pty Ltd. Melbourne, 1983.

Ingpen, Robert and Margaret Dunkle: *Conservation (A thoughtful way of explaining conservation to children)*. Hill of Content. Melbourne, 1987.

Macdonald, Caroline: *The Lake at the End of the World.* Viking Kestrel (Penguin Books Australia Ltd). Ringwood, Vic, 1988.

Mattingley, Christobel: *The Battle of the Galah Trees.* Illustrated by Gareth Floyd. Brockhampton Press. Leicester, UK, 1973.

Morimoto, Junko: *My Hiroshima.* Viking (Penguin Books Australia). Ringwood, Vic, 1987.

Noonuccal, Oodgeroo: see Kath Walker.

Novak, Jiri Tibor: *Birdman.* A Mark Macleod Book. Random House Australia. Milsons Point, NSW, 1994.

Oktober, Tricia (writer and illustrator): *Oil Spill.* A Hodder Children's Book. Rydalmere, NSW, 1996.

Oliver, Narelle (writer and illustrator): *The Best Beak in Boonaroo Bay.* A Lothian Book. Melbourne, 1993.

Oliver, Narelle (writer and illustrator): *High Above the Sea.* Jam Roll Press Pty Ltd. Nundah, Qld, 1991.

Oliver, Narelle (writer and illustrator): *Leaf Tail.* McCulloch Publishing Pty Ltd. Carlton, Vic, 1989.

Park, Ruth: *My Sister Sif.* Viking Kestrel (Penguin Books Australia Ltd). Ringwood, Vic, 1986.

Thiele, Colin: *Fight Against Albatross Two.* (Cover illustration by Donna Gynell; text illustration by Deborah Crooks.) Rigby Publishers. Dee Why West, NSW, 1986 (1974). (First published as *Albatross Two*, 1974).

Thiele, Colin: *The Sknuks.* Illustrated by Mary Milton. Rigby Ltd. Adelaide, 1977.

Thiele, Colin: *Little Tom Little.* (Photography by David Simpson.) Rigby Publishers Ltd. Adelaide, 1981.

Thiele, Colin: *Pinquo.* Illustrated by Mary Milton. Rigby Publishers. Adelaide, 1983.

Thiele, Colin: *Brahminy (The story of a boy and a sea eagle).* Walter McVitty Books. Montville, Qld, 1995.

Thiele, Colin: *High Valley.* Walter McVitty Books. Montville, Qld, 1996.

Wagner, Jenny: *Goanna.* (Illustrated by Noela Hills.) Puffin Books (Penguin Books Australia Ltd). Ringwood, Vic, 1989.

Walker, Kath (writer and illustrator): *Father Sky and Mother Earth.* The Jacaranda Press. Gladesville, NSW, 1981.

Wheatley, Nadia: *The Blooding.* Viking Kestrel. Ringwood, Vic, 1987.

Wrightson, Patricia: *The Nargun and the Stars.* Hutchinson Australia. Hawthorn, Vic, 1973 (1988).

Wrightson, Patricia: *Moon-Dark.* (Illustrated by Noela Young.) Margaret K. McElderry Books (Macmillan Publishing Company). New York, 1987 (originally published Century Hutchinson Ltd, London).

General

Anderson, Don: *Hot Copy (Reading and writing now)*. Penguin Books Australia Ltd. Ringwood, Vic, 1986.

Baker, Candida: *Yacker (Australian writers talk about their work)*. Pan Books (Australia) Pty Ltd (Picador). Sydney, 1986; *Yacker 2*. Pan Books. Sydney, 1987; *Yacker 3*. Pan. Sydney, 1989.

Bourke, Lawrence: *A Vivid Steady State (Les Murray and Australian poetry)*. University of NSW Press and New Endeavour Press. Kensington/Strawberry Hills, NSW, 1992.

Clancy, Laurie: *A Reader's Guide to Australian Fiction*. Oxford University Press Australia. Melbourne, 1992.

Cooper, Ross: see Andrew Pike, *Australian Film 1900–1977*.

Daniel, Helen: *The Good Reading Guide*. McPhee Gribble Publishers. Melbourne, 1989.

Davidson, Jim: *Sideways from the Page (The* Meanjin *interviews)*. Fontana/Collins. Melbourne, 1983.

De Groen, Frances: *Xavier Herbert: A Biography*. University of Queensland Press. St Lucia, Qld, 1998.

Dutton, Geoffrey (editor): *Country Childhoods*. University of Queensland Press. St Lucia, Qld, 1992.

Ellison, Jennifer: *Rooms of Their Own*. Penguin Books. Ringwood, Vic, 1986.

Fitzpatrick, Peter: After *'The Doll' (Australian drama since 1955)*. Edward Arnold (Australia) Pty Ltd. Melbourne, 1979. ('Studies in Australian Culture' series; general editor John Colmer.).

Goodwin, Ken: *A History of Australian Literature*. (Macmillan History of Literature series.) Macmillan Publishers Ltd. London, 1986.

Goodwin, Ken: *Adjacent Worlds (A literary life of Bruce Dawe)*. Longman Cheshire. Melbourne, 1988.

Green, Dorothy and David Headon (editors): *Imagining the Real: Australian Writing in the Nuclear Age*. ABC Enterprises. Sydney, 1987.

Hassall, Anthony J.: *Dancing on Hot Macadam (Peter Carey's fiction)*. University of Queensland Press. St Lucia, Qld, 1994.

Headon, David: see Dorothy Green, *Imagining the Real*.

Hergenhan, Laurie (general editor): *The Penguin New Literary History of Australia*. Penguin Books Australia Ltd. Ringwood, Vic, 1988.

Holloway, Peter (editor): *Contemporary Australian Drama*. Currency Press. Paddington, NSW, 1981 (wholly revised 1987).

Houbein, Lolo: *Wrong Face in the Mirror (An autobiography of race and identity)*. University of Queensland Press. St Lucia, Qld, 1990 (1988).

Kavanagh, Paul and Peter Kuch: *Conversations (Interviews with Australian writers)*. Angus & Robertson. North Ryde, NSW, 1991.

Kiernan, Brian: *David Williamson: A Writer's Career*. Currency Press Ltd. Paddington, NSW, 1996.

Kuch, Peter: *Bruce Dawe* (Oxford Australian Writers Series). Oxford University Press. Melbourne et al, 1995.

Lamb, Karen: *Peter Carey: The Genesis of Fame*. Angus & Robertson. Pymble, NSW, 1992.

Lees, Stella and Pam Macintyre (editors): *The Oxford Companion to Australian Children's Literature*. Oxford University Press (in association with ALIA Press). Melbourne, 1993.

McCallum, John: *Buzo*. Methuen Australia Pty Ltd. North Ryde, NSW, 1987 (monograph, part of Methuen Australian Drama series).

Macintyre, Pam: see Stella Lees, *The Oxford Companion to Australian Children's Literature*.

McVitty, Walter: *Innocence and Experience (Essays on contemporary Australian children's writers)*. Thomas Nelson Australia. Melbourne, 1981.

Marr, David: *Patrick White: A Life*. Random House Australia. Milsons Point, NSW, 1991.

Moffitt, Ian: *Deadlines*. Fontana Australia. Sydney, 1985.

Murray, Scott (compiler and editor): *Australian Film 1978–1992 (A survey of theatrical features)*. Oxford University Press Australia. Melbourne, 1993.

Nieuwenhuizen, Agnes: *No Kidding (Top writers for young people talk about their work)*. Pan Macmillan Publishers Australia. Chippendale, NSW, 1991.

Palmer, Jennifer (editor): *Contemporary Australian Playwrights*. Adelaide University Union Press. Adelaide, 1979.

Perkins, Elizabeth: *The Plays of Alma de Groen*. Rodopi. Amsterdam/Atlanta (Georgia), 1994. (Monograph No 6 in the series entitled 'Australian Playwrights', series editor Ortrun Zuber-Skerritt.)

Pike, Andrew and Ross Cooper (editors): *Australian Film 1900–1977*. (In association with the Australian Film Institute.) Oxford University Press. Melbourne, 1981.

Radic, Leonard: *The State of Play (The revolution in the Australian theatre since the 1960s)*. Penguin Books. Ringwood, Vic, 1991.

Rutherford, Anna (editor): *Aboriginal Culture Today*. Dangaroo Press. Sydney, 1988.

Shoemaker, Adam (editor): *Oodgeroo: a tribute*. University of Queensland Press. St Lucia, Qld, 1994. (A special issue of *Australian Literary Studies*, Vol 16 No 4.)

Taylor, Andrew: *Reading Australian Poetry*. University of Queensland Press. St Lucia, Qld, 1987.

Thiele, Colin: *With Dew on My Boots (A childhood revisited)*. Walter McVitty Books. Montville, Qld, 1997.

White, Patrick: *Patrick White Speaks*. (Edited by Paul Brennan and Christine Flynn.) Primavera Press. Leichhardt, NSW, 1989.

Willbanks, Ray: *Speaking Volumes (Australian writers and their work)*. Penguin Books Australia Ltd. Ringwood, Vic, 1991.

INDEX